Clinical Anatomy

An Illustrated Review with Questions and Explanations

Clinical Anatomy

An Illustrated Review with Questions and Explanations

Second Edition

Richard S. Snell, M.D., Ph.D.
Emeritus Professor of Anatomy
George Washington University
School of Medicine and Health Sciences
Washington, D.C.

Little, Brown and Company
Boston   New York   Toronto   London

Library of Congress Cataloging-in-Publication Data
Snell, Richard S.
    Clinical anatomy: a review with questions and explanations / Richard S. Snell. — 2nd ed.
      p. cm.
    Rev. ed. of: Gross anatomy / Richard S. Snell. 1st ed. ©1990.
    Includes index.
    ISBN 0-316-80307-3
    1. Human anatomy—Outlines, syllabi, etc. 2. Human anatomy—Examinations, questions, etc. I. Snell, Richard S.   Gross anatomy. II. Title.
    [DNLM: 1. Anatomy—examination questions. 2. Anatomy—outlines. QS 18.2 S671c 1995]
QM31.S7   1995
611'.0076—dc20
DNLM/DLC
for Library of Congress   95-20663
                                                   CIP

Printed in the United States of America

MV-NY

Editorial: Evan R. Schnittman
Production Editor: Marie A. Salter
Copyeditor: Libby Dabrowski
Indexer: Dorothy Hoffman
Production Supervisor/Cover Designer: Mike Burggren

*To those students who dislike examinations,
in the hope that this small book
will lighten their load*

Contents

Preface

This book is written for medical, dental, and allied health students who are preparing for examinations. The clinical anatomy is presented in a condensed form with simple diagrams, radiographs, CT scans, MRIs, and sonograms. At the end of each chapter are numerous National Board type questions, which are followed by answers and, where appropriate, brief explanations. Some of the questions in this edition are centered around a clinical problem that requires an anatomical answer. For a more extensive review, this book may be used in conjunction with *Clinical Anatomy for Medical Students,* Fifth Edition, also published by Little, Brown and Company.

The book begins with an introductory chapter dealing with basic anatomic structures. This is followed by chapters on regional anatomy, including thorax, abdomen, pelvis and perineum, upper limb, lower limb, head and neck, and back. The use of numerous tables simplifies the learning of muscles and their actions and nerve supplies, and assists in the memorization of the distribution of cranial and peripheral nerves.

The purpose of the National Board type questions is threefold: (1) They focus the student's attention on areas of importance; (2) they enable the student to find out his or her areas of weakness; and (3) when answered under examination conditions, with the factual material not in view, they provide the student with an effective form of self-evaluation. Apart from its use as a study guide for examinations, this book has been used extensively by students during their clinical clerkships for quick review of anatomical relationships.

I wish to express my sincere thanks to Ira Alan Grunther, AMI, for his excellent artwork. Finally, to the staff of Little, Brown and Company, I once again express my gratitude and appreciation for their assistance throughout the preparation of this book.

R.S.S.

Clinical Anatomy

An Illustrated Review with Questions and Explanations

Introduction to Basic Anatomic Structures

Bone

Bone is a living connective tissue that is hard because of its extracellular matrix; it possesses a degree of elasticity due to the presence of organic fibers.

Bone has the following functions:

1. **Protection.** For example, the skull and the vertebral column protect the brain and spinal cord, and the ribs and sternum protect the thoracic and upper abdominal viscera.
2. **Serves as a lever** on which muscles act to produce movements at joints.
3. **Storage area** for calcium and phosphorus.
4. **Contains bone marrow.** Bone houses and protects the delicate blood-forming bone marrow within its cavities.

Bone exists in two forms, **compact** and **cancellous.** Compact bone appears as a solid mass; cancellous bone consists of a branching network of trabeculae. The trabeculae are arranged to resist the stresses and strains to which the bone is exposed.

Classification of Bones

Bones may be classified according to their general shape (Fig. 1-1).

Long Bones

Long bones are found in the limbs (e.g., the humerus, the femur, the metacarpals, the metatarsals, and the phalanges). Their length is greater than their breadth. They have a tubular shaft, the **diaphysis,** and usually an **epiphysis** at each end. During the growing phase, the diaphysis is separated from the epiphysis by an **epiphyseal cartilage.** That part of the diaphysis that lies adjacent to the epiphyseal cartilage is called the **metaphysis.** The shaft has a central **marrow cavity** containing **bone marrow.** The outer part of the shaft is composed of compact bone that is covered by a connective tissue sheath, the **periosteum.**

The ends of long bones are composed of cancellous bone surrounded by a thin layer of compact bone. The articular surfaces of the ends of the bones are covered by hyaline cartilage.

Short Bones

Short bones are found in the hand and foot (e.g., the scaphoid, the lunate, the talus, and the calcaneum). They are roughly cuboidal in shape. They are composed of cancellous bone surrounded by a thin layer of compact bone. They are covered with periosteum, and the articular surfaces are covered by hyaline cartilage.

Flat Bones

Flat bones are found in the vault of the skull (e.g., the frontal and parietal bones). They are composed of thin inner and outer layers of compact bone, the **tables,** separated by a layer of cancellous bone, the **diploë.** The scapulae, although irregular, are included in this group.

Irregular Bones

Irregular bones include bones not assigned to the previous groups (e.g., the bones of the base of the skull, the vertebrae, and the pelvic bones). They are composed of a thin shell of compact bone and an interior made up of cancellous bone.

Sesamoid Bones

Sesamoid bones are small nodules of bone that are found in certain tendons where they rub over bony surfaces. The largest sesamoid bone is the **patella,** which is located in the tendon of the quadriceps femoris. Other examples are found in the tendons of flexor pollicis brevis and flexor hallucis brevis. The greater part of a sesamoid bone is buried in the tendon and the free surface is covered with cartilage. The function of a sesamoid bone is to reduce friction on the tendon; it may also alter the direction of pull of a tendon.

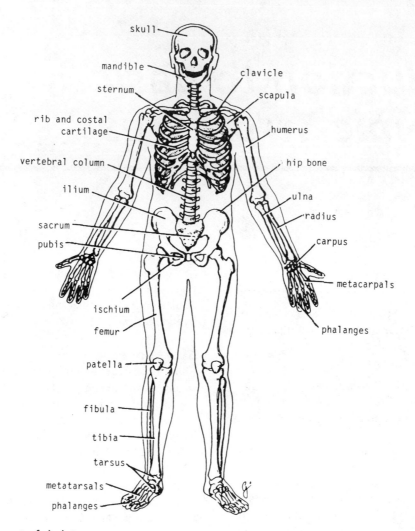

Fig. 1-1. Anterior aspect of skeleton.

Joints

The site where two or more bones come together, whether or not there is movement between them, is called a **joint.** Joints are classified according to the tissues that lie between the bones: fibrous joints, cartilaginous joints, and synovial joints (Fig. 1-2).

Fibrous Joints
The articulating surfaces of the bones are joined by fibrous tissue. The sutures of the skull and the inferior tibiofibular joints are examples. Very little movement is possible.

Cartilaginous Joints
Cartilaginous joints are of two types, primary and secondary. A **primary cartilaginous joint** is one in which the bones are united by a plate or bar of hyaline cartilage (e.g., the union between the **epiphysis** and the **diaphysis** of a growing bone and between the first rib and the manubrium sterni). No movement is possible.

A **secondary cartilaginous joint** (see Fig. 1-2) is one in which the bones are united by a plate of fibrocartilage, and the articular surfaces of the bones are covered by a thin layer of hyaline cartilage (e.g., **intervertebral joints** [joints between the vertebral bodies] and the **symphysis pubis**). A small amount of movement is possible.

Synovial Joints
The articular surfaces of the bones are covered by a thin layer of hyaline cartilage separated by a joint cavity (see Fig. 1-2). This arrangement permits a great degree of freedom of movement. The cavity of the joint is lined by **synovial membrane,** which extends from the margins of one articular surface to those of the other. The synovial membrane is protected on the outside by a tough fibrous membrane referred to as the **capsule** of the joint. The articular surfaces are lubricated by a viscous fluid, called **synovial fluid,** produced by the synovial membrane.

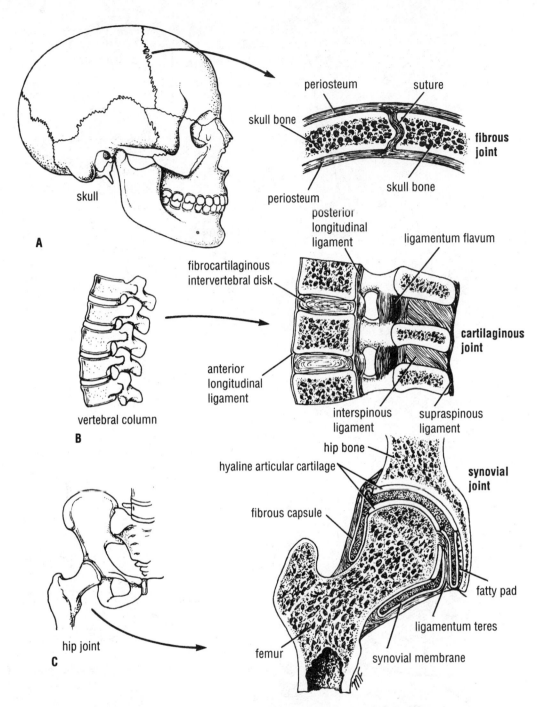

Fig. 1-2. Examples of three types of joints: (**A**) fibrous joint (coronal suture of skull), (**B**) cartilaginous joint (joint between two lumbar vertebral bodies), and (**C**) synovial joint (hip joint).

Types of Synovial Joints

Synovial joints are classified according to the shape of the articular surfaces and the movements that are possible (Fig. 1-3).

Plane Joints

The apposed articular surfaces and the bones slide upon one another (e.g., the sternoclavicular and acromioclavicular joints).

Hinge Joints

Because flexion and extension movements are possible, hinge joints resemble the hinge on a door (e.g., the elbow, knee, and ankle joints).

Pivot Joints

In pivot joints, there is a central bony pivot surrounded by a bony-ligamentous ring (e.g., the atlantoaxial and

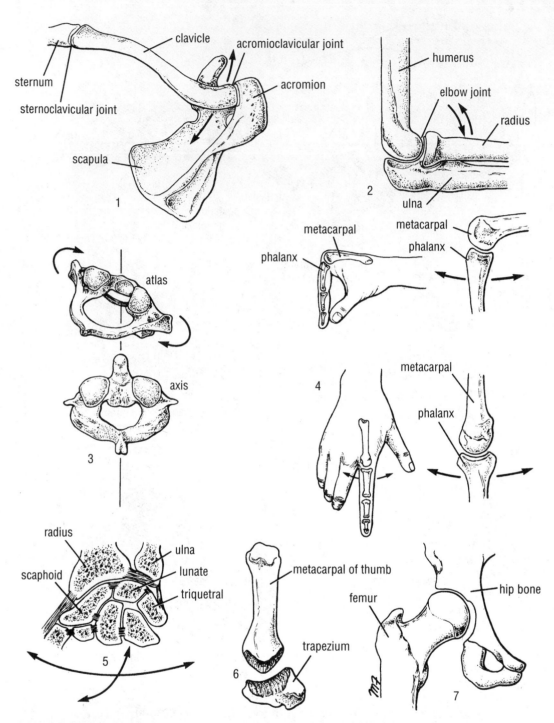

Fig. 1-3. Examples of the different types of synovial joints: (1) plane joints (sterno-clavicular and acromioclavicular joints), (2) hinge joint (elbow joint), (3) pivot joint (atlantoaxial joint), (4) condyloid joint (metacarpophalangeal joint), (5) ellipsoid joint (wrist joint), (6) saddle joint (carpometacarpal joint of the thumb), and (7) ball-and-socket joint (hip joint).

superior radioulnar joints). Rotation is the only movement possible.

Condyloid Joints

Condyloid joints have two distinct convex surfaces that articulate with two concave surfaces (e.g., the metacarpophalangeal and metatarsophalangeal joints). Flexion, extension, abduction, and adduction are possible with a small amount of rotation.

Ellipsoid Joints

Ellipsoid joints have an elliptical convex articular surface that fits into an elliptical concave articular surface (e.g., the wrist joint). Flexion, extension, abduction, and adduction can take place, but rotation is impossible.

Saddle Joints

In saddle joints, the articular surfaces are reciprocally concavoconvex and resemble a saddle on a horse's back (e.g., the carpometacarpal joint of the thumb). Flexion, extension, abduction, adduction, and rotation are possible.

Ball-and-Socket Joints

In ball-and-socket joints, a ball-shaped head of one bone fits into a socket-like concavity of another (e.g., shoulder and hip joints). Very free movements are possible, including flexion, extension, abduction, adduction, medial rotation, lateral rotation, and circumduction.

Muscle

There are three types of muscle: **skeletal, smooth,** and **cardiac.**

Skeletal (Voluntary) Muscle

A skeletal muscle has two or more attachments. The attachment that moves the least is referred to as the **origin,** and the one that moves the most as the **insertion.** The fleshy part of the muscle is referred to as its **belly.** The ends of a muscle are attached to bones, cartilage, or ligaments by cords of fibrous tissue called **tendons.** A thin, flat, expanded tendon is called an **aponeurosis.**

Nerve Supply of Skeletal Muscle

The nerve supply to a muscle is a mixed nerve (about 60% being motor and 40% sensory), and it also contains some sympathetic autonomic fibers.

Structures Associated with Skeletal Muscle
Bursae

A bursa is a lubricating device consisting of a closed fibrous sac lined with a smooth membrane. Its walls are separated by a film of viscous fluid. Bursae are found wherever tendons rub against bones, ligaments, or other tendons; they reduce friction.

Synovial Sheath

A synovial sheath is a tubular bursa that surrounds a tendon. Sheaths occur where tendons pass under ligaments and retinacula, and through osseofibrous tunnels. Their function is to reduce friction between the tendon and its surrounding structures.

Fasciae

The fasciae are of two types, **superficial** and **deep.** They lie between the skin and the underlying muscles and bones. The **superficial fascia,** subcutaneous tissue, is a mixture of areolar and adipose tissue that unites the skin to the underlying deep fascia. The **deep fascia** is a membranous layer of connective tissue that covers the muscles and other deep structures. In the limbs it forms a sheath around the muscles, holding them in place. In many places, septa of deep fascia pass between muscle groups dividing up the interior of the limbs into compartments.

Retinacula. Retinacula are thick restraining bands of deep fascia that occur in the region of joints. Their function is to hold underlying tendons in position or to serve as pulleys around which the tendons may move.

Smooth Muscle

Smooth muscle consists of long spindle-shaped cells arranged in bundles or sheets. In the tubes of the body the smooth muscle cells (fibers) are arranged in two layers, circular and longitudinal, and provide the motive power for propelling the contents through the lumen. In storage organs such as the urinary bladder, they are irregularly arranged and interlaced with one another. In blood vessels, the smooth muscle fibers are arranged circularly, and they serve to modify the diameter of the lumen.

Smooth muscle may be made to contract by local stretching of the fibers, by nerve impulses from autonomic nerves, or by hormonal stimulation.

Cardiac Muscle

Cardiac muscle consists of striated muscle fibers that branch and unite with each other. It forms the myocardium of the heart. Its fibers tend to be arranged in whorls and spirals, and they have the property of spontaneous and rhythmic contraction. Specialized cardiac muscle fibers form the **conducting system of the heart.**

Cardiac muscle is supplied by autonomic nerve fibers that terminate in the nodes of the conducting system and in the myocardium.

Vascular System

The vascular system consists of the heart and blood vessels (arteries, capillaries, and veins) that transport the blood through all regions of the body (Fig. 1-4).

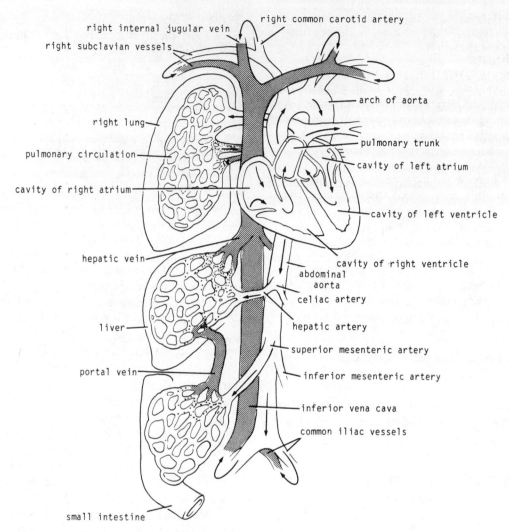

right internal jugular vein

right subclavian vessels

right common carotid artery

arch of aorta

right lung

pulmonary circulation

pulmonary trunk

cavity of left atrium

cavity of right atrium

cavity of left ventricle

cavity of right ventricle

hepatic vein

abdominal aorta

celiac artery

hepatic artery

liver

superior mesenteric artery

portal vein

inferior mesenteric artery

inferior vena cava

common iliac vessels

small intestine

Fig. 1-4. General plan of blood vascular system.

Heart

The heart is a hollow contractile muscular organ that lies within the pericardium in the mediastinum in the thorax. The heart is responsible for pumping blood through the pulmonary and systemic circulations.

Pulmonary Circulation

Venous blood from the body enters the right atrium of the heart and passes into the right ventricle. The blood is then pumped through the pulmonary arteries to the lungs for oxygenation; it returns through the pulmonary veins to the left atrium.

Systemic Circulation

The oxygenated blood in the left atrium passes into the left ventricle. The blood is then pumped through the aorta to all regions of the body and returns to the right atrium of the heart through the superior and inferior venae cavae and the cardiac veins.

Blood Vessels

Blood vessels are of three types: arteries, capillaries, and veins.

Arteries

The arteries transport blood from the heart and distribute it to the various tissues of the body by means of their branches. The smallest arteries, less than 0.1 mm in diameter, are referred to as **arterioles.** The joining of branches of arteries is called an **anastomosis.** There are no valves in arteries.

Capillaries

The capillaries are microscopic vessels in the form of a network connecting the arterioles to the venules. In some areas of the body, principally the tips of the fingers and toes, there are direct connections between the arteries and veins without the intervention of capillaries. Such connections are referred to as **arteriovenous anastomoses.**

Veins

The veins are vessels that transport blood back to the heart; many of them possess valves. The smallest veins are called **venules.** The small veins, or **tributaries,** unite to form larger veins, which commonly join with one another to form **venous plexuses.** Medium-sized deep arteries are often accompanied by two veins, one on each side, called **venae comitantes.**

Veins leaving the gastrointestinal tract do not go directly to the heart, but converge on the **portal vein;** this enters the liver and breaks up again into veins of diminishing size, which ultimately join capillary-like vessels, called **sinusoids,** in the liver. A **portal system** is thus a system of vessels interposed between two capillary beds.

Sinusoids

Sinusoids are thin-walled blood vessels having an irregular cross diameter; they are wider than capillaries. The sinusoids are lined by endothelium supported by a minimum of connective tissue. They are found in the bone marrow, spleen, liver, and some endocrine glands.

Lymphatic System

The lymphatic system consists of lymphatic tissues and lymphatic vessels (Fig. 1-5).

Lymphatic Tissues

Lymphatic tissues are a type of connective tissue that contains large numbers of lymphocytes. The lymphatic tissue is organized into the following organs or structures: the thymus, the lymph nodes, the spleen, and the lymphatic nodules. The lymphatic tissue is essential for the immunologic defenses of the body against bacteria and viruses.

Lymphatic Vessels

Lymphatic vessels are tubes that assist the cardiovascular system in the removal of tissue fluid from the tissue spaces of the body; the vessels then return the fluid to the blood. The lymphatic system is essentially a drainage system, and there is no circulation. Lymphatic vessels are found in all tissues and organs of the body except the central nervous system, the eyeball, the internal ear, the epidermis of the skin, cartilage, and bone.

Lymph

Lymph is the name given to tissue fluid once it has entered a lymphatic vessel. **Lymph capillaries** are a network of fine vessels that drain lymph from the tissues. The capillaries are in turn drained by **small lymph vessels,** which unite to form **large lymph vessels.** Lymph vessels have a beaded appearance due to the presence of numerous valves along their course.

Before lymph drains into the bloodstream, it passes through at least one lymph node and often through several.

The lymph reaches the bloodstream at the root of the neck by lymph vessels called the **right lymphatic duct** and the **thoracic duct** (see Fig. 1-5).

Nervous System

The nervous system is divided into two main parts, the **central nervous system,** which consists of the brain and spinal cord, and the **peripheral nervous system,** which consists of 12 pairs of cranial nerves and 31 pairs of spinal nerves and their associated ganglia.

Functionally, the nervous system can be further divided into the **somatic nervous system,** which controls voluntary activities, and the **autonomic nervous system,** which controls involuntary activities.

The nervous system, together with the endocrine system, controls and integrates the activities of the different parts of the body.

Central Nervous System

The central nervous system is composed of large numbers of nerve cells and their processes, supported by specialized tissue called **neuroglia.** The **neuron** is the name given to the nerve cell and all its processes. The nerve cell has two types of processes called **dendrites** and an **axon.** The dendrites are the short processes of the cell body; the axon is the name given to the longest process of the cell body.

The interior of the central nervous system is organized into gray and white matter. **Gray matter** consists of nerve cells embedded in neuroglia. **White matter** consists of nerve fibers (axons) embedded in neuroglia.

Peripheral Nervous System

The peripheral nervous system consists of the cranial and spinal nerves and their associated ganglia.

Cranial Nerves

There are 12 pairs of cranial nerves that leave the brain and pass through foramina in the skull. All the nerves are distributed in the head and neck except the tenth (vagus), which also supplies structures in the thorax and abdomen. The cranial nerves are described in Chapter 7.

Spinal Nerves

There are 31 pairs of spinal nerves that leave the spinal cord and pass through intervertebral foramina in the vertebral column. The spinal nerves are named according to the region of the vertebral column with which they are associated: 8 **cervical,** 12 **thoracic,** 5 **lumbar,** 5 **sacral,** and 1 **coccygeal. Note that there are 8 cervical nerves and only 7 cervical vertebrae and that there is 1 coccygeal nerve and 4 coccygeal vertebrae.**

Each spinal nerve is connected to the spinal cord by two **roots,** the **anterior root** and the **posterior root.** The ante-

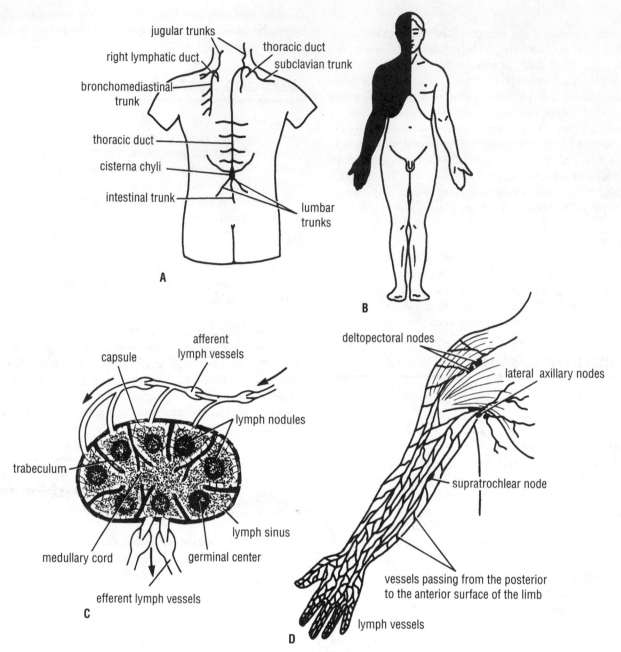

Fig. 1-5.   A. Thoracic duct and right lymphatic duct and their main tributaries.
B. The areas of body drained into thoracic duct (clear) and right lymphatic duct
(black). **C.** General structure of a lymph node. **D.** Lymph vessels and nodes of upper
limb.

rior root consists of bundles of nerve fibers carrying nerve
impulses away from the central nervous system. Such nerve
fibers are called **efferent motor fibers** and go to skeletal
muscle and cause them to contract. Their cells of origin are
located in the anterior gray horn of the spinal cord.

The posterior root consists of bundles of nerve fibers
that carry impulses to the central nervous system and are
called **afferent sensory fibers.** They are concerned with
conducting the sensations of touch, pain, temperature, and

vibrations. The cell bodies of these nerve fibers are situated in a swelling on the posterior root called the **posterior
root ganglion.**

During development, the spinal cord grows in length
more slowly than the vertebral column. In the adult, when
growth ceases, the lower end of the spinal cord only
reaches inferiorly as far as the lower border of the first
lumbar vertebra. To accommodate for this disproportionate
growth in length, the length of the roots increases progres-

sively from above downward. In the upper cervical region, the spinal nerve roots are short and run almost horizontally, but the roots of the lumbar and sacral nerves below the level of the termination of the cord form a vertical bundle of nerves that resemble a horse's tail and is called the **cauda equina.**

At each intervertebral foramen, the anterior and posterior roots unite to form a spinal nerve (Fig. 1-6). Here, the motor and sensory fibers become mixed together, so that a spinal nerve is made up of a mixture of motor and sensory fibers. After emerging from the intervertebral foramen, the spinal nerve divides into a large **anterior ramus** and a smaller **posterior ramus.** The posterior ramus passes posteriorly around the vertebral column to supply the muscles and skin of the back. The anterior ramus continues anteriorly to supply the muscles and skin over the anterolateral body wall and all the muscles and skin of the limbs.

In addition to the anterior and posterior rami, the spinal nerves give a small **meningeal branch** that supplies the vertebrae and the meninges. The spinal nerves also have branches called the **rami communicantes** that are associated with the sympathetic part of the autonomic nervous system (see p. 10).

Plexuses

At the root of the limbs, the anterior rami join one another to form complicated nerve plexuses. The **cervical** and **brachial plexuses** are found at the root of the upper limbs, and the **lumbar** and **sacral plexuses** are found at the root of the lower limbs.

Dermatomes

The area of skin supplied by a single spinal nerve, and therefore a single segment of the spinal cord, is called a **dermatome.** On the trunk, adjacent dermatomes overlap considerably, so that to produce a region of complete anesthesia, at least three contiguous spinal nerves have to be sectioned. Dermatomal charts for the anterior and posterior surfaces of the body are shown in Figures 1-7 and 1-8.

Autonomic Nervous System

The autonomic nervous system innervates involuntary structures such as the heart, smooth muscles, and glands. It is distributed throughout the central and peripheral nervous systems. The autonomic nervous system is divided into two parts, the **sympathetic** and **parasympathetic,** and in both parts there are afferent and efferent nerve fibers (Figs. 1-9 and 1-10).

Sympathetic Part

The sympathetic part of the autonomic nervous system prepares the body for an emergency. The heart rate is increased, the peripheral blood vessels are constricted, and the blood pressure is raised. The body blood is redistributed so that it leaves the skin and the gastrointestinal tract

Fig. 1-6. The association between spinal cord, spinal nerves, and sympathetic trunks.

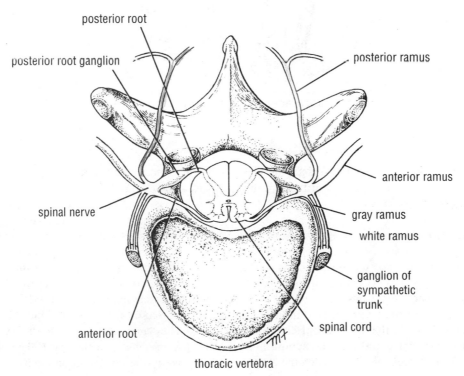

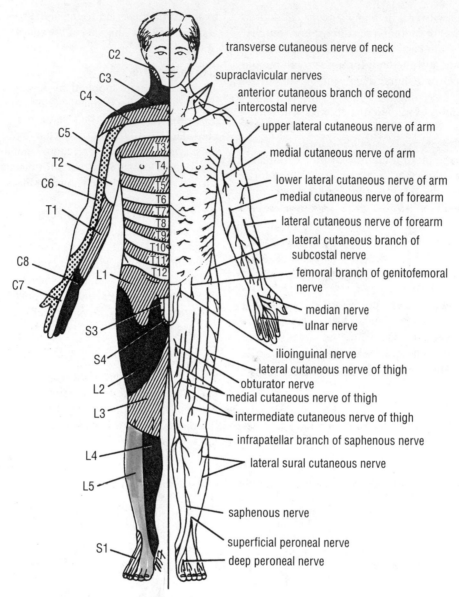

Fig. 1-7. Dermatomes and distribution of cutaneous nerves on the anterior aspect of the body.

and passes to the brain, the heart, and the skeletal muscles. The sympathetic nerves inhibit peristalsis of the gastrointestinal tract and close the sphincters.

Efferent Nerve Fibers

The lateral gray columns (horns) of the spinal cord from the first thoracic segment to the second lumbar segment (sometimes third lumbar segment) contain the cell bodies of the sympathetic connector neurons (see Fig. 1-9). The myelinated axons of these cells leave the cord in the anterior nerve roots and pass via the **white rami communicantes** to the **paravertebral ganglia** of the **sympathetic trunk.** Once these fibers (preganglionic) reach the ganglia

in the sympathetic trunk, they pass to the following destinations:

1. They may synapse with an excitor neuron in the ganglion (see Fig. 1-10). The postganglionic axons, which are nonmyelinated, leave the ganglion and pass to the thoracic spinal nerves as **gray rami communicantes.** These axons are distributed in the branches of the spinal nerves to smooth muscle in the blood vessel walls, the sweat glands, and the arrector pili muscles of the skin.

2. They may travel up in the sympathetic trunk to synapse in ganglia in the cervical region (see Fig. 1-10). Here again, the postganglionic nerve fibers pass via gray

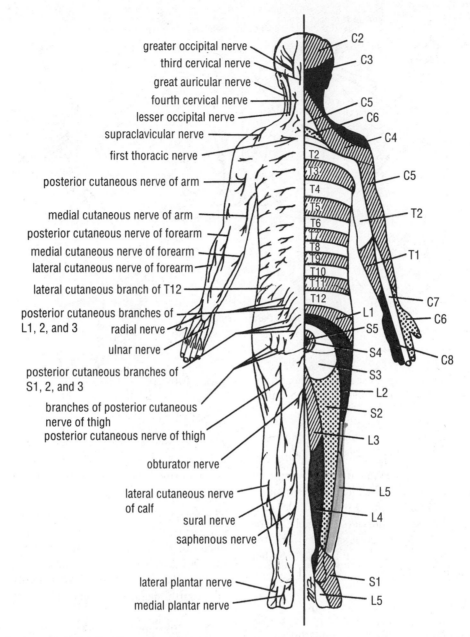

greater occipital nerve
third cervical nerve
great auricular nerve
fourth cervical nerve
lesser occipital nerve
supraclavicular nerve
first thoracic nerve
posterior cutaneous nerve of arm
medial cutaneous nerve of arm
posterior cutaneous nerve of forearm
medial cutaneous nerve of forearm
lateral cutaneous nerve of forearm
lateral cutaneous branch of T12
posterior cutaneous branches of L1, 2, and 3
radial nerve
ulnar nerve
posterior cutaneous branches of S1, 2, and 3
branches of posterior cutaneous nerve of thigh
posterior cutaneous nerve of thigh
obturator nerve
lateral cutaneous nerve of calf
sural nerve
saphenous nerve
lateral plantar nerve
medial plantar nerve

C2
C3
C5
C6
C4
C5
T2
T1
C7
C6
C8
L1
S5
S4
S3
L2
S2
L3
L5
L4
S1
L5

Fig. 1-8. Dermatomes and distribution of cutaneous nerves on the posterior aspect of the body.

rami communicantes to join the cervical spinal nerves. Many of the preganglionic fibers entering the part of the sympathetic trunk from the lower thoracic and upper two lumbar segments of the spinal cord travel down to synapse in ganglia in the lower lumbar and sacral regions. Once again, the postganglionic nerve fibers pass via gray rami communicantes to join the lumbar, sacral, and coccygeal spinal nerves (see Fig. 1-10).

3. They may pass through the ganglia of the sympathetic trunk without synapsing. These myelinated fibers leave the sympathetic trunk as the **greater splanchnic**

(T5–9), **lesser splanchnic** (T10–11), and **lowest** or **least splanchnic nerves** (T12). The greater and lesser splanchnic nerves pierce the diaphragm and synapse with excitor cells in ganglia of the **celiac plexus** (see Fig. 1-10). The lowest splanchnic nerve pierces the diaphragm and synapses with excitor cells in the **renal plexus.** The postganglionic fibers arising from the cells in the peripheral plexuses are distributed to the smooth muscles and glands of the viscera. A few preganglionic fibers traveling in the greater splanchnic nerve end directly on the cells of the suprarenal medulla.

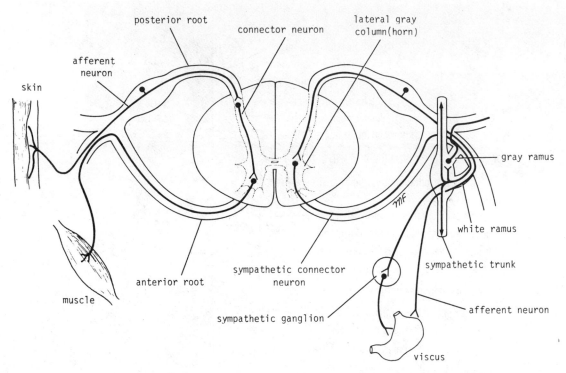

Fig. 1-9.   General arrangement of somatic part of nervous system (on left) compared with autonomic part of nervous system (on right).

Afferent Nerve Fibers

The afferent myelinated nerve fibers travel from the viscera through the sympathetic ganglia without synapsing. They pass to the spinal nerve via white rami communicantes and reach their cell bodies in the posterior root ganglion of the corresponding spinal nerve (see Fig. 1-9). The central axons then enter the spinal cord and may form the afferent component of a local reflex arc or ascend to higher autonomic centers in the brain.

Sympathetic Trunks

Sympathetic trunks are two ganglionated nerve trunks that extend the whole length of the vertebral column (see Fig. 1-10). There are 3 ganglia in each trunk in the neck, 11 or 12 ganglia in the thorax, 4 or 5 ganglia in the lumbar region, and 4 or 5 ganglia in the pelvis. The trunks lie anterior to the transverse processes of the cervical vertebrae in the neck, they are anterior to the heads of the ribs in the thorax, they are anterolateral to the bodies of the lumbar vertebrae in the abdomen, and they are anterior to the sacrum in the pelvis. The two trunks end below by joining together to form a single ganglion, the **ganglion impar.**

Parasympathetic Part

The parasympathetic part of the autonomic nervous system brings about the conservation and restoration of energy.

The heart rate is slowed, the peristalsis of the gastrointestinal tract is increased, the glandular secretions are increased, and the sphincters are opened.

Efferent Nerve Fibers

The cranial portion of the craniosacral outflow of the efferent part of the autonomic nervous system is located in the nuclei of the oculomotor (third), the facial (seventh), the glossopharyngeal (ninth), and the vagus (tenth) cranial nerves. The sacral outflow is situated in the second, third, and fourth sacral segments of the spinal cord (see Fig. 1-10).

The parasympathetic nucleus of the oculomotor nerve is called the **Edinger-Westphal nucleus,** those of the facial nerve are called the **lacrimatory** and **superior salivatory nuclei,** that of the glossopharyngeal nerve is called the **inferior salivatory nucleus,** and that of the vagus nerve, the **dorsal nucleus of the vagus.** The axons of these connector nerve cells are myelinated and emerge from the brain within the cranial nerves.

The parasympathetic sacral connector nerve cells give rise to myelinated axons that leave the spinal cord in the anterior nerve roots of the corresponding spinal nerves. They then leave the sacral nerves and form the **pelvic splanchnic nerves.**

The efferent fibers of the craniosacral outflow are preganglionic and synapse in peripheral ganglia located close to the viscera they innervate. The cranial parasympathetic

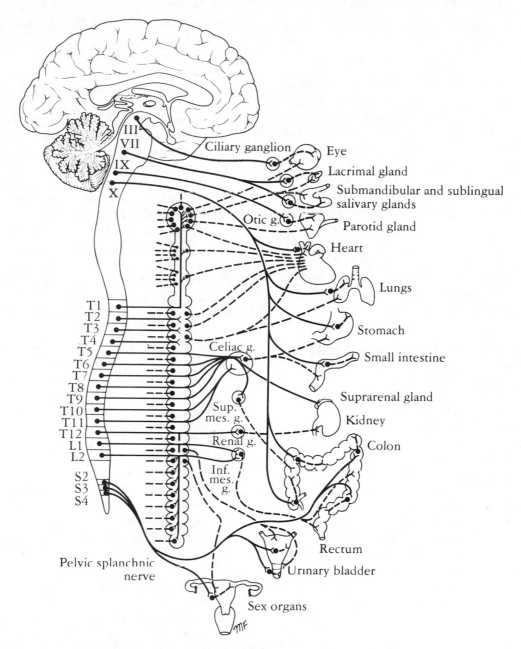

Fig. 1-10. Efferent part of autonomic nervous system. Preganglionic parasympathetic fibers are shown in solid black; postganglionic parasympathetic fibers are shown in interrupted black. Preganglionic sympathetic fibers are shown in solid black; postganglionic sympathetic fibers are shown in interrupted black.

ganglia are the **ciliary**, the **pterygopalatine**, the **submandibular**, and the **otic**. In certain locations, the ganglion cells are placed in nerve plexuses, such as the **cardiac plexus**, the **pulmonary plexus**, the **myenteric plexus** (**Auerbach's plexus**), and the **mucosal plexus** (**Meissner's plexus**); the last two plexuses are associated with the gastrointestinal tract. The pelvic splanchnic nerves synapse in ganglia in the pelvic plexuses. The postganglionic

parasympathetic fibers are nonmyelinated and short in length.

Afferent Nerve Fibers

The afferent myelinated fibers leave the viscera and reach their cell bodies in the sensory ganglia of cranial nerves or in posterior root ganglia of the sacral spinal nerves. The central axons then enter the central nervous system and form local reflex arcs or ascend to higher centers of the autonomic nervous system.

Control of the Autonomic Nervous System

The sympathetic outflow in the spinal cord (T1 to L2[3]) and the parasympathetic craniosacral outflow (cranial nerve nuclei 3, 7, 9, and 10; S2, 3, and 4) are controlled by the hypothalamus in the brain. Descending axons from nerve cells located in the hypothalamus synapse with neurons in the various cranial nerve nuclei in the brain stem and with neurons in the different segments of the spinal cord. The hypothalamus receives nerve input from all parts of the nervous system and afferent information from the viscera. It also receives information concerning the hormone levels of the blood. The hypothalamus appears to be able to integrate this nervous and hormonal input, and then, by appropriate nervous and hormonal output to the various body systems, preserves body homeostasis.

National Board Type Questions

Select the **best** response.

1. The normal functions of bone are:
   A. Storage of large quantities of lead.
   B. Provide the body with flexibility.
   C. Serve as levers and contain marrow.
   D. Form the main framework for the larynx.
   E. Provide the erect penis with rigidity.
2. Long bones have the following:
   A. The epiphysis forms the tubular shaft.
   B. The metaphysis forms the proximal end.
   C. The outer part of the shaft is formed of cancellous bone and the inner part of compact bone.
   D. The center of the shaft contains bone marrow.
   E. The epiphyseal cartilage (plate) lies between two epiphyseal areas.
3. The following bone or bones are classified as flat bones:
   A. Humerus.
   B. Clavicle.
   C. Ilium and pubis.
   D. Talus.
   E. Scapula and parietal.
4. A sesamoid bone has the following characteristics **except:**
   A. Reduces friction and may alter the direction of pull of the tendon.
   B. The greater part of the structure is composed of fibrocartilage.
   C. The largest sesamoid bone is the patella.
   D. Sesamoid bones are largely buried within a tendon.
   E. They are found in tendons that rub over bony surfaces.

5. Skeletal muscle has the following characteristics **except:**
   A. It is innervated by motor, sensory, and autonomic nerve fibers.
   B. The origin of a skeletal muscle is the attachment that moves the least.
   C. All skeletal muscle is under voluntary control.
   D. An aponeurosis of a skeletal muscle is a thin, flat, expanded tendon.
   E. Many skeletal muscles in the limbs are compartmentalized by septa formed of deep fascia.
6. Which of the following statements concerning smooth muscle is **incorrect?**
   A. Smooth muscle is found in the wall of the intestinal tract as an outer longitudinal and an inner circular layer.
   B. Smooth muscle can be made to contract by stretching its fibers.
   C. Smooth muscle in the wall of blood vessels is commonly arranged circularly.
   D. Smooth muscle in the wall of the urinary bladder is arranged in three definite layers.
   E. Smooth muscle is often sensitive to hormonal stimulation.
7. Which of the following statements concerning cardiac muscle is **incorrect?**
   A. Cardiac muscle fibers are branched.
   B. Cardiac muscle is innervated by autonomic nerves.
   C. Cardiac muscle has the property of spontaneous rhythmic contraction.
   D. Cardiac muscle does not contribute to the structure of the conducting system of the heart.
   E. Cardiac muscle fibers are arranged in whorls and spirals.
8. The vascular system has the following characteristics **except:**
   A. In the systemic circulation, the venous blood returns to the right atrium of the heart.
   B. In the pulmonary circulation, the right ventricle pumps blood into the pulmonary arteries.
   C. Oxygenated blood returns to the heart in the pulmonary veins that drain into the left atrium.
   D. The smallest arteries, less than 0.1 mm in diameter, are called arterioles.
   E. The cardiac veins drain into the left atrium.

Match each bone on the left with an appropriate class on the right. Each lettered answer may be used more than once.

9. Femur.
10. Proximal phalanx of thumb.
11. Calcaneum.
12. Atlas.

A. Short bone.
B. Irregular bone.
C. Long bone.
D. Flat bone.
E. Sesamoid bone.

For each joint on the left, give the most appropriate classification from the list on the right. Each lettered answer may be used more than once.

13. Sutures between bones of vault of skull.
14. Joint between the first rib and the manubrium sterni.
15. Symphysis pubis.
16. Ankle joint.

A. Synovial joint.
B. Cartilaginous joint.
C. Fibrous joint.
D. None of the above.

For each type of synovial joint on the left, give an appropriate example from joints on the right.

17. Saddle joint.
18. Condyloid joint.
19. Hinge joint.
20. Plane joint.

A. Wrist joint.
B. Carpometacarpal joint of thumb.
C. Metatarsophalangeal joint.
D. Knee joint.
E. Acromioclavicular joint.

In each of the following questions, answer:

A. If only (1) is correct
B. If only (2) is correct
C. If both (1) and (2) are correct
D. If neither (1) nor (2) is correct

21. (1) A retinaculum is a thick restraining band of deep fascia that holds tendons in place.
(2) A synovial sheath surrounds a tendon and provides it with added strength.

22. (1) Venae comitantes are small arteries that sometimes accompany veins.
(2) A portal system is a system of blood vessels interposed between two capillary beds.

23. (1) Lymph is the name given to tissue fluid once it has entered a lymphatic capillary.
(2) The right lymphatic duct drains the lymph from the right side of the head and neck, the right upper limb, and the right side of the thorax.

24. (1) A bursa consists of a closed fibrous sac whose walls are separated by a film of viscous fluid.
(2) All bursae communicate with joint cavities.

25. (1) Before lymph enters the bloodstream, it usually passes through several lymph nodes.
(2) Medium-sized and large lymph vessels possess numerous valves that ensure that the lymph flows toward the larger lymph vessels that drain into the large veins at the root of the neck.

26. (1) There are 12 pairs of cranial nerves.
(2) There are 42 pairs of spinal nerves.

27. (1) Each spinal nerve is connected to the spinal cord by an anterior root and a posterior root.
(2) Motor nerve fibers to skeletal muscle leave the spinal cord in an anterior root.

28. (1) All the cranial nerves are distributed in the head and neck except the glossopharyngeal nerve, which also supplies structures in the thorax.

(2) There are 7 cervical spinal nerves and 7 cervical vertebrae.

29. (1) Afferent sensory fibers in a spinal nerve have cell bodies located in the specialized sensory nerve endings.
(2) The cauda equina consists of the anterior and posterior roots of the lumbar and sacral spinal nerves below the level of the first lumbar vertebra.

30. (1) The posterior ramus of a spinal nerve supplies the muscles of the limbs.
(2) The roots of a spinal nerve unite at an intervertebral foramen to form a spinal nerve.

31. (1) The sympathetic and parasympathetic parts of the autonomic nervous system have both afferent and efferent nerve fibers.
(2) The sympathetic nerves stimulate peristalsis of the gastrointestinal tract.

32. (1) The sympathetic efferent nerves originate from the first thoracic segment to the second lumbar segment (sometimes third lumbar segment) of the spinal cord.
(2) Preganglionic sympathetic nerve fibers run from the anterior nerve roots to the sympathetic trunk via gray rami communicantes.

33. (1) Postganglionic sympathetic nerve fibers travel in spinal nerves to be distributed to smooth muscle in blood vessel walls, sweat glands, and the arrector pili muscles of the skin.
(2) The greater splanchnic nerve is composed of preganglionic parasympathetic nerve fibers.

34. (1) Afferent sympathetic nerve fibers travel through sympathetic ganglia without synapsing and join the spinal nerves via white rami communicantes.
(2) Afferent parasympathetic nerve fibers traveling with cranial nerves have their cell bodies located in ganglia along these nerves.

35. (1) The parasympathetic part of the autonomic nervous system brings about the conservation and restoration of energy within the body.
(2) The cranial part of the parasympathetic outflow takes place in cranial nerves 3, 7, 9, and 10.

36. (1) The spinal part of the parasympathetic outflow is situated in the five sacral segments of the spinal cord.
(2) The efferent fibers of the parasympathetic craniosacral outflow tend to have short preganglionic nerve fibers and long postganglionic fibers.

In the next section, study each case history. Select the **best** answer to each question following it.

An 8-year-old girl was playing in the garden when she fell and grazed her right elbow. She had no other signs or symptoms. Her mother thoroughly washed the injured area of skin and applied a Band-Aid. The next morning the child complained of pain and tenderness over the right elbow and discomfort in the right armpit; there were also red streaks passing up from the damaged skin area into the upper arm.

37. Why did the patient experience discomfort in the right armpit?
    A. Did she injure her right shoulder joint when she fell?
    B. Did she tear a muscle in her right axilla?
    C. Was the skin area infected at the time of injury and subsequently organisms ascended to the lymph nodes in the right axilla?
    D. Was she imagining the discomfort and trying to miss going to school?
38. Can you explain the red skin streaks that extended from the damaged skin area up into the arm?
    A. Were the skin streaks merely fine scratch marks passing up the arm that the mother had not noticed the previous day?
    B. Were the streaks caused by pressure from her night-clothes or teddy bear that she cuddled while asleep?
    C. Did the red lines indicate that the lymphatic vessels lying in the subcutaneous tissue were inflamed by the spread of infection from the injured skin area?

A 10-year-old boy was examined by a pediatrician because he was complaining of weakness and lassitude. His parents had noticed that he bruised easily after minor trauma and often suffered from nosebleeds. Abdominal examination revealed an enlarged liver and spleen. Because of the strong possibility that this child might have acute leukemia, the physician requested a complete laboratory examination of his blood and a bone marrow biopsy.

39. The bone marrow biopsy would determine if the formation of red and white blood cells was taking place normally. Can you explain why the marrow biopsy was performed at the crest of the iliac bone?
    A. The iliac crest is the only part of the skeletal system in a child that contains bone marrow.
    B. The physician was wrong to choose the iliac crest since it consists entirely of compact bone and does not possess bone marrow.
    C. In a normal child the bone marrow present within the iliac crest is of the yellow variety and does not have a blood-forming function.
    D. The bone marrow in the iliac crest is of the red type and is ideally situated to obtain a biopsy.
40. Name the structures that the needle passes through to reach the marrow when a marrow biopsy is being performed at the iliac crest.
    A. Skin, superficial fascia, deep fascia, periosteum, compact bone, cancellous bone.
    B. Skin, superficial fascia, deep fascia, and periosteum.
    C. Skin, superficial fascia, deep fascia, and compact bone.

Answers and Explanations

1. C Bones form rigid levers permitting muscles to act with mechanical advantage. Many bones contain red or yellow marrow within their cavities. In normal individuals, lead is only found in minimal amounts in bone. However, should lead be ingested (lead paint) or inhaled (lead-polluted air), it can be absorbed into the body and stored in bones. Nerve paralysis and severe brain dysfunction will follow its slow release from bone into the bloodstream. Cartilages form the main framework of the larynx (see p. 254). Bone is not found in the penis of man; however, it is present in the penis of many vertebrates.
2. D The center of the shaft of a long bone contains a cavity (the marrow cavity) that is filled with marrow. At birth, all the marrow of long bones is red and forming blood cells. At 7 years of age, the red marrow begins to be replaced with yellow marrow in the distal bones of the limbs. This process continues during childhood, so that by the time the person reaches adulthood, the red marrow is restricted to the heads of the humerus and femur, as well as other bones including the skull, vertebral column, thoracic cage, and girdle bones.
3. E The scapula, although having some bony projections, is nevertheless classified as a flat bone.
4. B Sesamoid bones are composed of bone. They are covered by cartilage on their free surface that rubs against another bone.
5. C Not all skeletal muscle is under voluntary control. For example, skeletal muscle fibers present in the wall of the upper two-thirds of the esophagus are certainly not under voluntary control.
6. D Smooth muscle fibers found in storage organs such as the urinary bladder are irregularly arranged and interlaced with one another.
7. D Specialized cardiac muscle fibers form the conducting system of the heart.
8. E The cardiac veins, which drain the wall of the heart, open into the right atrium of the heart.
9. C
10. C
11. A
12. B
13. C
14. B
15. B
16. A
17. B
18. C
19. D
20. E

21. A   A synovial sheath is a delicate structure that provides a lubricating mechanism for a tendon as it passes beneath a restraining fascial band.
22. B   Venae comitantes are small veins, commonly two in number, that often accompany deeply placed arteries.
23. C
24. A   Most bursae do not communicate with a joint cavity. However, two bursae around the knee joint communicate with the synovial cavity. They are the suprapatellar bursa and the popliteal bursa; the semimembranosus bursa also frequently communicates with the joint cavity.
25. C
26. A   There are 31 pairs of spinal nerves.
27. C
28. D   The vagus (tenth cranial nerve) is the only cranial nerve to be distributed outside the head and neck. Having supplied structures in the head and neck, it also supplies viscera in the thorax and abdomen. There are 8 cervical spinal nerves and 7 cervical vertebrae.
29. B   Afferent sensory fibers in a spinal nerve have their cell bodies in the posterior root ganglion.
30. B   The anterior ramus is the largest of the two rami and supplies the muscles and skin over the anterolateral body wall and all the muscles and skin of the limbs.
31. A   The sympathetic nerves inhibit peristalsis of the gastrointestinal tract and cause the sphincters to contract.
32. A   Preganglionic sympathetic nerve fibers run from the anterior spinal nerve roots to the sympathetic trunk via white rami communicantes.
33. A   The greater splanchnic nerve is composed of preganglionic sympathetic nerve fibers.
34. C
35. C
36. D   The spinal part of the parasympathetic outflow is situated in segments S2, 3, and 4 only. Efferent parasympathetic nerve fibers have long preganglionic fibers and short postganglionic fibers; the ganglia tend to be located close to the organ they supply.
37. C   The skin abrasion was contaminated with dirt containing pathogenic bacteria. These organisms ascended in the lymph to the lymph nodes in the axilla, where they multiplied and produced inflammation of the nodes (lymphadinitis). The infected nodes became swollen and tender causing discomfort. The little girl did not injure her right shoulder joint nor did she damage a muscle in her right axilla, as seen by the absence of signs or symptoms immediately following the fall.
38. C   As the pathogenic bacteria ascended in the lymph to the axillary lymph nodes, they inflamed the walls of the lymphatic vessels conducting the lymph, a condition known as lymphangitis. The blood vessels accompanying the lymph vessels underwent a vasodilatation causing the red streaks beneath the skin.
39. D   The iliac crest is a common site for obtaining a sample of red bone marrow at any age. It is easy to palpate just beneath the skin.
40. A   Following the introduction of a local anesthetic at the site of the biopsy, the needle is gently but forcibly pushed through the skin, superficial fascia, deep fascia, periosteum, and compact bone. It finally enters the red marrow located between the trabeculae of the cancellous bone.

Thorax

Thoracic Wall

Bones of the Thoracic Wall

These bones consist of the thoracic part of the vertebral column, the sternum, the ribs, and the costal cartilages (Fig. 2-1).

Thoracic Part of Vertebral Column

The thoracic part of the vertebral column is concave forward and consists of 12 vertebrae, together with their intervertebral discs.

Sternum

The sternum is a flat bone that may be divided into three parts: the manubrium, the body, and the xiphoid process.

Manubrium

The manubrium forms the upper part of the sternum. It articulates with the body of the sternum at the manubriosternal joint. It also articulates with the clavicles and the first and upper part of the second costal cartilage on each side.

The **sternal angle (angle of Louis)** is formed by the articulation of the manubrium with the body of the sternum (see Fig. 2-1). It can be felt on the anterior aspect of the sternum as a transverse ridge. This ridge is an important surface landmark and lies at the level of

1. The second costal cartilage
2. The intervertebral disc between the fourth and fifth thoracic vertebrae
3. The junction of the ascending aorta and the aortic arch, and the junction of the aortic arch and the descending thoracic aorta
4. The bifurcation of the trachea
5. The junction of the superior mediastinum and the inferior mediastinum (see p. 25).

Body of the Sternum

The body of the sternum articulates above with the manubrium by means of a fibrocartilaginous joint, the **manubri-osternal joint,** and below by means of a cartilaginous joint with the xiphoid process at the **xiphisternal joint.** On each side, it articulates with the lower part of the second costal cartilage and the third to the seventh costal cartilages (Fig. 2-2).

Xiphoid Process

The xiphoid process is a thin plate of hyaline cartilage that becomes ossified at its proximal end in adult life; no ribs or costal cartilages are attached to it.

Ribs

There are 12 pairs of ribs, all of which are attached posteriorly to the thoracic vertebrae. **True ribs:** The upper seven pairs are attached to the sternum by their costal cartilages. **False ribs:** The eighth, ninth, and tenth ribs are attached anteriorly to each other and to the seventh rib by means of their costal cartilages and small synovial joints. **Floating ribs:** The eleventh and twelfth pairs have no anterior attachment.

Typical Rib

The typical rib is a long, twisted flat bone with a rounded superior border and a grooved inferior border, the **costal groove,** which accommodates the intercostal vessels and nerve. The anterior end of each rib is attached to the corresponding costal cartilage (see Fig. 2-2).

A rib has a **head, neck, tubercle, shaft,** and **angle.** The **head** has two facets for articulation with the numerically corresponding vertebral body and with the vertebra immediately above. The **tubercle** has a facet for articulation with the transverse process of the numerically corresponding vertebra. The **angle** is where the shaft bends sharply forward.

Atypical Rib

The **first rib** is atypical. It is important because of its close relationship to the nerves of the brachial plexus and the subclavian artery and vein. The rib is small and flattened from above downward. The scalenus anterior is attached to its upper surface and inner border. Anterior to the scalenus

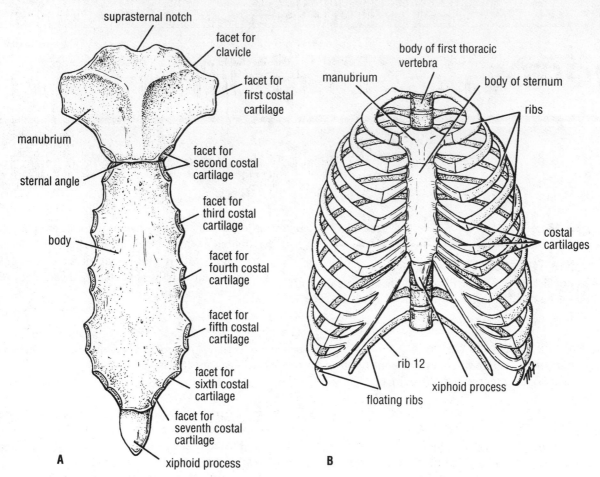

Fig. 2-1. **A.** Anterior view of sternum. **B.** Sternum, ribs, and costal cartilage forming thoracic skeleton.

anterior, the subclavian vein crosses the rib; posterior to the muscle attachment, the subclavian artery and the lower trunk of the brachial plexus lie in contact with the bone.

Cervical Rib
A cervical rib occurs in about 0.5 percent of persons. It arises from the transverse process of the seventh cervical vertebra. It may have a free anterior end, may be connected to the first rib by a fibrous band, or may articulate with the first rib. It may cause pressure on the lower trunk of the brachial plexus or the subclavian artery leading to symptoms and signs referred to as the **thoracic outlet syndrome.**

Costal Cartilages
The costal cartilages are bars of hyaline cartilage connecting the upper seven ribs to the lateral edge of the sternum, and the eighth, ninth, and tenth ribs to the cartilage immediately above (see Fig. 2-1). The cartilages of the eleventh and twelfth ribs end in the abdominal musculature.

Joints of the Thoracic Wall
Manubriosternal Joint
The manubriosternal joint is a secondary cartilaginous joint. The bony surfaces are covered with hyaline cartilage and joined by a disc of fibrocartilage. A small amount of movement is possible during respiration.

Xiphisternal Joint
The xiphisternal joint is a secondary cartilaginous joint. The xiphoid process usually fuses with the body of the sternum during middle age.

Costovertebral Joints
From the second to the ninth ribs, the head articulates by means of a synovial joint with the corresponding vertebral body and that of the vertebra above. There is a strong **intra-articular ligament** that connects the head to the intervertebral disc. The head of the first and that of the lower three ribs has a simple synovial joint with its corresponding vertebral body.

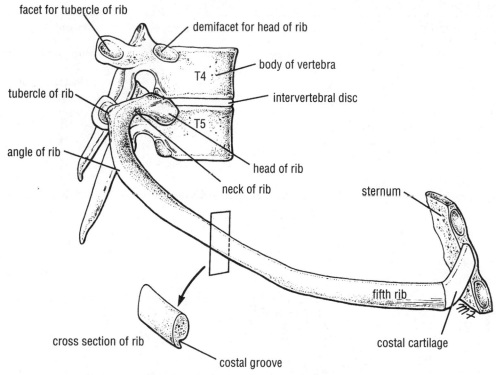

Fig. 2-2. Fifth right rib as it articulates with vertebral column posteriorly and with sternum anteriorly. Note that rib head articulates with vertebral body of its own number and that of vertebra immediately above. Note also presence of costal groove along inferior border of the rib.

The tubercle of a rib articulates by a synovial joint with the transverse process of the corresponding vertebra. (This joint is absent on the eleventh and twelfth ribs.)

Costochondral Joints
Costochondral joints are primary cartilaginous joints. No movements are possible.

Joints of the Costal Cartilages with the Sternum
The first costal cartilages articulate with the manubrium by means of primary cartilaginous joints that permit no movement. The second costal cartilages articulate with the manubrium and body of the sternum by a synovial joint. The third to the seventh costal cartilages articulate with the lateral border of the body of the sternum by synovial joints. The sixth, seventh, eighth, ninth, and tenth costal cartilages articulate with one another along their borders by small synovial joints. The cartilages of the eleventh and twelfth ribs are embedded in the abdominal musculature.

The raising and lowering of the ribs during respiration are accompanied by gliding movements in both the joints of the head and the tubercle, permitting the neck of each rib to rotate around its own axis.

Muscles of the Thoracic Wall
The muscles of the thoracic wall are summarized in Table 2-1.

Intercostal Spaces
The spaces between the ribs are called intercostal spaces. Each space contains three muscles of respiration: the external intercostal, the internal intercostal, and the transversus thoracis muscle. The transversus thoracis muscle is lined internally by the **endothoracic fascia** and parietal pleura. The intercostal nerves and blood vessels run between the intermediate and deepest layer of muscles (Fig. 2-4). They are arranged in the following order from above downward: intercostal vein, intercostal artery, and intercostal nerve (i.e., VAN).

Diaphragm
The diaphragm is the most important muscle of respiration (Fig. 2-3). It is dome-shaped and consists of a peripheral muscular part and a centrally placed tendon. The origin of the diaphragm may be divided into three parts:

1. A **sternal part** arising from the posterior surface of the xiphoid process

Table 2-1   Muscles of the Thorax

Name of muscle	Origin	Insertion	Nerve supply	Action
Diaphragm	Xiphoid process; lower 6 costal cartilages; 1–3 lumbar vertebrae by crura and medial and lateral arcuate ligaments	Central tendon	Phrenic nerve	Most important muscle of inspiration, increases vertical diameter of thorax by pulling down central tendon; assists in raising lower ribs
Intercostal muscles				
External intercostal (fibers pass downward and forward)	Inferior border of rib above	Superior border of rib below	Intercostal nerves	With first rib fixed, they raise ribs during inspiration and thus increase anteroposterior and transverse diameters of thorax; with last rib fixed by abdominal muscles, they lower ribs during expiration
Internal intercostal (fibers pass downward and backward)	Inferior border of rib above	Superior border of rib below	Intercostal nerves	Assist external intercostal muscles
Transversus thoracis (fibers pass transversely, forms incomplete layer of muscle)	Adjacent ribs	Adjacent ribs	Intercostal nerves	Assist external and internal intercostal muscles
Levatores costarum (12 in number)	Transverse processes of seventh cervical to eleventh thoracic vertebra	Superior border of ribs	Posterior rami of thoracic spinal nerves	Elevate ribs
Serratus posterior superior	Ligamentum nuchae and upper thoracic spines	Upper ribs	Intercostal nerves	Elevates ribs
Serratus posterior inferior	Lower thoracic and upper lumbar spines	Lower ribs	Intercostal nerves	Lowers ribs

2. A **costal part** arising from the deep surfaces of the lower six ribs and their costal cartilages
3. A **vertebral part** arising by means of vertical columns or **crura** and from the arcuate ligaments

The **right crus** arises from the sides of the bodies of the first three lumbar vertebrae and the intervertebral discs; it splits to enclose the esophagus. The **left crus** arises from the sides of the bodies of the first two lumbar vertebrae and the intervertebral disc. Lateral to the crura, the diaphragm arises from the medial and lateral arcuate ligaments. The **medial arcuate ligament** extends from the side of the body of the second lumbar vertebra to the transverse process of the first lumbar vertebra. The **lateral arcuate ligament** extends from the transverse process of the first lumbar vertebra to the twelfth rib. The diaphragm is inserted into a **central tendon.**

As seen from in front, the diaphragm curves up into right and left domes. Note that the domes support the right and left lungs whereas the central tendon supports the heart.

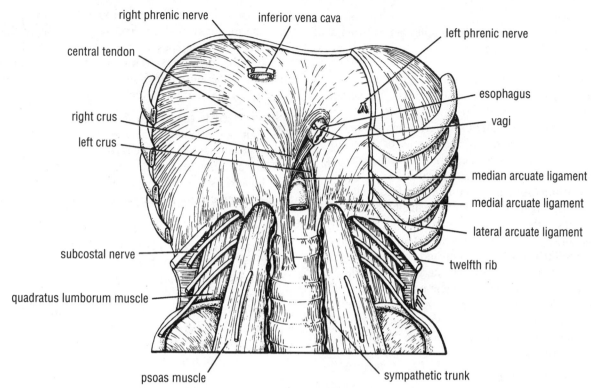

Fig. 2-3. Diaphragm as seen from below. The anterior portion of the right side has been removed. Note sternal, costal, and vertebral origins of muscle and important structures that pass through it.

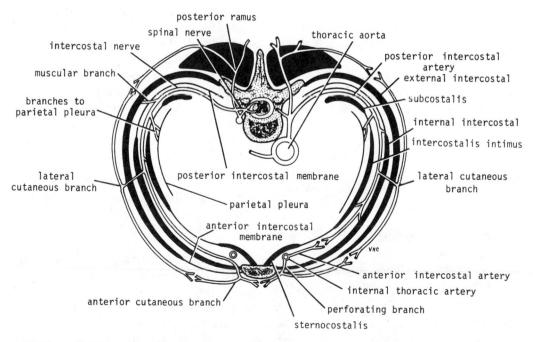

Fig. 2-4. Cross section of thorax, showing distribution of a typical intercostal nerve and a posterior and an anterior intercostal artery.

Nerve Supply

The motor nerve supply of the diaphragm is from the phrenic nerve (C3, 4, and 5).

Action

1. **Muscle of inspiration.** On contraction, the diaphragm pulls down its central tendon and increases the vertical diameter of the thorax.
2. **Muscle of abdominal straining.** The contraction of the diaphragm assists that of the muscles of the anterior abdominal wall in raising the intra-abdominal pressure for micturition, defecation, and parturition.
3. **Weight-lifting muscle.** By taking a deep breath and holding it (fixing the diaphragm), the diaphragm assists the muscles of the anterior abdominal wall in raising the intra-abdominal pressure to such an extent that it will help support the vertebral column and prevent its flexion.
4. **Thoracoabdominal pump.** The descent of the diaphragm decreases the intrathoracic pressure and increases the intra-abdominal pressure. This mechanism assists in the return of venous blood in the inferior vena cava to the right atrium and the passage of lymph upward in the thoracic duct.

Openings in the Diaphragm

Aortic Opening.   The aortic opening lies anterior to the body of the twelfth thoracic vertebra between the crura. It transmits the aorta, the thoracic duct, and the azygos vein.

Esophageal Opening.   The esophageal opening lies at the level of the tenth thoracic vertebra in a sling of muscle fibers derived from the right crus. It transmits the esophagus, the right and left vagus nerves, the esophageal branches of the left gastric vessels, and the lymphatic vessels from the lower one-third of the esophagus.

Caval Opening.   The caval opening lies at the level of the eighth thoracic vertebra in the central tendon. It transmits the inferior vena cava and terminal branches of the right phrenic nerve.

In addition to these structures, the splanchnic nerves pierce the crura, the sympathetic trunk passes posterior to the medial arcuate ligament on each side, and the superior epigastric vessels pass between the sternal and costal origins of the diaphragm on each side.

Blood Vessels of the Thoracic Wall

Intercostal Arteries and Veins

Each intercostal space possesses a large single **posterior intercostal artery** and two small **anterior intercostal arteries.**

Posterior Intercostal Arteries

The posterior intercostal arteries (see Fig. 2-4) of the first two spaces are branches of the superior intercostal artery, a branch of the costocervical trunk of the subclavian artery. The posterior intercostal arteries of the lower nine spaces are branches of the thoracic aorta.

Anterior Intercostal Arteries

The anterior intercostal arteries (see Fig. 2-4) of the first six spaces are branches of the internal thoracic artery. The anterior intercostal arteries of the lower spaces are branches of the musculophrenic artery, one of the terminal branches of the internal thoracic artery.

Branches of Intercostal Arteries

The branches of intercostal arteries pass to the parietal pleura, muscles, skin, and breast tissue.

The corresponding **posterior intercostal veins** drain into the azygos or hemiazygos veins, and the **anterior intercostal veins** drain into the internal thoracic and musculophrenic veins.

Internal Thoracic Artery and Vein

Internal Thoracic Artery

The internal thoracic artery (see Fig. 2-4) arises from the first part of the subclavian artery. It descends directly behind the first six costal cartilages and in front of the parietal pleura; it lies about a fingerbreadth lateral to the sternum. It terminates in the sixth intercostal space by dividing into the superior epigastric artery and musculophrenic arteries.

Branches

1. **Anterior intercostal arteries** for the upper six intercostal spaces
2. **Perforating arteries** pierce the intercostal muscles and supply the skin and the mammary gland
3. **Pericardiophrenic artery** supplies the pericardium and the diaphragm
4. **Mediastinal arteries** to the mediastinum, including the thymus gland
5. **Superior epigastric artery** enters the rectus sheath and supplies the upper part of the rectus muscle
6. **Musculophrenic artery** follows the costal margin on the upper surface of the diaphragm; it supplies the diaphragm and the lower intercostal spaces

Internal Thoracic Vein

The internal thoracic vein begins as venae comitantes of the internal thoracic artery. The venae eventually join to form a single vessel, which drains into the brachiocephalic vein on each side.

Lymphatic Drainage of the Thoracic Wall

The **skin** of the anterior chest wall drains to the anterior axillary lymph nodes; that of the posterior chest wall drains to the posterior axillary nodes. (The lymphatic drainage of the breast is described on p. 123.)

The **intercostal spaces** drain forward to the **internal thoracic nodes,** situated along the internal thoracic artery, and posteriorly to the **posterior intercostal nodes,** close to the heads of the ribs and the **para-aortic nodes** in the posterior mediastinum.

Nerves of the Thoracic Wall
Intercostal Nerves
The intercostal nerves are the anterior rami of the first 11 thoracic spinal nerves (Fig. 2-5). Each nerve enters an intercostal space and runs forward inferiorly to the intercostal vessels in the subcostal groove of the corresponding rib, between the transversus thoracis and internal intercostal muscle (see Fig. 2-4).

The first six nerves are distributed within their intercostal spaces. The seventh, eighth, and ninth intercostal nerves leave the anterior ends of their intercostal spaces by passing deep to the costal cartilages to enter the abdominal wall. The tenth and eleventh nerves pass forward directly into the abdominal wall.

Branches
1. **Collateral branch** runs forward below the main nerve.
2. **Lateral cutaneous branch** divides into anterior and posterior branches that supply the skin.
3. **Anterior cutaneous branch** is the terminal part of the main nerve. It divides into a medial and a lateral branch and supplies the skin near the midline.
4. **Muscular branches** to the intercostal muscles.
5. **Pleural branches** to the parietal pleura and **peritoneal branches (7–11 intercostal nerves only)** to parietal peritoneum. These are sensory nerves.

It should be noted that the seventh to eleventh intercostal nerves supply: (1) the skin and parietal peritoneum covering the outer and inner surfaces of the anterior abdominal wall, respectively, and (2) the anterior abdominal muscles (external and internal oblique, transversus abdominis, and rectus abdominis muscles).

The **first and second intercostal nerves** are exceptions. The **first intercostal nerve** gives rise to a large branch (equivalent to the lateral cutaneous branch of typical intercostal nerves) that joins the anterior ramus of the eighth cervical nerve to form the lower trunk of the brachial plexus (see p. 153). The remainder of the first intercostal nerve is small.

The **second intercostal nerve** is joined to the medial cutaneous nerve of the arm by the **intercostobrachial nerve.** The second intercostal nerve therefore supplies the skin of the armpit and the upper medial side of the arm. In coronary heart disease, pain is referred along this nerve to the medial side of the arm.

Thoracic Cavity

The thoracic cavity may be divided into a median partition, called the mediastinum, and the laterally placed pleurae and lungs.

The thoracic cavity communicates with the root of the neck through an opening called the **thoracic inlet** (clinicians often call it the thoracic outlet since important vessels and nerves emerge from the thorax here to enter the neck and upper limbs). The opening is bounded posteriorly by the first thoracic vertebra, laterally by the medial borders of the first ribs and costal cartilages, and anteriorly by the superior border of the manubrium sterni. Through this opening pass the esophagus and trachea and many vessels and nerves. Because of the obliquity of the opening, the apices of the lungs and pleurae project upward into the neck.

Below, the thoracic cavity communicates with the abdomen through a large opening called the **thoracic outlet.** The opening is bounded posteriorly by the twelfth thoracic vertebra, laterally by the curving costal margin, and anteriorly by the xiphisternal joint. Through this large opening, which is closed by the diaphragm, pass the esophagus and many large vessels and nerves, all of which pierce the diaphragm.

Mediastinum
The mediastinum is an interpleural partition that extends superiorly to the thoracic inlet and the root of the neck and inferiorly to the diaphragm. It extends anteriorly to the sternum and posteriorly to the 12 thoracic vertebrae. It may be divided into the **superior and inferior mediastina** by an imaginary plane passing from the sternal angle anteriorly to the lower border of the body of the fourth thoracic vertebra posteriorly (Fig. 2-6). The inferior mediastinum is further subdivided into the middle mediastinum, which consists of the pericardium and heart; the **anterior mediastinum,** which is a space between the pericardium and the sternum; and the **posterior mediastinum,** which lies between the pericardium and the vertebral column.

Superior Mediastinum
The contents of the superior mediastinum, **from anterior to posterior,** include the following: remains of thymus, brachiocephalic veins, upper part of superior vena cava, brachiocephalic artery, left common carotid artery, left subclavian artery, arch of aorta, both phrenic and vagus nerves, left recurrent laryngeal and cardiac nerves, trachea and lymph nodes, esophagus and thoracic duct, and sympathetic trunks.

Anterior Mediastinum
The contents of the anterior mediastinum include the following: sternopericardial ligaments, lymph nodes, and remains of thymus.

Middle Mediastinum
The contents of the middle mediastinum include the following: pericardium, heart and roots of great blood vessels, phrenic nerves, bifurcation of trachea, and lymph nodes.

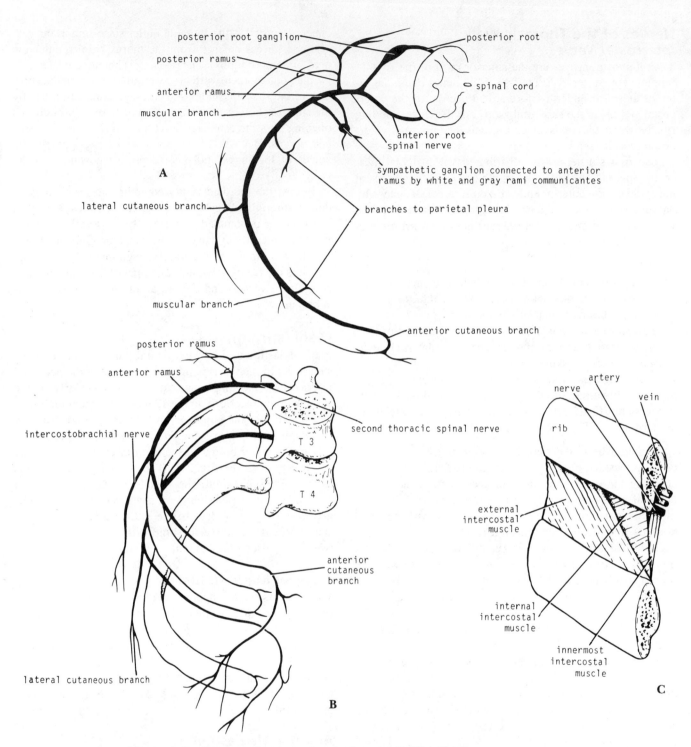

posterior root ganglion

posterior root

posterior ramus

spinal cord

anterior ramus

muscular branch

anterior root
spinal nerve

A

sympathetic ganglion connected to anterior
ramus by white and gray rami communicantes

lateral cutaneous branch

branches to parietal pleura

muscular branch

anterior cutaneous branch

posterior ramus

anterior ramus

artery

nerve

vein

intercostobrachial nerve

rib

second thoracic spinal nerve

T 3

external
intercostal
muscle

T 4

anterior
cutaneous
branch

internal
intercostal
muscle

innermost
intercostal
muscle

lateral cutaneous branch

C

B

Fig. 2-5. **A.** The origin and distribution of a thoracic spinal nerve. **B.** The distribution of two intercostal nerves relative to the rib cage. **C.** Section through an intercostal space, showing the positions of the intercostal nerve, artery, and vein relative to the intercostal muscles.

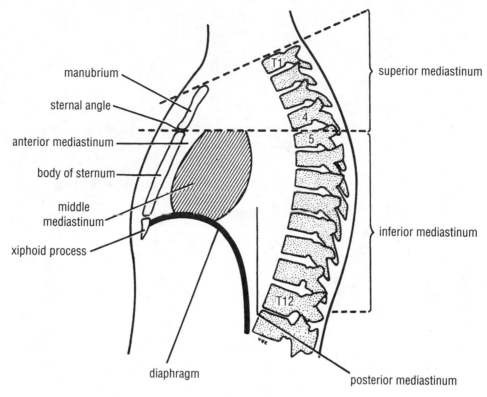

Fig. 2-6. Subdivisions of mediastinum.

Posterior Mediastinum

The contents of the posterior mediastinum include the following: descending thoracic aorta, esophagus, thoracic duct, azygos and hemiazygos veins, vagus nerves, splanchnic nerves, sympathetic trunks, and lymph nodes.

Trachea and Bronchi
Trachea

The trachea is a mobile cartilaginous and membranous tube (Fig. 2-7). It begins as a continuation of the larynx at the lower border of the cricoid cartilage (level of sixth cervical vertebra). It descends in the midline of the neck and ends in the thorax by dividing into right and left principal (main) bronchi at the level of the sternal angle (disc between T4–5 vertebrae).

The trachea is about 4½ in. (11.25 cm) long and 1 in. (2.5 cm) in diameter in the adult. The fibroelastic tube is kept from collapsing by the presence of U-shaped cartilaginous rings embedded in its wall.

Tracheal Relations in Superior Mediastinum

Anteriorly: Sternum, thymus, left brachiocephalic vein, origins of brachiocephalic and left common carotid arteries, and arch of aorta

Posteriorly: Esophagus and left recurrent laryngeal nerve

Right side: Azygos vein, right vagus nerve, and pleura

Left side: Arch of aorta, left common carotid, left subclavian arteries, left vagus nerve, left phrenic nerve, and pleura

Bronchi
Right Principal

The **right principal (main) bronchus** is wider, shorter, and more vertical than the left. Before entering the hilum of the right lung, it gives off the **superior lobar bronchus.** On entering the hilum, it divides into a **middle and inferior lobar bronchus.**

Left Principal

The **left principal (main) bronchus** is narrower, longer, and more horizontal than the right. It passes to the left below the arch of the aorta and in front of the esophagus. On entering the hilum of the left lung, it divides into a **superior and an inferior lobar bronchus.**

Pleurae

The pleurae are two serous sacs surrounding and covering the lungs (Fig. 2-8). Each pleura has two parts: (1) a **parietal pleura,** which lines the thoracic wall, covers the thoracic surface of the diaphragm and the lateral surface of the mediastinum, and (2) a **visceral pleura,** which covers the outer surfaces of the lungs and extends into the interlobar fissures. The parietal pleura becomes continuous with the

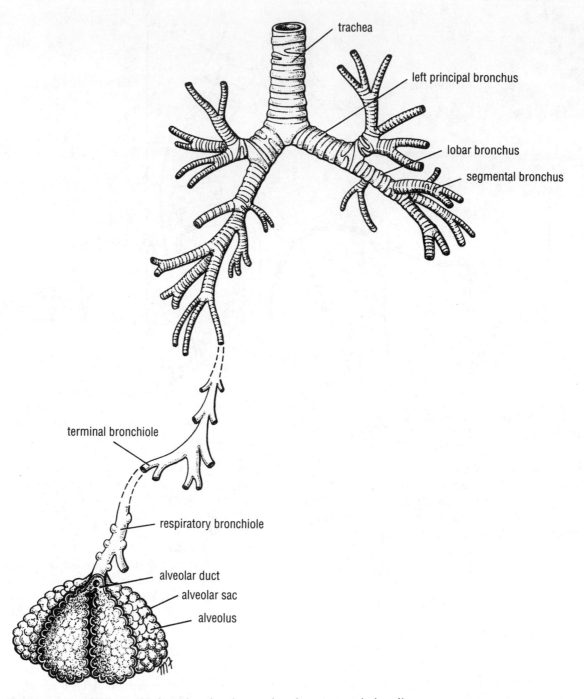

Fig. 2-7.   Trachea, bronchi, bronchioles, alveolar ducts, alveolar sacs, and alveoli. Note the path taken by inspired air from the trachea to the alveoli.

visceral pleura at the **hilum** of each lung. Here, they form a cuff that surrounds the structures entering and leaving the lung at the **lung root.** The **pulmonary ligament** is a loose extension of the cuff below the lung root to allow for movement during respiration.

The **pleural cavity** is a slitlike space that separates the parietal and visceral pleurae. It normally contains a small amount of **pleural fluid** that lubricates the apposing

pleural surfaces. The **costodiaphragmatic recess** is the lowest area of the pleural cavity into which the lungs expand on deep inspiration.

Nerve Supply of the Pleura
Parietal Pleura

The parietal pleura is sensitive to pain, temperature, touch, and pressure. The costal pleura is supplied by intercostal

Thorax

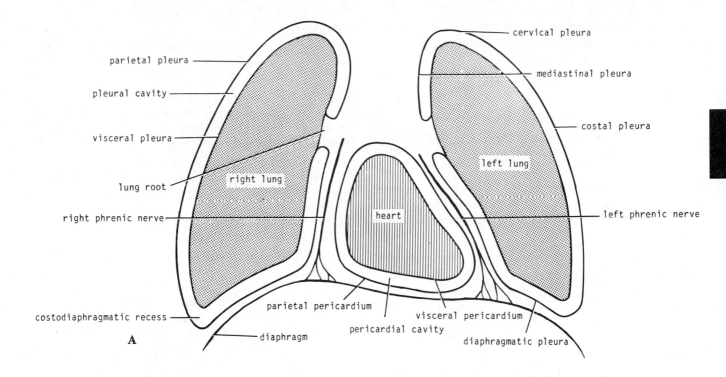

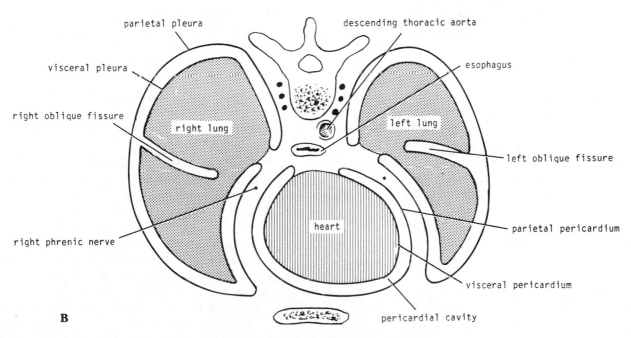

Fig. 2-8.   **A.** Coronal section of the thorax, showing the arrangement of the visceral and parietal layers of pleura and the visceral and parietal layers of serous pericardium. **B.** Horizontal section of the thorax, showing the arrangement of the pleura and pericardium.

nerves; the mediastinal pleura is supplied by the phrenic nerve; and the diaphragmatic pleura is supplied over the domes by the phrenic nerve and around the periphery by the lower intercostal nerves.

Visceral Pleura

The visceral pleura receives an autonomic nerve supply from the pulmonary plexus. It is sensitive to stretch but is insensitive to common sensations such as pain and touch.

Endothoracic Fascia

The endothoracic fascia is a thin layer of loose connective tissue that separates the parietal pleura from the thoracic wall.

Suprapleural Membrane

The suprapleural membrane is a thickening of the endothoracic fascia that covers the dome of the parietal pleura projecting into the root of the neck.

Lungs

The lungs, right and left, are situated on each side of the mediastinum (see Fig. 2-8). Between them in the mediastinum lie the heart and great vessels. The lungs are conical in shape and are covered with visceral pleura. Although the lungs are freely suspended, they are attached by their roots to the mediastinum.

Each lung has a blunt **apex,** which projects upward into the neck (Fig. 2-9) for about 1 in. (2.5 cm) above the clavicle; a concave **base** that sits on the diaphragm; a convex costal surface that corresponds to the concave chest wall; and a concave **mediastinal surface** that is molded to the pericardium and other mediastinal structures. About the middle of this surface is the **hilum,** a depression where the bronchi, vessels, and nerves enter the lung to form the **root.**

The **anterior border** is thin and overlaps the heart; it is here on the left lung that there is a notch called the **cardiac notch.** The **posterior border** is thick and lies beside the vertebral column.

Lobes and Fissures
Right Lung

The right lung is slightly larger than the left and divided into the **upper, middle,** and **lower lobes** by the oblique and horizontal fissures (see Fig. 2-9).

The **oblique fissure** runs from the inferior border upward and backward across the medial and costal surfaces until it cuts the posterior border. The **horizontal fissure** runs horizontally across the costal surface to meet the oblique fissure. The middle lobe is thus a small triangular lobe bounded by the horizontal and oblique fissures.

Left Lung

The left lung is divided by only one fissure, the oblique fissure, into two lobes, the **upper** and **lower lobes.**

Bronchopulmonary Segments

Bronchopulmonary segments are the anatomic, functional, and surgical units of the lungs. Each lobar (secondary) bronchus, which passes to a lobe of the lung, gives off branches called **segmental (tertiary) bronchi** (see Fig. 2-7). Each segmental bronchus then enters a **bronchopulmonary segment.** A bronchopulmonary segment has the following characteristics:

1. It is a subdivision of a lung lobe.
2. It is pyramidal in shape with its apex toward the lung root.
3. It is surrounded by connective tissue.
4. It has a segmental bronchus, a segmental artery, lymph vessels, and autonomic nerves.
5. The segmental vein lies in the connective tissue between adjacent bronchopulmonary segments.
6. A diseased segment, since it is a structural unit, can be removed surgically.

Blood Supply of the Lungs

The bronchi, connective tissue, and visceral pleura are supplied by the bronchial arteries, branches of the descending thoracic aorta. The bronchial veins drain into the azygos and hemiazygos veins.

The alveoli receive deoxygenated blood from the pulmonary arteries. Two pulmonary veins leave each lung root.

Lymph Drainage of the Lungs

The lymph vessels originate in superficial and deep plexuses; they are not present in the alveolar walls. The **superficial plexus** lies beneath the visceral pleura and drains over the surface of the lung toward the hilum, where the lymph vessels enter the **bronchopulmonary nodes.** The **deep plexus** travels along the bronchi and pulmonary vessels toward the hilum of the lung, passing through **pulmonary nodes** located within the lung substance; the lymph then enters the bronchopulmonary nodes in the hilum of the lung. All the lymph from the lung leaves the hilum and drains into the **tracheobronchial nodes** and then into the **bronchomediastinal lymph trunks.**

Nerve Supply of the Lungs

Each lung is supplied by the **pulmonary plexus.** The plexus is formed from branches of the sympathetic trunk and receives parasympathetic fibers from the vagus nerve.

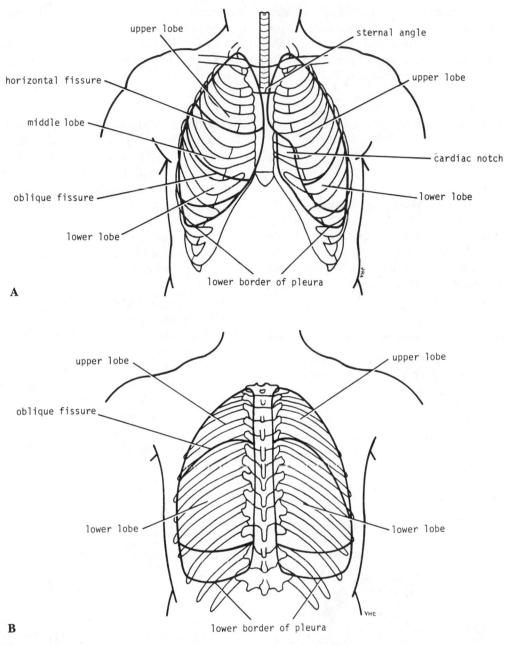

Fig. 2-9. **A.** Surface markings of lungs and parietal pleura on anterior thoracic wall. **B.** Surface markings of lungs and parietal pleura on posterior thoracic wall.

Mechanics of Respiration

Inspiration and expiration are accomplished by the alternate increase and decrease of the capacity of the thoracic cavity (16–20 times/minute in normal resting adults).

Quiet Inspiration

The vertical diameter of the thoracic cavity is increased by the contraction and descent of the diaphragm. The antero-posterior diameter is increased by raising the ribs and thrusting forward the sternum by the contraction of the intercostal muscles. The transverse diameter is increased by raising the ribs (like bucket handles) by contracting the intercostal muscles.

Forced Inspiration

In addition to the muscles used in quiet inspiration, the maximum increase in thoracic capacity is brought about by the contraction of the scalenus anterior and medius (raises first rib), sternocleidomastoid (raises the sternum), and serratus anterior and pectoralis minor (raise the ribs). If the upper limb is fixed, the sternal origin of pectoralis major may also assist in elevating the sternum and ribs.

Quiet Expiration

Quiet expiration is a passive process brought about by the elastic recoil of the lungs and the relaxation of the intercostal muscles and diaphragm.

Forced Expiration

Forced expiration is an active process accomplished by the contraction of the muscles of the anterior abdominal wall (forcing upward the relaxed diaphragm by raising intra-abdominal pressure) and the contraction of the quadratus lumborum (pulling downward the twelfth rib). The latissimus dorsi muscle may also assist in pulling down the lower ribs.

Pericardium

The pericardium is a fibroserous sac that encloses the heart and the roots of the great blood vessels (Fig. 2-10). It lies within the middle mediastinum (see Fig. 2-6).

Fibrous Pericardium

The fibrous pericardium is the fibrous part of the sac. It is strong and limits unnecessary movements of the heart. It

Fig. 2-10. Different layers of the pericardium.

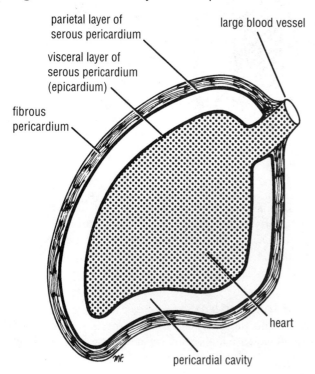

parietal layer of serous pericardium

large blood vessel

visceral layer of serous pericardium (epicardium)

fibrous pericardium

heart

pericardial cavity

fuses above with the walls of the great blood vessels (ascending aorta, pulmonary trunk, superior and inferior venae cavae, and the pulmonary veins). It is firmly attached below to the central tendon of the diaphragm. It is attached anteriorly to the sternum by the **sternopericardial ligaments.**

Serous Pericardium

The serous pericardium has parietal and visceral layers (see Fig. 2-10). The **parietal layer** lines the fibrous pericardium and is reflected around the roots of the great vessels to become continuous with the **visceral layer** that closely covers the heart (**epicardium**). The **pericardial cavity** is the slitlike space between the parietal and visceral layers. The **pericardial fluid** is a small amount of fluid normally present in the pericardial cavity to act as a lubricant to facilitate cardiac movements.

Pericardial Sinuses

The **transverse sinus** is a passage on the posterior surface of the heart that lies between the reflection of serous pericardium around the ascending aorta and pulmonary trunk and the reflection around the great veins (Fig. 2-11). The **oblique sinus** is a recess formed by the reflection of the serous pericardium around the venae cavae and the four pulmonary veins (see Fig. 2-11).

Nerve Supply of Pericardium

The nerve supply of the pericardium consists of the phrenic nerves.

Heart

The heart is a hollow muscular organ that lies within the pericardium in the middle mediastinum. It is somewhat pyramidal in shape and has three surfaces: (1) sternocostal (anterior), (2) diaphragmatic (inferior), and (3) a base (posterior). It also has an **apex** that is directed downward, forward, and to the left (see Fig. 2-11). The apex of the heart lies at the level of the fifth left intercostal space, about 3½in. (9 cm) from the midline.

Structure

The heart is divided by vertical septa into four chambers, the right and left atria and the right and left ventricles. The right atrium lies anterior to the left atrium and the right ventricle lies anterior to the left ventricle (Fig. 2-12). The walls of the heart consist of three layers: (1) outer visceral layer of serous pericardium, the **epicardium;** (2) the middle thick layer of cardiac muscle, the **myocardium;** and (3) the inner thin layer, the **endocardium.** The **skeleton of the heart** consists of fibrous rings that surround the atrioventricular, pulmonary, and aortic orifices and are contin-

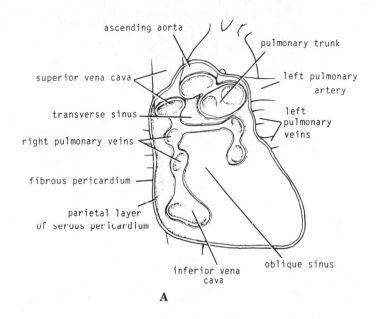

A

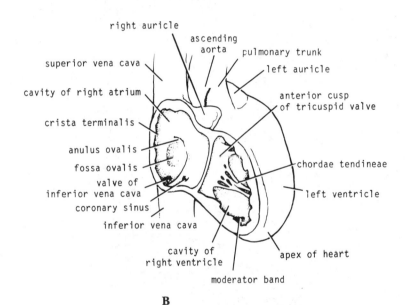

B

Fig. 2-11. **A.** Great blood vessels and interior of pericardium. **B.** Interior of right atrium and right ventricle.

uous with the membranous upper part of the ventricular septum.

Chambers
Right Atrium
The right atrium lies anterior to the left atrium and consists of a main cavity and an **auricle** (see Fig. 2-11). At the junction of the parts is an external vertical groove, the **sulcus terminalis,** which on the inside forms a ridge, the **crista terminalis** (the junction between sinus venosus and right atrium proper). The part of the atrium that lies posterior to

the ridge is smooth walled, whereas the interior of the auricle is roughened by bundles of muscle fibers, the **musculi pectinati.**

Openings
The **superior vena cava** opens into the upper part of the right atrium; there is no valve. The **inferior vena cava** (larger than the superior vena cava) opens into the lower part of the right atrium; there is a rudimentary valve.

The **coronary sinus** opens into the right atrium (see Fig. 2-11) between the inferior vena cava and the atrioventricu-

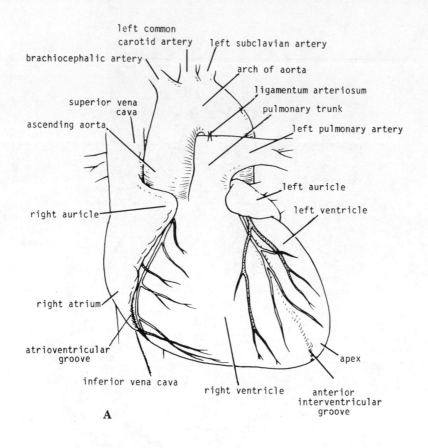

left common
carotid artery   left subclavian artery

brachiocephalic artery   arch of aorta

ligamentum arteriosum

superior vena
cava   pulmonary trunk

ascending aorta   left pulmonary artery

left auricle

right auricle   left ventricle

right atrium

atrioventricular
groove   apex

inferior vena cava   right ventricle   anterior
interventricular
groove

A

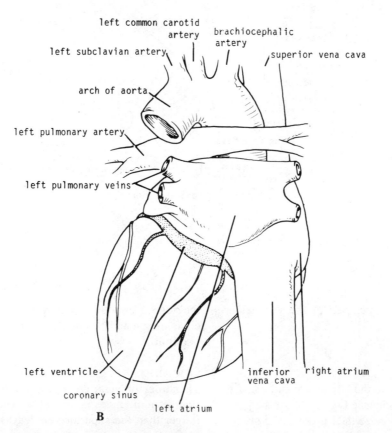

left common carotid
artery   brachiocephalic
artery

left subclavian artery   superior vena cava

arch of aorta

left pulmonary artery

left pulmonary veins

left ventricle   inferior
vena cava   right atrium

coronary sinus   left atrium

B

Fig. 2-12. **A.** Anterior surface of heart and great blood vessels. **B.** Posterior surface
of heart.

lar orifice (see Fig. 2-12); it is guarded by a rudimentary valve.

The **right atrioventricular orifice** lies anterior to the inferior vena caval opening and is guarded by the tricuspid valve (see Fig. 2-11).

There are also many small orifices of small veins that drain the wall of the heart and open directly into the right atrium.

Fetal Remnants

In addition to the rudimentary valve of the inferior vena cava, there are the **fossa ovalis** and **anulus ovalis** (see Fig. 2-11). These structures lie on the atrial septum that separates the right atrium from the left atrium. The fossa ovalis is a shallow depression that is the site of the **foramen ovale** in the fetus. (Before birth oxygenated blood passed through this foramen from the right atrium into the left atrium.) The anulus ovalis forms the upper margin of the fossa.

Right Ventricle

The right ventricle forms the greater part of the anterior surface of the heart; it lies anterior to the left ventricle (see Fig. 2-11). The right ventricle communicates with the right atrium through the atrioventricular orifice and with the pulmonary trunk through the pulmonary orifice. The approach to the pulmonary orifice is funnel-shaped and known as the **infundibulum.**

The walls of the right ventricle are much thicker than those of the right atrium. The internal surface shows projecting ridges called **trabeculae carneae.** There are three types of ridges.

1. **Papillary muscles.** These are attached by their bases to the ventricular wall, and their apices are connected by fibrous chords (the **chordae tendineae**) to the cusps of the tricuspid valve.
2. The ridges are attached at their ends to the ventricular wall and are free in the middle. The **moderate band** is a large ridge. It is attached at its ends to the septal and anterior ventricular walls. It conveys within it the right branch of the atrioventricular bundle, part of the conducting system of the heart.
3. Simple prominent projections.

Tricuspid Valve

The tricuspid valve guards the atrioventricular orifice (see Fig. 2-11). It consists of three cusps whose bases are attached to the fibrous ring of the skeleton of the heart. To their free edges and ventricular surfaces are attached the chordae tendineae. The chordae tendineae connect the cusps to the papillary muscles. The cusps are **anterior, septal,** and **inferior.** The anterior cusp lies anteriorly, the septal cusp lies against the ventricular septum, and the inferior cusp lies inferiorly.

Pulmonary Valve

The pulmonary valve guards the pulmonary orifice. The three semilunar cusps are attached by their curved lower margin to the arterial wall. The open mouths of the cusps are directed upward into the pulmonary trunk. There is one posterior and two anterior semilunar cusps. The **pulmonary sinuses** are three dilatations found at the root of the pulmonary trunk, one being situated external to each cusp (see the section on the aortic valve).

Left Atrium

The left atrium (see Fig. 2-12) lies posterior to the right atrium. It consists of a main cavity and an auricle. The interior of the auricle possesses muscular ridges as on the right side.

Openings

The four **pulmonary veins,** two from each lung, open through the posterior wall; there are no valves. The left **atrioventricular orifice** is guarded by the mitral valve.

Left Ventricle

The left ventricle is situated largely behind the right ventricle (see Fig. 2-12). A small portion, however, projects to the left, forming the left margin of the heart and the heart apex. The left ventricle communicates with the left atrium through the atrioventricular orifice and with the aorta through the aortic orifice. The walls of the left ventricle are three times thicker than those of the right ventricle. There are **trabeculae carneae** and two **papillary muscles;** however, there is no moderator band. The **aortic vestibule** is the part of the ventricle below the aortic orifice.

Mitral Valve

The mitral valve guards the atrioventricular orifice. It consists of two cusps, one anterior and one posterior. Attached to the cusps are chordae tendineae and papillary muscles similar to the tricuspid valve.

Aortic Valve

The aortic valve guards the aortic orifice and, as with the pulmonary valve, the aortic valve consists of three semilunar cusps. One cusp is located on the anterior wall and two are located on the posterior wall. Behind each cusp, the aortic wall bulges to form an **aortic sinus.** The anterior aortic sinus gives origin to the right coronary artery, and the left posterior sinus gives origin to the left coronary artery.

Heart Sounds

The heart makes two sounds: LŪB—DŬP. The first sound is produced by the contraction of the ventricles and the closure of the tricuspid and mitral valves. The second shorter sound is produced by the sharp closure of the aortic and pulmonary valves.

The **tricuspid valve** is best heard over the right half of the lower end of the body of the sternum.

The **mitral valve** is best heard over the apex beat, that is, at the level of the fifth left intercostal space, about 3½ in. (9 cm) from the midline.

The **pulmonary valve** is best heard over the medial end of the second left intercostal space.

The **aortic valve** is best heard over the medial end of the second right intercostal space.

Conducting System

The conducting system of the heart is composed of modified cardiac muscle.

Sinoatrial Node (Pacemaker)

The sinoatrial node initiates the heartbeat. It is situated at the upper part of the sulcus terminalis close to the opening of the superior vena cava (Fig. 2-13). It is usually supplied by the right but sometimes the left coronary artery.

Atrioventricular Node

The atrioventricular node is situated in the lower part of the atrial septum just above the attachment of the septal cusp of the tricuspid valve (see Fig. 2-13). It receives its blood supply from the right coronary artery.

Atrioventricular Bundle

The atrioventricular bundle is continuous with the atrioventricular node above and with the fibers of the Purkinje plexus below. It descends behind the septal cusp of the tricuspid valve on the membranous part of the ventricular septum. On reaching the muscular part of the septum, it divides into two branches, one for each ventricle (see Fig. 2-13). The atrioventricular bundle is the only muscular connection between the myocardium of the atria and the myocardium of the ventricles.

The terminal branches of the atrioventricular bundle spread out into the ventricular walls and become continuous with the fibers of the **Purkinje plexus.**

The atrioventricular bundle is supplied by the right coronary artery; the right terminal branch is supplied by the left coronary artery, and the left terminal branch of the bundle is supplied by the right and left coronary arteries.

Blood Supply of the Heart
Arterial Supply
Right Coronary Artery

The right coronary artery arises from the anterior aortic sinus of the ascending aorta (Fig. 2-14). It descends in the

Fig. 2-13. The heart. The arrangement and position of the different parts of the conducting system are indicated.

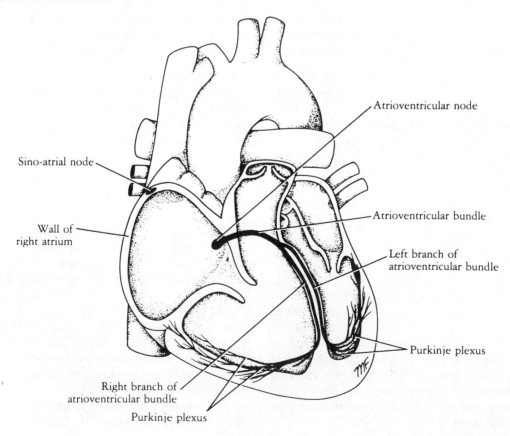

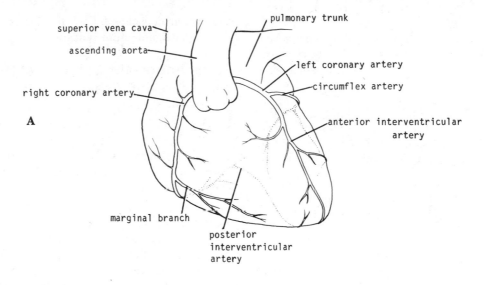

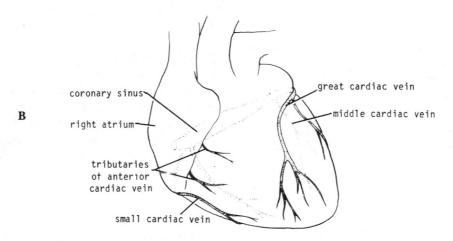

Fig. 2-14. **A.** Coronary arteries. **B.** Cardiac veins.

right atrioventricular groove, giving branches to the right atrium and right ventricle. It ends by anastomosing with the left coronary artery in the posterior interventricular groove.

Branches

1. The **marginal branch** supplies the right ventricle.
2. The **posterior interventricular branch** supplies both ventricles; it ends by anastomosing with the anterior interventricular branch of the left coronary artery.

Left Coronary Artery

The left coronary artery arises from the left posterior aortic sinus of the ascending aorta (see Fig. 2-14). The left coronary artery is larger than the right coronary artery. In the atrioventricular groove, it divides into an anterior interventricular branch and a circumflex branch.

Branches

1. The **anterior interventricular branch** descends to the apex of the heart in the anterior interventricular groove, supplying both ventricles. It passes around the apex

and anastomoses with the posterior interventricular branch of the right coronary artery.
2. The **circumflex branch** winds around the left margin of the heart supplying the left atrium and the left ventricle. It ends by anastomosing with the right coronary artery.

The above arrangement is subject to variation. In the case of "right dominance," the posterior interventricular artery is a large branch of the right coronary artery. In the case of "left dominance," the posterior interventricular artery is a branch of the left coronary artery.

Venous Drainage

Most of the venous blood from the heart wall drains into the right atrium through the coronary sinus (see Fig. 2-14). The remainder drains directly into the right atrium through the **anterior cardiac vein** and small veins, the **venae cordis minimae.**

Coronary Sinus

The coronary sinus lies in the posterior part of the atrioventricular groove and is the largest vein draining the heart wall (see Fig. 2-14). It is a continuation of the great cardiac vein.

Great Cardiac Vein

The great cardiac vein ascends from the apex of the heart in the anterior interventricular groove (see Fig. 2-14). It then enters the atrioventricular groove curving to the left side and back of the heart to empty into the coronary sinus.

Middle Cardiac Vein

The middle cardiac vein runs from the apex of the heart in the posterior interventricular groove and empties into the coronary sinus (see Fig. 2-14).

Small Cardiac Vein

The small cardiac vein accompanies the marginal artery along the inferior border of the heart and empties into the coronary sinus (see Fig. 2-14).

Anterior Cardiac Vein

The anterior cardiac vein drains the anterior surface of the right atrium and the right ventricle and empties directly into the right atrium (see Fig. 2-14).

Nerve Supply of the Heart

The heart is innervated by sympathetic and parasympathetic fibers of the autonomic nervous system via the **cardiac plexuses.** The postganglionic sympathetic fibers terminate on the sinoatrial and atrioventricular nodes, on cardiac muscle fibers, and on coronary arteries. Activation of these nerves results in cardiac acceleration, increased force of contraction of the cardiac muscle, and dilatation of the coronary arteries. The parasympathetic fibers reach the cardiac plexuses in the vagus nerves. Postganglionic fibers terminate on the sinoatrial and atrioventricular nodes and on the coronary arteries. Activation of the parasympathetic nerves results in a reduction in the rate and force of contraction of the heart and a constriction of the coronary arteries.

Gross Anatomic Changes in the Fetal Circulation at Birth

When the umbilical cord is tied, there is an immediate fall in blood pressure in the inferior vena cava (Fig. 2-15). This fact, coupled with the increased left atrial pressure from the increased pulmonary blood flow, causes the **foramen ovale** to close.

The diminished pulmonary vascular resistance associated with inflation of the lungs causes the direction of flow (from right to left) through the ductus arteriosus to be changed to the neonatal route of left to right. The ductus arteriosus constricts as a reaction of its muscle to the raised oxygen tension. It later closes and becomes the **ligamentum arteriosum.** In addition, the wall of the ductus venosus contracts and the lumen is closed. Later the ductus becomes fibrosed to form the **ligamentum venosum.**

Some Important Congenital Defects of the Heart and Great Vessels

Atrial Septal Defect

In 25 percent of individuals, the foramen ovale does not completely close. When the opening is small, it is of no clinical significance. Occasionally the opening is large and results in oxygenated blood from the left atrium passing over into the right atrium (Fig. 2-16).

Ventricular Septal Defects

The ventricular septum is normally formed by a fusion of the small membranous upper part with the larger lower muscular part. Ventricular septal defects are found in the membranous part of the septum. Oxygenated blood passes through the defect from left to right, causing enlargement of the right ventricle.

Tetralogy of Fallot

The following four defects occur with tetralogy of Fallot: (1) large ventricular septal defeat, (2) stenosis of the pulmonary trunk, and (3) exit of the aorta from the heart immediately above the ventricular septal defect. The resulting high blood pressure in the right ventricle causes (4) hypertrophy of the right ventricle (see Fig. 2-16).

Patent Ductus Arteriosus

Normally, by the end of the first month after birth, the ductus arteriosus has closed. The failure to close results in aortic blood passing into the pulmonary artery, which raises the pressure in the pulmonary circulation and causes hypertrophy of the right ventricle (see Fig. 2-16).

Coarctation of the Aorta

Coarctation of the aorta is a narrowing of the aorta just proximal to, opposite, or distal to the site of attachment of the ligamentum arteriosum (see Fig. 2-16). It arises after birth and is believed to be due to the contraction of ductus arteriosus muscle tissue that has been incorporated in the wall of the aorta. When the ductus arteriosus normally contracts, the aortic wall also contracts and the aortic lumen is narrowed. Later fibrosis causes permanent narrowing.

Large Veins of the Thorax

Brachiocephalic Veins

The **right brachiocephalic vein** is formed at the root of the neck by the union of the right subclavian and right internal jugular veins. The **left brachiocephalic vein** has a similar origin on the left side of the root of the neck. It

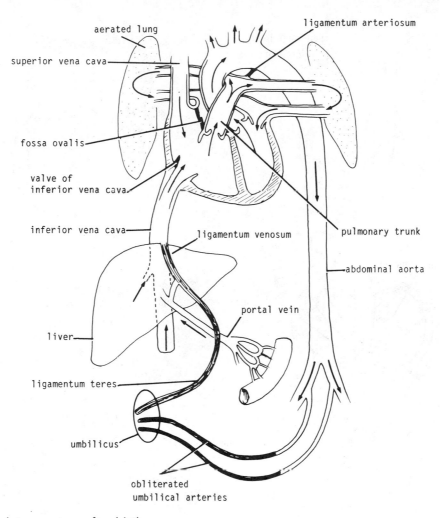

Fig. 2-15. The circulatory system after birth.

passes downward and to the right and joins the right brachiocephalic vein to form the superior vena cava.

Superior Vena Cava

The superior vena cava is a large vein formed by the union of the two brachiocephalic veins. It descends vertically to drain into the right atrium of the heart (see Fig. 2-12). The azygos vein joins the posterior aspect of the superior vena cava.

Azygos Vein

The azygos vein has a variable origin. Commonly, it is formed by the union of the **right ascending lumbar vein** and the **right subcostal vein.** It ascends through the aortic opening in the diaphragm, and at the level of the fifth thoracic vertebra it arches forward to join the superior vena cava. The azygos vein has numerous tributaries that include the eight lower right intercostal veins, the right superior intercostal vein, the superior and inferior hemiazygos veins, and numerous mediastinal veins.

Inferior Hemiazygos Vein

The inferior hemiazygos vein may be formed by the union of the left ascending lumbar vein and the left subcostal vein. It ascends through the left crus of the diaphragm to join the azygos vein at the level of the eighth thoracic vertebra.

Superior Hemiazygos Vein

The superior hemiazygos vein is formed by the union of the fourth to the eighth intercostal veins. It joins the azygos vein at the level of the seventh thoracic vertebra.

Inferior Vena Cava

The inferior vena cava is formed in the abdomen (see p. 75). It perforates the central tendon of the diaphragm, the pericardium, and opens into the right atrium of the heart.

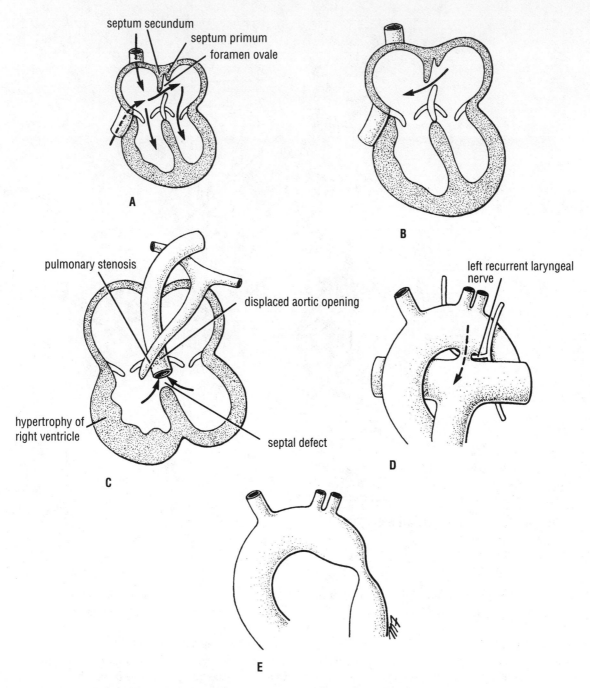

Fig. 2-16. **A.** Normal fetal heart. **B.** Atrial septal defect. **C.** Tetralogy of Fallot. **D.** Patent ductus arteriosus (note close relationship to left recurrent laryngeal nerve). **E.** Coarctation of aorta.

Pulmonary Veins

There are four pulmonary veins, two from each lung (see Fig. 2-12). They carry oxygenated blood from the lungs and open into the left atrium of the heart.

Large Arteries of the Thorax

Aorta

The aorta may be divided into four parts: the ascending aorta, arch of aorta, descending aorta, and abdominal aorta. The first three parts are in the thorax.

Ascending Aorta

The ascending aorta is an artery that arises from the left ventricle and ascends behind the sternum to the level of the sternal angle where it becomes continuous with the arch of the aorta (see Fig. 2-12). At its root it possesses three bulges, the **sinuses of the aorta,** one behind each aortic cusp.

Branches

1. The **right coronary artery** arises from the anterior aortic sinus.
2. The **left coronary artery** arises from the left posterior aortic sinus.

Arch of Aorta

A continuation of the ascending aorta, the arch of the aorta arches upward, backward, and to the left behind the manubrium sterni and in front of the trachea. At the level of the sternal angle it becomes continuous with the descending aorta. The arch is related inferiorly to the root of the left lung, the ligamentum arteriosum, the left recurrent laryngeal nerve, and the bifurcation of the pulmonary trunk.

Branches

Three branches arise from the convex surface of the aortic arch (see Fig. 2-12):

1. The **brachiocephalic artery**
2. The **left common carotid artery**
3. The **left subclavian artery**

Descending Aorta

The descending thoracic aorta is an artery that begins as a continuation of the arch of the aorta on the left side of the lower border of the body of the fourth thoracic vertebra (level of sternal angle). It descends through the posterior mediastinum and reaches the anterior surface of the twelfth thoracic vertebra. Here, it enters the abdomen by passing behind the diaphragm (through the aortic opening) in the midline and becomes continuous with the abdominal aorta.

Branches

1. The **posterior intercostal arteries** pass to the lower nine intercostal spaces on each side.
2. The **subcostal arteries** are given off on each side and run along the lower border of the twelfth rib to enter the abdominal wall.
3. The **pericardial arteries.**
4. The **esophageal arteries.**
5. The **bronchial arteries.**

Pulmonary Trunk

The pulmonary trunk conveys deoxygenated blood from the right ventricle to the lungs. It ascends from the right ventricle and terminates in the concavity of the aortic arch by dividing into right and left pulmonary arteries (see Fig. 2-12).

The **ligamentum arteriosum** is a fibrous band that connects the bifurcation of the pulmonary trunk to the lower surface of the aortic arch (see Fig. 2-12). It is the remains of the **ductus arteriosus** (see p. 39).

Branches

Right and left pulmonary arteries enter the root of their respective lungs where they divide into branches for each lobe.

Lymph Nodes and Vessels of the Thorax

Lymph Nodes
Internal Thoracic Nodes

Five in number, the internal thoracic nodes lie alongside the internal thoracic artery. They drain lymph from the medial quadrants of the breast, the deep structures of the anterior thoracic and abdominal walls down as far as the umbilicus, and the upper surfaces of the liver.

Intercostal Nodes

The intercostal nodes lie close to the heads of the ribs. They receive lymph from the intercostal spaces and from the breast.

Diaphragmatic Nodes

Located on the upper surface of the diaphragm, the diaphragmatic nodes drain lymph from the diaphragm and the upper surface of the liver.

Brachiocephalic Nodes

The brachiocephalic nodes lie alongside the brachiocephalic veins and drain lymph from the thyroid and the pericardium.

Posterior Mediastinal Nodes

The posterior mediastinal nodes lie alongside the descending aorta. They drain lymph from the esophagus, the pericardium, and the diaphragmatic nodes.

Tracheobronchial Nodes

The tracheobronchial nodes lie alongside the trachea and bronchi. They drain lymph from the lungs, the trachea, and the heart.

Lymph Vessels
Thoracic Duct

In the root of the left side of the neck, the thoracic duct conveys lymph to the blood from the following structures: the lower limbs, pelvic cavity, abdominal cavity, left side of the thorax, and left side of the head, neck, and left upper limb.

The thoracic duct begins in the abdomen as a dilated sac, the **cisterna chyli.** It ascends through the aortic opening in the diaphragm on the right side of the descending aorta. It eventually reaches the left border of the esophagus and follows the esophagus to the root of the neck. Here, it turns laterally behind the carotid sheath and enters the beginning of the left brachiocephalic vein.

At its termination, the thoracic duct receives the left jugular, subclavian, and mediastinal lymph trunks, although these trunks may drain independently into neighboring large veins in this region.

Right Lymphatic Duct

In the root of the right side of the neck, the right lymphatic duct conveys lymph to the blood from the following structures: the right side of the head and neck, the right upper limb, and the right side of the thorax. This duct is about ½ in. (1.3 cm) long and opens into the beginning of the right brachiocephalic vein.

Sometimes the right lymphatic duct is absent and the right jugular, right subclavian, and right bronchomediastinal trunks open independently into the great veins at the root of the neck.

Large Nerves of the Thorax

Vagus Nerves

The **right vagus nerve** crosses the anterior surface of the subclavian artery and descends laterally to the trachea and medially to the azygos vein. It runs behind the root of the right lung contributing to the **pulmonary plexus.** It then passes onto the posterior surface of the esophagus and contributes to the **esophageal plexus.** It then leaves the thorax to enter the abdomen behind the esophagus, passing through the esophageal opening of the diaphragm (see Fig. 7-18).

The **left vagus nerve** descends into the thorax between the left common carotid and the left subclavian arteries. It crosses the left side of the aortic arch and descends **behind** the root of the left lung, contributing to the **pulmonary plexus.** The left vagus then passes down on the anterior surface of the esophagus, contributing to the **esophageal plexus.** It enters the abdomen through the esophageal opening of the diaphragm lying in front of the esophagus. (The abdominal course is shown on p. 243.)

Branches

1. **Recurrent laryngeal nerves.** The left recurrent laryngeal nerve arises from the vagus as the latter crosses the arch of the aorta. It hooks beneath the arch behind the ligamentum arteriosum and ascends into the neck between the trachea and the esophagus. (The right recurrent laryngeal nerve arises from the right vagus in the neck and hooks around the subclavian artery.)
2. **Cardiac branches.** Two or three branches arise from the vagus in the neck and descend into the thorax to end in the **cardiac plexuses.**
3. **Pulmonary branches.**
4. **Esophageal branches.**

The vagus nerves thus supply the thorax, the heart, the trachea, the bronchi, the lungs, and the esophagus with parasympathetic and sensory nerve fibers.

Phrenic Nerves

The phrenic nerves arise in the neck from the anterior rami of the third, fourth, and fifth cervical nerves.

The **right phrenic nerve** descends in the thorax along the right side of the superior vena cava and **in front of** the root of the right lung. It then passes over the pericardium to the diaphragm.

The **left phrenic nerve** descends along the left side of the left subclavian artery and crosses the left side of the aortic arch and the left vagus nerve. It passes **in front of** the root of the left lung and descends on the pericardium to the diaphragm.

The phrenic nerve is the **only** motor nerve supply to the diaphragm. It also gives off sensory branches to the pericardium, mediastinal parietal pleura, and the pleura and peritoneum covering the upper and lower surfaces of the central part of the diaphragm.

Thoracic Part of Sympathetic Trunk

The most laterally placed structure in the mediastinum, the thoracic part of the sympathetic trunk runs downward on the heads of the ribs. It leaves the thorax by passing behind the medial arcuate ligament to become continuous with the lumbar part of the sympathetic trunk. The trunk has 11 or 12 segmentally arranged ganglia (see Fig. 1-10). The first ganglion is often fused with the inferior cervical ganglion to form the **stellate ganglion.**

Branches
1. The **white rami communicantes** join each ganglion to a corresponding thoracic spinal nerve. A white ramus contains preganglionic nerve fibers and afferent sensory nerve fibers.
2. The **gray rami communicantes** join each ganglion to a corresponding thoracic spinal nerve. A gray ramus contains postganglionic nerve fibers.
3. The **cardiac, aortic, pulmonary,** and **esophageal branches** arise from the first five ganglia.
4. The **splanchnic nerves** descend and pierce the crura of the diaphragm to supply abdominal viscera. The **greater splanchnic nerve** arises from ganglia 5 to 9, the **lesser splanchnic nerve** arises from ganglia 10 and 11, and the **lowest splanchnic nerve** arises from the last thoracic ganglion.

Esophagus

The esophagus is a muscular tube about 10 in. (25 cm) long that is continuous above with the pharynx opposite the sixth cervical vertebra. It passes through the diaphragm at the level of the tenth thoracic vertebra to join the stomach. It has three constrictions: (1) where it begins, (2) where it is crossed by the left bronchus, and (3) where it pierces the diaphragm.

The esophagus descends through the thorax behind the trachea, the left bronchus, and the left atrium of the heart (see Fig. 2-8).

Blood Supply
Upper third: inferior thyroid artery
Middle third: branches from descending thoracic aorta
Lower third: left gastric artery

The veins from the upper third drain into the inferior thyroid veins, from the middle third into the azygos veins, and from the lower third into the left gastric vein, a tributary of the portal vein.

Lymphatic Drainage
Upper third: deep cervical lymph nodes
Middle third: superior and posterior mediastinal lymph nodes
Lower third: left gastric nodes and celiac nodes in the abdomen

Nerve Supply
The nerve supply is parasympathetic and sympathetic, and includes branches from the recurrent laryngeal nerves, vagus nerves, sympathetic trunks, and greater splanchnic nerves.

Thymus

The thymus is a flattened, bilobed structure lying between the sternum and the pericardium. It continues to grow until puberty. It is an important source of T-lymphocytes.

National Board Type Questions

Match the numbered structures shown in the postero-anterior radiograph of the thorax with the appropriate lettered structures on the right.

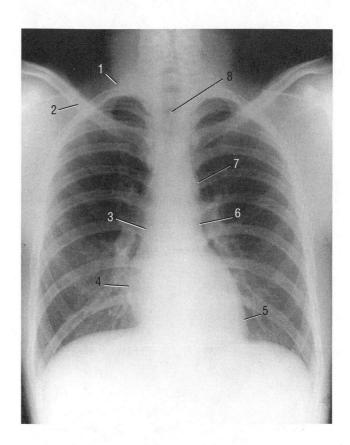

1. Structure 1.	A. Trachea.
2. Structure 2.	B. Pulmonary trunk.
3. Structure 3.	C. Right atrium.
4. Structure 4.	D. Clavicle.
5. Structure 5.	E. Left ventricle.
6. Structure 6.	F. Superior vena cava.
7. Structure 7.	G. Aortic arch.
8. Structure 8.	H. First rib.
	I. Right ventricle.
	J. None of the above.

Match the numbered structures shown in the computed tomographic (CT) scan of the thorax with the appropriate lettered structures on the right.

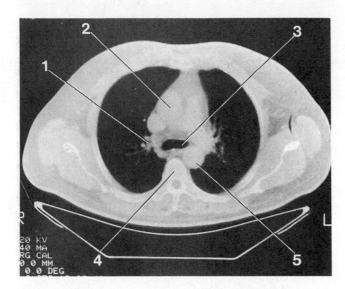

9. Structure 1.	A. Body of thoracic vertebra.
10. Structure 2.	B. Bifurcation of trachea.
11. Structure 3.	C. Descending thoracic aorta.
12. Structure 4.	D. Right pulmonary artery.
13. Structure 5.	E. Ascending aorta.
	F. None of the above.

Match the numbered structures shown in the aortic arch an-
giogram with the appropriate lettered structures on the
right.

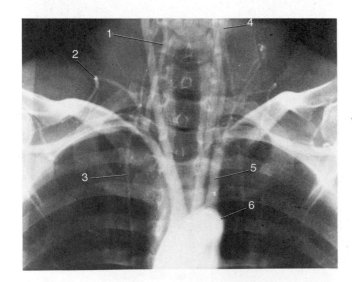

14. Structure 1.	A. Right internal thoracic artery.
15. Structure 2.	B. Left subclavian artery.
16. Structure 3.	C. Right vertebral artery.

17. Structure 4.	D. Suprascapular artery.
18. Structure 5.	E. Left common carotid artery.
19. Structure 6.	F. Arch of aorta.
	G. Brachiocephalic artery.

Match the numbered structures shown in the oblique radiograph of the thorax with the appropriate lettered structures on the right (note barium swallow).

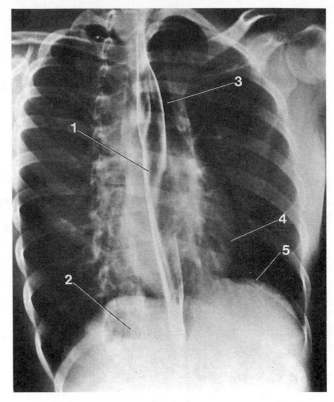

20. Structure 1.	A. Gas bubble in fundus of stomach.
21. Structure 2.	B. Right dome of diaphragm.
22. Structure 3.	C. Esophagus.
23. Structure 4.	D. Trachea.
24. Structure 5.	E. Right ventricle.
	F. Superior vena cava.

Match the missing structure with an appropriate structure
below.

25. The _____ is located on the heart surface immediately alongside the interventricular branch of
the left coronary artery.
26. The _____ runs within the posterior interventricular groove on the heart surface.

A. Great cardiac vein.
B. Middle cardiac vein.
C. Small cardiac vein.
D. Coronary sinus.
E. Anterior cardiac vein.

Match the valves of the heart on the left with the areas on the chest wall where they are best heard with a stethoscope on the right.

27. Tricuspid valve.
28. Mitral valve.
29. Pulmonary valve.
30. Aortic valve.

A. Second right intercostal space.
B. Lower end of sternum.
C. Fifth left intercostal space 3½ in. (9 cm) from the midline.
D. Second left intercostal space.
E. Sixth left intercostal space 3½ in. (9 cm) from the midline.

Match the structures on the left with the regions of the heart on the right. Each lettered region may be used more than once.

31. Moderator band.
32. Sinoatrial node.
33. Bicuspid valve.
34. Aortic vestibule.

A. Left ventricle.
B. Right ventricle.
C. Left atrium.
D. Right atrium.
E. Right side of membranous part of interventricular septum.

Match the structures on the left with the appropriate sympathetic ganglia on the right.

35. Lowest splanchnic nerve.
36. Sympathetic innervation to the head and neck.
37. Lesser splanchnic nerve.
38. Greater splanchnic nerve.

A. T1–4.
B. T5–9.
C. T12.
D. T10–11.
E. None of the above.

Match the nerve on the left with the appropriate statement on the right.

39. Right vagus nerve.
40. Left vagus nerve.
41. Right phrenic nerve.
42. Left phrenic nerve.

A. Passes through aortic hiatus in diaphragm.
B. Lies lateral to trachea and medial to azygos vein.
C. Descends in thorax lateral to superior vena cava.
D. Gives off recurrent branch in the thorax, which passes around the ligamentum arteriosum.
E. Pierces muscular part of diaphragm.

In each of the following questions, answer:

A. If only (1) is correct
B. If only (2) is correct
C. If both (1) and (2) are correct
D. If neither (1) nor (2) is correct

43. Enlarged tracheobronchial lymph nodes on the right side could result from the invasion of malignant tumor cells spreading from the:
    (1) Middle lobe of the right lung.
    (2) Inferior lobe of the right lung.

44. The following structure(s) may impress the esophagus during the passage of a barium meal:
    (1) Left ventricle.
    (2) Left principal (main) bronchus.

45. Coarctation (narrowing) of the distal end of the aortic arch will result in a direct diminution of the amount of blood flow into the:
    (1) Third to eleventh posterior intercostal arteries.
    (2) First and second posterior intercostal arteries.

46. Which of the following statements regarding structures in the intercostal space is (are) correct?
    (1) The anterior intercostal arteries of the lower five intercostal spaces are branches of the musculophrenic artery.
    (2) The sensory fibers in the lower five intercostal nerves supply the skin of the lateral thoracic and anterior abdominal walls.

47. (1) The trachea bifurcates opposite the manubriosternal angle in the midrespiratory position.
    (2) The lower margin of the left lung on full inspiration could extend down in the midclavicular line to the eighth costal cartilage.

48. (1) The apex of the heart can normally be felt in the fifth left intercostal space 3½ in. (9 cm) from the midline.
    (2) The internal thoracic vessels run vertically downward posterior to the costal cartilages 2 in. (5 cm) lateral to the edge of the sternum.

49. (1) The trabeculae carneae are internal surface structures of both the left and right ventricles.
    (2) The pericardial cavity is the potential space between the fibrous and serous pericardia.

50. (1) The coronary arteries are functional end arteries.
    (2) The sinoatrial (SA) node is supplied by both the left and right coronary arteries.

51. With respect to the peripheral nervous system:
    (1) The lung and visceral pleura are **not** sensitive to sensations of temperature, touch, or pressure.
    (2) Motor innervation to the diaphragm is provided by the third, fourth, and fifth cervical spinal nerves and the lower six intercostal nerves.

52. (1) The lymphatic drainage of the posterior surface of the visceral pleura is into the posterior intercostal nodes.
    (2) The lymphatic drainage of the heart is to the tracheobronchial nodes.

53. (1) The thymus lies in the superior mediastinum and may extend superiorly into the neck.
    (2) The thymus receives its arterial supply mainly from the internal thoracic arteries.

54. Concerning the trachea:
    (1) The carina is the name given to the site of bifurcation of the trachea.
    (2) The sensory nerve supply of the mucous mem-

brane lining the lower part of the trachea is from the phrenic nerves.

55. Concerning the ligamentum arteriosum:
    (1) It is the remains of the ductus arteriosus.
    (2) It is formed from the sixth left pharyngeal arch artery.

Select the **best** response.

56. All of the following commonly occur upon inhalation **except:**
    A. The diaphragm descends.
    B. The external intercostal muscles contract.
    C. The abdominal muscles contract pushing abdominal viscera cranially.
    D. The ribs are raised.
    E. The vertical dimension of the thoracic cavity increases.

57. With the patient in the standing position, fluid in the left pleural cavity tends to gravitate down to the:
    A. Oblique fissure.
    B. Cardiac notch.
    C. Costomediastinal recess.
    D. Costodiaphragmatic recess.
    E. Horizontal fissure.

58. When passing a needle through the chest wall into the pleural cavity in the midaxillary line, the following structures will be pierced **except** the:
    A. External intercostal muscle.
    B. Skin.
    C. Parietal pleura.
    D. Levator costarum.
    E. Internal intercostal muscle.

59. Which of the following statements is (are) **correct** concerning the bronchopulmonary segments?
    A. The right lung has three segments.
    B. The arteries supplying each segment are segmental in position.
    C. Each pyramid-shaped segment has its base pointing toward the hilus.
    D. The veins draining each segment are segmental in position.
    E. There are no autonomic nerves within a segment.

60. Which structure(s) compress(es) the posterior surface of the heart during cardiopulmonary resuscitation?
    A. Body of the sternum.
    B. Heads of the ribs.
    C. Tracheal bifurcation.
    D. Inferior vena cava.
    E. Bodies of the thoracic vertebrae.

61. Pericardiocentesis is best achieved by passing a needle through the:
    A. Fourth intercostal space.
    B. Sixth intercostal space at the left paravertebral border.

C. Second intercostal space at the midclavicular line.
    D. Subcostal angle.
    E. Second intercostal space at the right sternal angle.

62. The following statements concerning the main bronchi are true **except:**
    A. The right main bronchus is wider than the left main bronchus.
    B. The right main bronchus is shorter than the left main bronchus.
    C. The right main bronchus is more vertical in position than the left main bronchus.
    D. The left main bronchus passes to the left in front of the esophagus.
    E. The left main bronchus gives off the superior lobar bronchus before entering the hilum of the lung.

63. The following statements concerning the diaphragm are true **except:**
    A. All the muscle of the diaphragm is supplied by the phrenic nerves.
    B. On contraction, the diaphragm increases the intrathoracic pressure.
    C. The right crus provides a sphincter-like mechanism for the esophagus.
    D. On contraction, the diaphragm assists in the return of venous blood to the right atrium and lymph to the thoracic duct.
    E. When viewed from in front the central tendon of the diaphragm lies behind the xiphisternal joint.

64. Which of the following statements is (are) **incorrect** with regard to the root of the right lung?
    A. The right phrenic nerve descends anterior to the lung root.
    B. The right pulmonary artery lies superior to the principal bronchus in the lung root.
    C. The right vagus nerve descends posterior to the lung root.
    D. The azygos vein passes forward below the inferior margin of the lung root.
    E. The lung root is made up of the bronchi, vessels, and nerves that enter or leave the lung.

65. Which statement concerning the lungs is **incorrect?**
    A. Each lung is very elastic, and should the thoracic cavity be opened by a stab wound, it shrinks to one-third or less in volume.
    B. The cardiac notch lies in the lower lobe of the left lung.
    C. The visceral pleura covering each lung lines the fissures that are situated between the lobes.
    D. The apex of each lung extends up into the root of the neck and lies anterior to the lower roots of the brachial plexus.

E. The bronchi, connective tissue, and visceral pleura of the lungs are supplied by the bronchial arteries.

66. With aging the following detrimental changes take place in the thorax **except:**
    A. The ribs and costal cartilages become more rigid.
    B. The elastic tissue in the lungs tends to degenerate.
    C. The manubriosternal joint becomes more mobile.
    D. The thoracic and abdominal muscles tend to atrophy.
    E. The xiphoid process becomes ossified.

67. The following anatomic events take place at the level of the sternal angle (angle of Louis) **except:**
    A. The right and left pulmonary arteries enter the lungs.
    B. The right recurrent laryngeal nerve arises from the right vagus.
    C. The trachea bifurcates.
    D. The ascending aorta becomes continuous with the arch of the aorta.
    E. The second costal cartilages articulate with the sternum.

68. The following structures open into the right atrium **except:**
    A. Superior vena cava.
    B. Coronary sinus.
    C. Anterior cardiac vein.
    D. Inferior vena cava.
    E. Right pulmonary veins.

69. Sudden occlusion of the anterior interventricular branch of the left coronary artery will result in infarction (tissue death) of one of the following areas:
    A. The entire diaphragmatic surface of the left ventricle.
    B. The anterior part of the ventricular septum and the anterior (septal) papillary muscle of the left ventricle.
    C. The posterior wall of the right atrium.
    D. The atrioventricular bundle.
    E. The right auricle.

70. The conducting system of the heart is made up of the following structures **except:**
    A. The Purkinje plexus.
    B. The deep cardiac plexus.
    C. The sinoatrial node.
    D. The atrioventricular bundle.
    E. The atrioventricular node.

71. The following anatomic facts concerning the right coronary artery are true **except:**
    A. Gives rise to a marginal branch.
    B. Passes forward between the right auricle and the pulmonary trunk.
    C. Gives rise to an anterior interventricular branch.
    D. Arises from the anterior aortic sinus.
    E. Descends in the right atrioventricular groove.

72. During fetal life the following facts concerning the circulatory system are true **except:**
    A. The blood passing through the foramen ovale is more oxygenated than that passing through the right atrioventricular orifice.
    B. The direction of blood flow through the ductus arteriosus is from right to left.
    C. The ductus venosus permits blood to bypass the liver.
    D. The blood in the abdominal aorta is relatively more oxygenated than that in the internal carotid arteries.
    E. The valve of the inferior vena cava deflects blood toward the foramen ovale.

73. In the tetralogy of Fallot, there are four cardiac anomalies. Which of the following is characteristic of the condition?
    A. Pulmonary stenosis with hypertrophy of the right ventricle.
    B. Large atrial septal defect.
    C. Hypertrophy of the left ventricle.
    D. Stenosis of the aorta.
    E. High blood pressure in the left ventricle.

74. Pain arising in the heart is commonly referred to the following skin areas **except:**
    A. Up into the neck and jaw.
    B. Down the medial side of the arm.
    C. The point of the shoulder.
    D. The epigastric area.
    E. Over the sternum.

In the next section, study each case history. Select the **best** answer to each question following it.

A 22-year-old man was seen in the emergency department after a street shoot-out. The patient showed signs of severe hemorrhagic shock. A small entrance wound was found in the third left intercostal space about 1 in. (2.5 cm) from the lateral margin of the sternum. There was no exit wound. The left side of his thorax was dull on percussion, and breath sounds were absent on that side of the chest. It was decided to open the chest (thoracotomy) through the fourth left intercostal space. It was found that the left atrium had been perforated by the bullet.

75. When the thoracotomy incision was made to enter the pleural cavity the following structures were incised **except:**
    A. Skin and subcutaneous tissue.
    B. Pectoral muscles and the serratus anterior.
    C. Latissimus dorsi muscle.
    D. External intercostal muscle and the anterior intercostal membrane.
    E. Internal intercostal and transversus thoracis muscles.
    F. Endothoracic fascia and parietal pleura.

76. The following important structures are situated in the region of the thoracotomy incision **except:**
    A. Internal thoracic artery.
    B. Intercostal nerve.
    C. Superior epigastric artery.
    D. Intercostal artery.
    E. Intercostal vein.

A 35-year-old woman was seen in the emergency department following an automobile accident. She had been driving her car with a lap belt on but without the shoulder strap and she hit a utility pole head-on. Examination of the thoracic cage revealed the presence of a fracture of the body of the sternum and the third and fourth left ribs.

77. In this case the following structure(s) located behind the body of the sternum and the left ribs could have been injured **except:**
    A. Pericardium.
    B. Right ventricle of the heart.
    C. Right atrium of the heart.
    D. Left ventricle of the heart.
    E. Phrenic nerves.
    F. Esophagus.
78. The patient's age in such an accident may play a large role in the extent of anatomical injury. Which of the following statements is likely to be correct?
    A. The increased elasticity of fibrous structures in the elderly.
    B. The highly flexible rib cage present in children.
    C. The ossification of the xiphoid cartilage with age.

Answers and Explanations

1. H
2. D
3. F
4. C
5. E
6. B
7. G
8. A
9. D
10. E
11. B
12. A
13. C
14. C
15. D
16. A
17. E
18. B
19. F
20. C
21. A
22. D
23. E
24. B
25. A
26. B
27. B
28. C
29. D
30. A

In answers 27–30, the sites chosen are those where each valve is most clearly heard with the minimum noise from other valves.

31. B
32. D
33. A
34. A
35. C   Sometimes this nerve is absent.
36. A   The preganglionic fibers ascend in the sympathetic trunk and synapse in the superior, middle, or inferior cervical ganglia.
37. D
38. B
39. B
40. D
41. C
42. E
43. C
44. B   The left principal (main) bronchus crosses the anterior surface of the esophagus and causes it to be narrowed at this site.
45. A   Only the third to the eleventh posterior intercostal arteries arise distal to the point of narrowing of the aorta.
46. C
47. C   The bifurcation of the trachea in the midrespiratory position occurs opposite the manubriosternal angle, which is the level of the disc between the fourth and fifth thoracic vertebrae. On full inspiration it may descend to the level of the sixth thoracic vertebra, and on full expiration the bifurcation may rise to the third thoracic vertebra. On full inspiration the lower margin of the lung fills the costodiaphragmatic recess and may reach the eighth costal cartilage in the midclavicular line.
48. A   The internal thoracic vessels lie only a fingerbreadth lateral to the margin of the sternum.
49. A   The pericardial cavity lies between the parietal and visceral layers of serous pericardium.
50. C   Coronary arteries are functional end arteries. Functional end arteries are those that do anastomose, but the

caliber of the anastomosis is insufficient to keep the tissue alive should one of the arteries become occluded.

51. A   The lung and visceral pleura are supplied by autonomic afferent nerves and they are only sensitive to stretch. The motor innervation of the diaphragm is only the phrenic nerve (C3, 4, and 5).

52. B   The lymphatic drainage of the visceral pleura is into the superficial pulmonary plexus, which drains into the bronchopulmonary nodes.

53. C

54. A   The sensory nerve supply of the mucous membrane of the lower part of the trachea is the pulmonary plexus.

55. C

56. C   On inspiration, the abdominal muscles relax to accommodate the abdominal viscera as the diaphragm descends.

57. D   The costodiaphragmatic recess is the most dependent part of the pleural cavity.

58. D   The levator costarum muscles are small accessory muscles of inspiration found on the back of the chest wall alongside the vertebral column.

59. B   The right and left lungs have 10 pulmonary segments each. Each segment has its base pointing toward the pleural surface. The veins draining each segment are intersegmental in position. Each segment has autonomic nerves.

60. E

61. D   Pericardial fluid may be aspirated from the pericardial cavity by inserting the needle to the left of the xiphoid process in an upward and backward direction at an angle of 45 degrees to the skin. Due to the presence of the cardiac notch, the needle misses the pleura and lungs and pierces the pericardium.

62. E   The right principal (main) bronchus gives off the superior lobar bronchus before entering the hilum of the right lung. The left principal bronchus gives off the superior lobar bronchus after it enters the left lung.

63. B   The diaphragm on contraction descends and thus increases the vertical diameter of the thoracic cavity and reduces the intrathoracic pressure.

64. D   The azygos vein runs forward above the right lung root to enter the superior vena cava.

65. B   The cardiac notch of the left lung lies in the upper lobe.

66. C   The manubriosternal joint becomes less mobile with age and eventually the manubrium becomes joined to the body of the sternum by bone.

67. B   At the level of the sternal angle, the left recurrent laryngeal nerve arises from the left vagus, hooks beneath the arch of the aorta, and ascends to the neck. The right recurrent laryngeal nerve arises from the right vagus in the neck and hooks beneath the right subclavian artery; the right recurrent laryngeal nerve does not enter the thorax.

68. E   The two right pulmonary veins and the two left pulmonary veins return oxygenated blood to the heart by entering the left atrium.

69. B   The diaphragmatic surface of the left ventricle receives its blood supply from the posterior interventricular branch of the right coronary artery as well as from the anterior interventricular branch of the left coronary artery. The posterior wall of the right atrium, the right auricle, and the atrioventricular bundle are all supplied by the right coronary artery.

70. B   The deep cardiac plexus is an autonomic nerve plexus that lies outside the heart below the arch of the aorta.

71. C   The anterior interventricular artery is a branch of the left coronary artery.

72. D   The most richly oxygenated blood reaches the brain via the internal carotid arteries. The abdominal aorta contains blood that has come from the right ventricle, the pulmonary trunk, and the ductus arteriosus; this blood is poorly oxygenated.

73. A   The tetralogy of Fallot has pulmonary stenosis, ventricular septal defect, exit of the aorta from the heart immediately above the ventricular septal defect, and hypertrophy of the right ventricle. The stenosis of the pulmonary trunk and the leakage of left ventricular blood through the ventricular septal defect are responsible for the raised blood pressure in the right ventricle and the resulting right ventricular hypertrophy.

74. C   Pain arising from the myocardium ascends to the central nervous system through the cardiac branches of the sympathetic trunk and enters the spinal cord through the posterior roots of the upper four thoracic nerves. The pain is not felt in the heart, but is referred to skin areas supplied by the upper four intercostal nerves and by the intercostobrachial nerve (T2). The intercostobrachial nerve communicates with medial cutaneous nerves of the upper part of the arm. A certain amount of spread of nervous information occurs within the central nervous system, for the pain is sometimes felt in the neck and jaw and the epigastrium. Cardiac pain is not referred to the point of the shoulder.

75. C   The latissimus dorsi is a sheet of muscle that covers the back of the thoracic cage and is inserted into the floor of the bicipital groove of the humerus. It is located too far posteriorly to be damaged by this exploratory thoracotomy.

76. C   The internal thoracic artery terminates in the sixth intercostal space by dividing into the musculophrenic artery and the superior epigastric artery. The superior

epigastric artery quickly enters the anterior abdominal wall.

77. F   The heart can be squeezed between the sternum and the vertebral column when the thorax is subjected to a severe frontal impact. The right atrium, right ventricle, or part of the left ventricle is commonly injured. The pericardium may be punctured by a fractured rib or rarely may be ruptured by a sudden blow.

If pericardial rupture should occur it tends to take place along the lateral margins with involvement of the phrenic nerves.

78. B   The highly flexible rib cage present in children makes myocardial bruising a common occurrence. In the elderly toughness of the pericardium and diminished elasticity may make rupture more common.

Abdomen

Anterior Abdominal Wall

Skin
Nerve Supply

The cutaneous nerve supply to the anterior abdominal wall is derived from the anterior rami of the lower six thoracic and the first lumbar nerves (Figs. 3-1 and 3-2). The thoracic nerves are the lower five intercostal and the subcostal nerves, and the first lumbar nerve is represented by the iliohypogastric and ilioinguinal nerves. The dermatome of T7 is located in the epigastrium over the xiphoid process, that of T10 includes the umbilicus, and that of L1 lies just above the inguinal ligament and the symphysis pubis.

Blood Supply
Arteries

The skin near the midline is supplied by branches of the superior and inferior epigastric arteries; the skin of the flanks is supplied by branches from the intercostal, lumbar, and deep circumflex iliac arteries (see Fig. 3-1).

Veins

The venous drainage passes above into the axillary vein via the lateral thoracic vein and below into the femoral vein via the superficial epigastric and great saphenous veins (note the important indirect connection between the superior and inferior venae cavae). The small **paraumbilical veins** connect the veins of the umbilical region along the ligamentum teres to the portal vein (note the important portal-systemic anastomosis).

Lymph Drainage

The cutaneous lymph vessels above the level of the umbilicus drain upward into the anterior axillary lymph nodes. The vessels below this level drain downward into the superficial inguinal nodes.

Superficial Fascia

The superficial fascia may be divided into the superficial **fatty layer (fascia of Camper)** and a deep **membranous layer (Scarpa's fascia).**

The fatty layer is continuous with the superficial fascia over the rest of the body and may be extremely thick in obese individuals. The membranous layer fades out laterally and above. Inferiorly, it passes over the inguinal ligament to fuse with the deep fascia of the thigh (fascia lata) about one fingerbreadth below the inguinal ligament. In the midline it is not attached to the pubis, but forms a tubular sheath for the penis (clitoris). In the perineum it is attached on each side to the margins of the pubic arch and is known as **Colles' fascia.** Posteriorly, it fuses with the perineal body and the posterior margin of the perineal membrane.

In the scrotum the fatty layer of superficial fascia is represented as a thin layer of smooth muscle, the **dartos muscle.**

Deep Fascia

In the anterior abdominal wall, the deep fascia is merely a thin layer of areolar tissue covering the muscles.

Muscles of the Anterior Abdominal Wall

The muscles of the anterior abdominal wall consist mainly of three broad thin sheets that are aponeurotic in front; they are the **external oblique, internal oblique, and transversus** from exterior to interior (Fig. 3-3). On either side of the midline anteriorly there is, in addition, a wide vertical muscle, the **rectus abdominis** (Fig. 3-4). As the aponeuroses of the three sheets pass forward, they enclose the rectus abdominis to form the **rectus sheath.**

In the lower part of the rectus sheath, there may be present a small muscle called the **pyramidalis.**

The **cremaster muscle,** which is derived from the lower fibers of the internal oblique, passes inferiorly as a covering of the spermatic cord and enters the scrotum.

The muscles of the anterior abdominal wall are shown in Table 3-1.

Rectus Sheath

The rectus sheath (Fig. 3-5) is a long fibrous sheath that encloses the rectus abdominis muscle and pyramidalis muscle (if present) and contains the anterior rami of the lower

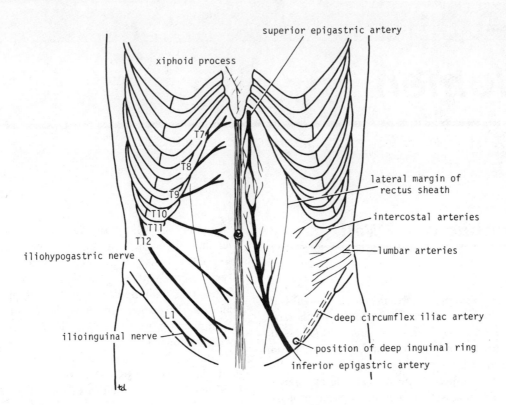

Fig. 3-1. On left, segmental innervation of anterior abdominal wall. On right, arterial supply to anterior abdominal wall.

six thoracic nerves and the superior and inferior epigastric vessels and lymph vessels. It is formed mainly by the aponeuroses of the three lateral abdominal muscles. The internal oblique aponeurosis splits at the lateral edge of the rectus abdominis to form two laminae: one passes anteriorly to the rectus and one passes posteriorly. The aponeurosis of the external oblique fuses with the anterior lamina, and the transversus aponeurosis fuses with the posterior lamina. At the level of the anterior superior iliac spines, all three aponeuroses pass anteriorly to the rectus muscle leaving the sheath deficient posteriorly below this level. The lower crescent-shaped edge of the posterior wall of the sheath is called the **arcuate line.** All three aponeuroses fuse with each other and with their fellows of the opposite side in the midline between the right and left recti muscles to form a fibrous band called the **linea alba.** The linea alba extends from the xiphoid process above to the pubic symphysis below.

Note that the anterior wall of the retus sheath is firmly attached to the tendinous intersections of the rectus abdominis muscle, whereas the posterior wall of the sheath has no attachment to the muscle. The transverse **tendinous intersections,** which divide the rectus abdominis muscle into segments, are usually three in number: one at the level of the xiphoid process, one at the level of the umbilicus, and one between these two.

Linea Semilunaris

The linea semilunaris is the lateral edge of the rectus abdominis muscle; it crosses the costal margin at the tip of the ninth costal cartilage.

Conjoint Tendon

The internal oblique muscle has a lower free border that arches over the spermatic cord (or round ligament of the uterus) and then descends behind it to be attached to the pubic crest and the pectineal line. Near their insertion, the lowest tendinous fibers are joined by similar fibers from the transversus abdominis to form the **conjoint tendon.** The conjoint tendon strengthens the medial half of the posterior wall of the inguinal canal.

Inguinal Ligament

The inguinal ligament (see Fig. 3-3) connects the anterior superior iliac spine to the pubic tubercle. It is formed by the lower border of the aponeurosis of the external oblique muscle, which is folded back upon itself. From the medial end of the ligament, the **lacunar ligament** extends backward and upward to the pectineal line on the superior ramus of the pubis where it becomes continuous with the **pectineal ligament** (thickening of periosteum). The lower border of the inguinal ligament is attached to the deep fascia of the thigh, the **fascia lata.**

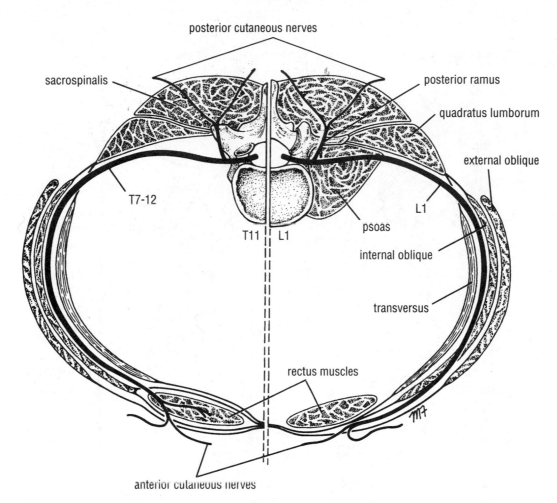

posterior cutaneous nerves

sacrospinalis

posterior ramus

quadratus lumborum

external oblique

T7-12

L1

psoas

T11   L1

internal oblique

transversus

rectus muscles

anterior cutaneous nerves

Fig. 3-2. Cross section of abdomen, showing courses of lower thoracic and first lumbar nerves.

Fascia Transversalis

The fascia transversalis is a thin layer of fascia that lines the transversus muscle and is continuous with a similar layer lining the diaphragm and the iliacus muscle. The **femoral sheath** for the femoral vessels is formed from the fascia transversalis and the fascia iliaca.

Inguinal Canal

The inguinal canal (Fig. 3-6) is an oblique passage through the lower part of the anterior abdominal wall and is present in both sexes. It allows structures to pass to and from the testis to the abdomen in the male. In the female it permits the passage of the round ligament of the uterus from the uterus to the labium majus.

The canal is about 1½ in. (4 cm) long in the adult and extends from the deep inguinal ring, a hole in the fascia transversalis, downward and medially to the superficial inguinal ring, a hole in the aponeurosis of the external oblique muscle. It lies parallel to and immediately above the inguinal ligament.

The **deep inguinal ring,** an oval opening in the fascia transversalis, lies about ½ in. (1.3 cm) above the inguinal ligament. The margins of the ring give attachment to the **internal spermatic fascia.**

The **superficial inguinal ring** is a triangular-shaped defect in the aponeurosis of the external oblique muscle and lies immediately above and medial to the pubic tubercle. The margins of the ring give attachment to the **external spermatic fascia.**

Walls

Anterior wall: External oblique aponeurosis, reinforced laterally by origin of internal oblique from inguinal ligament (see Fig. 3-6)

Posterior wall: Conjoint tendon medially, fascia transversalis laterally (see Fig. 3-6)

Roof or superior wall: Arching fibers of internal oblique and transversus (see Fig. 3-6)

Floor or inferior wall: Inguinal ligament and lacunar ligament

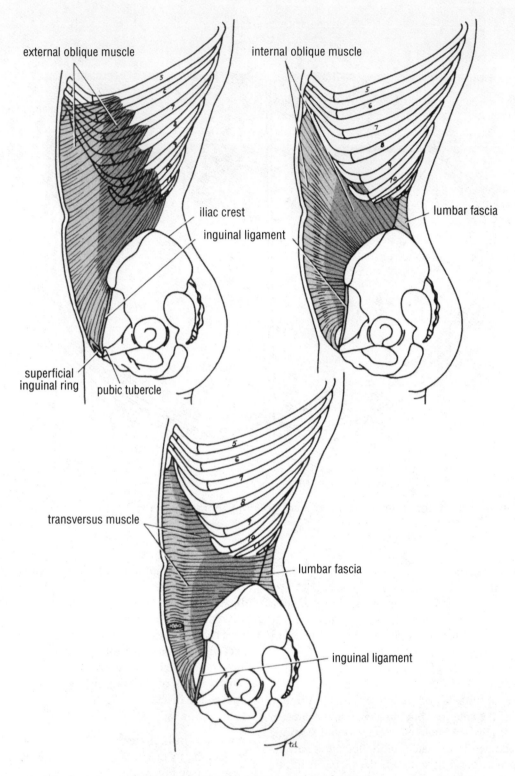

Fig. 3-3.   External oblique, internal oblique, and transversus muscles of anterior abdominal wall.

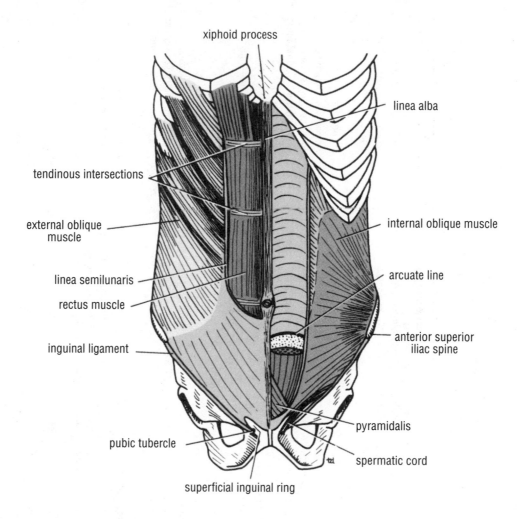

Fig. 3-4.   Anterior view of rectus abdominis muscle and rectus sheath. On left, anterior wall of sheath has been partly removed, revealing rectus muscle with its tendinous intersections. On right, posterior wall of rectus sheath is shown. Edge of arcuate line is shown at level of anterior superior iliac spine.

Function of the Inguinal Canal

The inguinal canal allows structures to pass to and from the testis to the abdomen in the male. (Normal spermatogenesis only takes place if the testis leaves the abdominal cavity to enter a cooler environment in the scrotum.) In the female, the smaller canal permits the passage of the round ligament of the uterus from the uterus to the labium majus. In both sexes, the canal also transmits the ilioinguinal nerve.

Mechanics of the Inguinal Canal

The inguinal canal is a site of potential weakness in both sexes. On coughing and straining, as in micturition, defecation, and parturition, the arching lowest fibers of the internal oblique and transversus abdominis muscles contract, flattening out the arch so that the roof of the canal is lowered toward the floor and the canal is virtually closed.

Inguinal Hernia

The inguinal hernia occurs above the inguinal ligament, whereas a femoral hernia occurs below the inguinal ligament. Inguinal hernias are of two types: indirect and direct.

Indirect Inguinal Hernia

1. The hernial sac is the remains of the processus vaginalis.
2. It is more common than a direct inguinal hernia.
3. It is much more common in the male as compared with the female.
4. It is more common on the right side.
5. It is most common in children and young adults.
6. The hernial sac enters the inguinal canal through the deep inguinal ring and **lateral** to the inferior epigastric vessels. The neck of the sac is narrow.
7. The hernial sac may extend through the superficial inguinal ring above and medial to the pubic tubercle

Table 3-1   Muscles of the Anterior Abdominal Wall

Name of muscle	Origin	Insertion	Nerve supply	Action
External oblique	Lower eight ribs	Xiphoid process, linea alba, pubic crest, pubic tubercle, iliac crest	Lower six thoracic nerves, iliohypogastric and ilioinguinal nerves (L1)	Compresses abdominal contents; assists in flexing and rotation of trunk; pulls down ribs in forced expiration
Internal oblique	Lumbar fascia, iliac crest, lateral two-thirds of inguinal ligament	Lower three ribs and costal cartilages, xiphoid process, linea alba, symphysis pubis; forms conjoint tendon with transversus	Lower six thoracic nerves, iliohypogastric and ilioinguinal nerves (L1)	Compresses abdominal contents; assists in flexing and rotation of trunk; pulls down ribs in forced expiration
Transversus	Lower six costal cartilages, lumbar fascia, iliac crest, lateral third of inguinal ligament	Xiphoid process, linea alba, symphysis pubis; forms conjoint tendon with internal oblique	Lower six thoracic nerves, iliohypogastric and ilioinguinal nerves (L1)	Compresses abdominal contents
Rectus abdominis	Symphysis pubis and pubic crest	Fifth, sixth, and seventh costal cartilages and xiphoid process	Lower six thoracic nerves	Compresses abdominal contents and flexes vertebral column; accessory muscle of expiration
Pyramidalis (often absent)	Anterior surface of pubis	Linea alba	Twelfth thoracic nerve	Tenses the linea alba
Cremaster	Lower margin of internal oblique muscle	Pubic crest	Genital branch of genito-femoral nerve (L1, 2)	Retracts testis

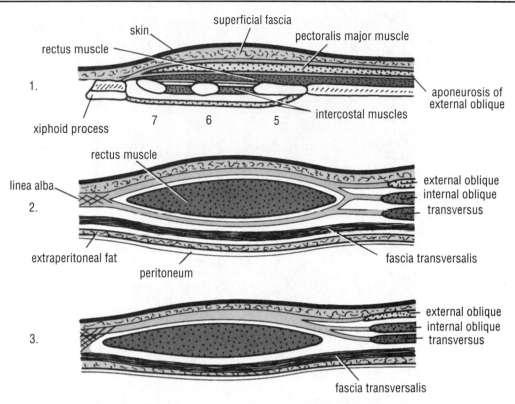

Fig. 3-5. Transverse sections of rectus sheath seen at three levels: (1) above costal margin, (2) between costal margin and level of anterior superior iliac spine, (3) below level of anterior superior iliac spine and above pubis.

(femoral hernia below and lateral to the pubic tubercle).

8. The hernial sac may extend down into the scrotum or labium majus.

Direct Inguinal Hernia

1. This hernia is common in elderly men with weak abdominal muscles. It is rare in women.
2. The hernial sac bulges forward through the posterior wall of the inguinal canal **medial** to the inferior epigastric vessels.
3. The neck of the hernial sac is wide.

Spermatic Cord

The spermatic cord is a collection of the following structures that pass through the inguinal canal to and from the testis:

1. Vas deferens
2. Testicular artery
3. Testicular veins (pampiniform plexus)
4. Testicular lymph vessels
5. Autonomic nerves
6. Processus vaginalis (remains of)
7. Cremasteric artery
8. Artery of the vas deferens
9. Genital branch of the genitofemoral nerve, which supplies the cremaster muscle.

Coverings of the Spermatic Cord. There are three concentric layers of fascia derived from the layers of the anterior abdominal wall.

1. **External spermatic fascia** derived from the external oblique muscle and attached to the margins of the superficial inguinal ring
2. **Cremasteric fascia** derived from the internal oblique muscle
3. **Internal spermatic fascia** derived from the fascia transversalis and attached to the margins of the deep inguinal ring

Processus Vaginalis

The processus vaginalis is a peritoneal diverticulum formed in the fetus that passes through the layers of the lower part of the anterior abdominal wall to form the inguinal canal. The **tunica vaginalis** is the lower expanded part of the processus vaginalis. Normally, the cavity of the tunica vaginalis becomes shut off from the upper part of the processus and the peritoneal cavity just before birth. The tunica vaginalis is thus a closed sac, invaginated from behind by the testis.

Gubernaculum Testis

The gubernaculum testis is a muscular-ligamentous cord that connects the fetal testis to the floor of the developing

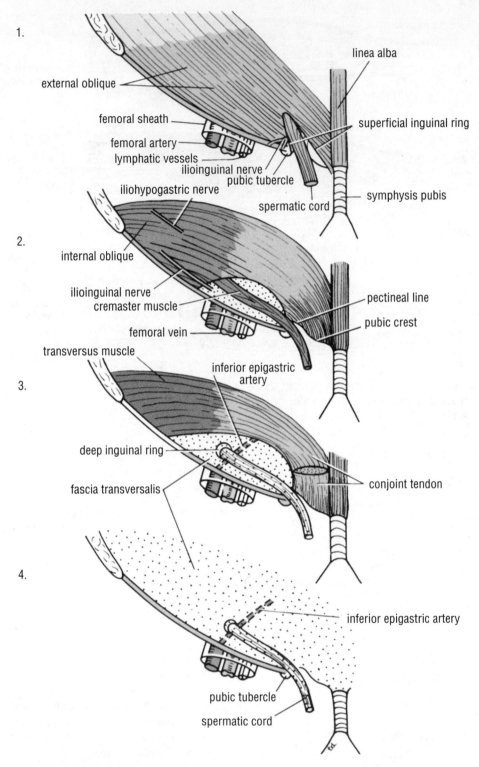

Fig. 3-6. Inguinal canal, showing arrangement of (1) external oblique muscle, (2) internal oblique muscle, (3) transversus muscle, and (4) fascia transversalis. Note that anterior wall of canal is formed by external oblique and internal oblique, and posterior wall is formed by fascia transversalis and conjoint tendon. Deep inguinal ring lies lateral to inferior epigastric artery.

scrotum. It plays an important role in the descent of the testis. It is homologous to the female round ligament of the ovary and the round ligament of the uterus.

Scrotum

The scrotum is an outpouching of the lower part of the anterior abdominal wall and contains the testes, the epididymides, and the lower ends of the spermatic cords.

The wall of the scrotum has the following layers:

1. Skin
2. Superficial fascia: dartos muscle (smooth muscle) replaces fatty layer
3. External spermatic fascia from the external oblique
4. Cremasteric fascia from the internal oblique (cremasteric muscle supplied by genital branch of genitofemoral nerve)
5. Internal spermatic fascia from the fascia transversalis
6. Tunica vaginalis: a closed sac that covers the anterior, medial, and lateral surfaces of each testis

Testes

The testes are paired ovoid organs responsible for the production of spermatozoa and testosterone. Normal spermatogenesis will only take place at a temperature lower than that of the abdominal cavity, hence the descent of the testes into the scrotum. The **tunica albuginea** is the outer fibrous capsule of the testis.

Epididymides

The epididymis on each side lies posterior to the testis and has a **head, body,** and **tail.** It is a coiled tube about 20 ft. (6 m) long. The vas deferens emerges from the tail.

Blood Supply of the Testis and Epididymis

The testicular artery is a branch of the abdominal aorta. The testicular vein emerges from the testis and the epididymis as a venous network, the **pampiniform plexus.** This becomes reduced to a single vein as it ascends through the inguinal canal. The right testicular vein drains into the inferior vena cava, and the left vein joins the left renal vein.

Lymph Drainage of the Testis and Epididymis

Para-aortic lymph nodes on the side of the aorta at the level of the first lumbar vertebra.

Nerves of the Anterior Abdominal Wall

The nerves of the anterior abdominal wall are the anterior rami of the lower six thoracic and the first lumbar nerves (see Fig. 3-2). They run downward and forward between the internal oblique and transversus muscles. They supply the skin, the muscles, and the parietal peritoneum of the anterior abdominal wall. Note that the lower six thoracic nerves pierce the posterior wall of the rectus sheath. The first lumbar nerve is represented by the **iliohypogastric** and **ilioinguinal nerves,** which do not enter the rectus sheath. The iliohypogastric nerve pierces the external oblique aponeurosis above the superficial inguinal ring, and the ilioinguinal nerve passes through the inguinal canal to emerge through the ring.

Blood Supply of the Anterior Abdominal Wall
Arteries

The **superior epigastric artery** arises from the internal thoracic artery and enters the rectus sheath. It descends behind the rectus muscle and supplies the upper central part of the anterior abdominal wall. It anastomoses with the inferior epigastric artery.

The **inferior epigastric artery** arises from the external iliac artery above the inguinal ligament. It runs medial to the deep inguinal ring and enters the rectus sheath. It ascends behind the rectus muscle and supplies the lower central part of the anterior abdominal wall. It anastomoses with the superior epigastric artery.

The **deep circumflex iliac artery** is a branch of the external iliac artery. It runs upward and laterally toward the anterior superior iliac spine. It supplies the lower lateral part of the abdominal wall.

The **lower two posterior intercostal arteries** from the descending thoracic aorta and the **four lumbar arteries** from the abdominal aorta supply the lateral part of the anterior abdominal wall. The **superficial epigastric artery,** the **superficial circumflex iliac artery,** and the **superficial external pudendal artery** branches of the femoral artery also supply the lower part of the anterior abdominal wall.

Veins

The **superior and inferior epigastric veins** and the **deep circumflex iliac veins** follow the arteries and drain into the internal thoracic and external iliac veins. The **posterior intercostal veins** drain into the azygos veins, and the **lumbar veins** drain into the inferior vena cava. The **superficial epigastric,** the **superficial circumflex iliac,** and the **superficial external pudendal veins** drain into the great saphenous vein and thence to the femoral vein. The **thoracoepigastric vein** is the name given to the anastomoses between the lateral thoracic vein and the superficial epigastric vein, a tributary of the great saphenous vein. This vein provides an alternative path for the venous blood should obstruction occur in the superior or inferior vena cava.

Lymph Drainage of the Anterior Abdominal Wall

The cutaneous lymph vessels above the level of the umbilicus drain upward into the anterior axillary lymph nodes. The vessels below this level drain downward into the superficial inguinal nodes. The deep lymph vessels follow the arteries and drain into the internal thoracic, external iliac, posterior mediastinal, and para-aortic (lumbar) nodes.

Peritoneum

The peritoneum is the serous membrane lining the abdominal and pelvic cavities and clothing the viscera (Fig. 3-7). It may be regarded as a balloon into which organs are pressed into from the outside. The **parietal layer** lines the walls of the abdominal and pelvic cavities, and the **visceral layer** covers the organs. The potential space between the parietal and visceral layers of peritoneum is called the **peritoneal cavity.** In the male this is a closed cavity, but in the female there is a communication with the exterior through the uterine tubes, the uterus, and the vagina.

The peritoneal cavity may be divided into two parts, the greater sac and the lesser sac (see Fig. 3-7). The **greater sac** is the main compartment of the peritoneal cavity and extends from the diaphragm down into the pelvis. The **lesser sac** is smaller and lies behind the stomach. The greater and lesser sacs are in free communication with one another through the **epiploic foramen.** The peritoneum secretes a small amount of serous fluid, which lubricates the surfaces of the peritoneum and facilitates free movement between the viscera.

Fig. 3-7.   **A.** Sagittal section of female abdomen, showing arrangement of peritoneum. **B.** Transverse section of abdomen, showing arrangement of peritoneum. Note that this section is viewed from below.

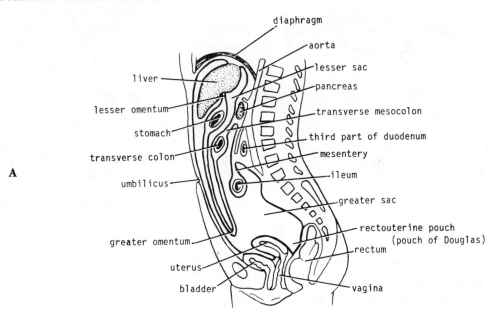

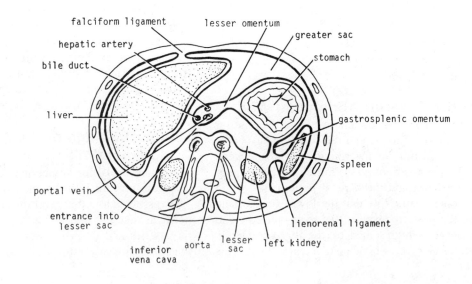

Peritoneal Ligaments, Omenta, and Mesenteries

The peritoneal ligaments, omenta, and mesenteries permit blood, lymph vessels, and nerves to reach the viscera.

Peritoneal Ligaments

Peritoneal ligaments are two-layered folds of peritoneum that connect solid viscera to the abdominal walls. The liver, for example, is connected to the diaphragm by the **falciform ligament,** the **coronary ligament,** and the **right and left triangular ligaments** (see Fig. 3-12).

Omenta

Omenta are two-layered folds of peritoneum that connect the stomach to another viscus. The **greater omentum** connects the greater curvature of the stomach to the transverse colon (see Fig. 3-7). It hangs down like an apron in front of the coils of the small intestine and is folded back on itself. The **lesser omentum** suspends the lesser curvature of the stomach to the fissure for the ligamentum venosum and the porta hepatis of the liver (see Fig. 3-7; see also Fig. 3-12). The **gastrosplenic omentum** (ligament) connects the stomach to the hilus of the spleen.

Mesenteries

Mesenteries are two-layered folds of peritoneum connecting parts of the intestines to the posterior abdominal wall (e.g., the **mesentery of the small intestine,** the **transverse mesocolon,** and the **sigmoid mesocolon**) (see Fig. 3-7).

Lesser Sac

The lesser sac lies behind the stomach and lesser omentum (see Fig. 3-7). It extends upward as far as the diaphragm and downward between the layers of the greater omentum. The left margin is formed by the spleen, the gastrosplenic omentum, and the lienorenal ligament. The right margin of the sac opens into the greater sac (i.e., the main part of the peritoneal cavity) through the **epiploic foramen.**

Boundaries of Epiploic Foramen

Anteriorly: free border of lesser omentum, bile duct, hepatic artery, and portal vein

Posteriorly: inferior vena cava

Superiorly: caudate process of caudate lobe of liver

Inferiorly: first part of duodenum

Peritoneal Fossae, Spaces, and Gutters

Duodenal Fossae, Cecal Fossae, Spaces, and Gutters

Duodenal Fossae

Close to the duodenojejunal junction, there **may** be four small pouches of peritoneum called the **superior duodenal fossa,** the **inferior duodenal fossa,** the **paraduodenal fossa,** and the **retroduodenal fossa.**

Cecal Fossae

Folds of peritoneum close to the cecum produce three peritoneal fossae called the **superior ileocecal, the inferior ileocecal, and the retrocecal fossae.**

Subphrenic Spaces

Subphrenic spaces lie between the diaphragm and the liver and are called the **right** and **left anterior** and **posterior subphrenic spaces.** Clinically, these spaces are important since they may provide sites for the accumulation of pus.

Paracolic Gutters

Paracolic gutters lie on the lateral and medial sides of the ascending and descending colons respectively; they provide channels for the movement of infected fluid in the peritoneal cavity.

Nerve Supply of Peritoneum

The **parietal peritoneum** (for pain, temperature, touch, and pressure) is supplied by the lower six thoracic and first lumbar nerves. The parietal peritoneum in the pelvis is mainly supplied by the obturator nerve.

The **visceral peritoneum** (for stretch only) is supplied by autonomic nerves that supply the viscera or are traveling in the mesenteries.

Gastrointestinal Viscera

Stomach

The stomach is a dilated portion of the alimentary canal and is situated in the upper part of the abdomen (Fig. 3-8). It is roughly J-shaped and has two openings, the **cardiac and pyloric orifices;** two curvatures, the **greater and lesser curvatures;** and two surfaces, an **anterior** and a **posterior surface.**

The stomach may be divided into the following parts:

Fundus: This is dome-shaped and projects upward and to the left of the cardiac orifice. It is usually full of gas.

Body: This extends from the cardiac orifice to the **incisura angularis,** a constant notch in the lower part of the lesser curvature.

Pyloric antrum: This extends from the incisura angularis to the pylorus.

Pylorus: This is the most tubular part of the stomach. It has a thick muscular wall called the **pyloric sphincter.** The cavity of the pylorus is the **pyloric canal.**

The **lesser curvature** forms the right border of the stomach and is connected to the liver by the lesser omentum. The **greater curvature** is much longer than the lesser curvature and extends from the left of the cardiac orifice over the dome of the fundus and along the left border of the stomach. The gastrosplenic omentum (ligament) extends from

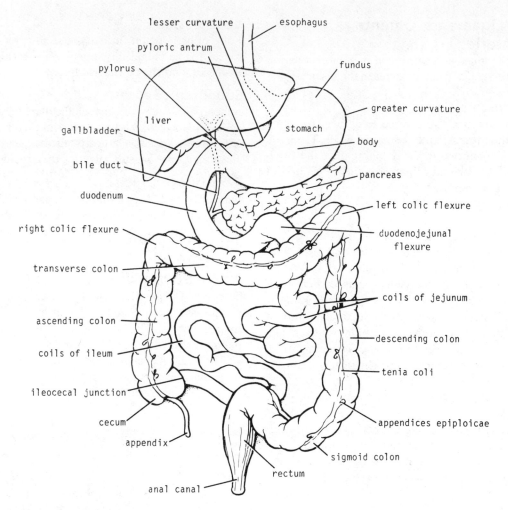

Fig. 3-8. General arrangement of abdominal viscera.

the upper part of the greater curvature to the spleen, and the greater omentum extends from the lower part of the greater curvature to the transverse colon.

The **cardiac orifice** is where the esophagus enters the stomach. Although no anatomic sphincter can be demonstrated here, a physiologic mechanism exists that prevents regurgitations of stomach contents into the esophagus. The **pyloric orifice** is formed by the pyloric canal. The circular muscle coat of the stomach is much thicker here and forms the anatomic and physiologic **pyloric sphincter.**

Blood Supply
Arteries
Right and let gastric arteries supply the lesser curvature. Right and left gastroepiploic arteries supply the greater curvature. Short gastric arteries, from the splenic artery, supply the fundus (Fig. 3-9).

Veins
The veins drain into the portal circulation. The right and left gastric veins drain into the portal vein; the short gastric

and left gastroepiploic veins drain into the splenic vein; and the right gastroepiploic vein drains into the superior mesenteric vein.

Lymph Drainage
The lymph vessels follow the arteries into the left and right gastric nodes, the left and right gastroepiploic nodes, and the short gastric nodes. All lymph from the stomach eventually passes to the celiac nodes.

Nerve Supply
Sympathetic from the celiac plexus and parasympathetic from the vagus nerves.

Small Intestine
The greater part of digestion and food absorption takes place in the small intestine, which extends from the pylorus of the stomach to the ileocecal junction (see Fig. 3-8). The small intestine is divided into three parts: the duodenum, the jejunum, and the ileum.

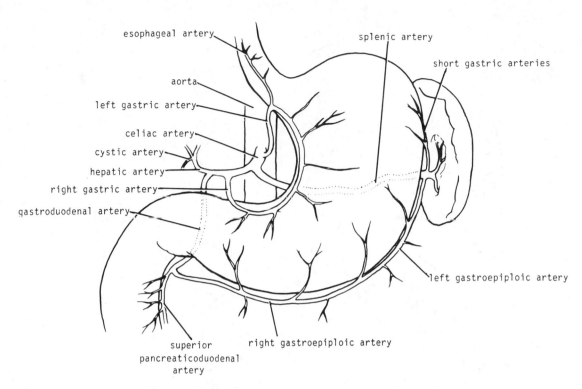

Fig. 3-9. Arterial supply to the stomach. Note that all the arteries are branches of the celiac artery.

Duodenum

The duodenum is a C-shaped tube about 10 in. (25 cm) long, which curves around the head of the pancreas (see Fig. 3-8; Fig. 3-10). It begins at the pyloric sphincter of the stomach and ends by becoming continuous with the jejunum. Although the first inch of the duodenum is covered on its anterior and posterior surfaces with peritoneum, and it has the lesser omentum attached to its upper border and the greater omentum attached to its lower border, the remainder of the duodenum is retroperitoneal.

The duodenum is divided into four parts:

First part: This runs upward and backward on the transpyloric plane at the level of the first lumbar vertebra.

Second part: This runs vertically downward, and the bile and main pancreatic ducts pierce the medial wall about halfway down. They unite to form an ampulla that opens on the summit of a **major duodenal papilla** (see Fig. 3-10). The accessory pancreatic duct, if present, opens into the duodenum on a **minor duodenal papilla** about ¾ in. (1.9 cm) above the major duodenal papilla.

Third part: This passes horizontally in front of the vertebral column. It is crossed anteriorly by the root of the mesentery of the small intestine and the superior mesenteric vessels.

Fourth part: This runs upward and to the left to the **duodenojejunal flexure.** The flexure is held in position by the **ligament of Treitz,** which is attached to the right crus of the diaphragm.

Blood Supply

Arteries.   The upper half of the duodenum is supplied by the superior pancreaticoduodenal artery, a branch of the gastroduodenal artery. The lower half is supplied by the inferior pancreaticoduodenal artery, a branch of the superior mesenteric artery.

Veins.   The superior pancreaticoduodenal vein joins the portal vein; the inferior vein joins the superior mesenteric vein.

Lymph Drainage

The lymph vessels drain upward via pancreaticoduodenal nodes to gastroduodenal nodes and celiac nodes; they drain downward via pancreaticoduodenal nodes to superior mesenteric nodes.

Nerve Supply

Sympathetic and vagus nerves via the celiac and superior mesenteric plexuses.

Jejunum and Ileum

The jejunum measures about 8 ft. (2.5 m) and the ileum about 12 ft. (3.6 m) long. The jejunum begins at the duodenojejunal flexure (see Fig. 3-8) in the upper part of the abdominal cavity to the left of the midline. It is wider in diameter, thicker walled, and redder in color than the ileum.

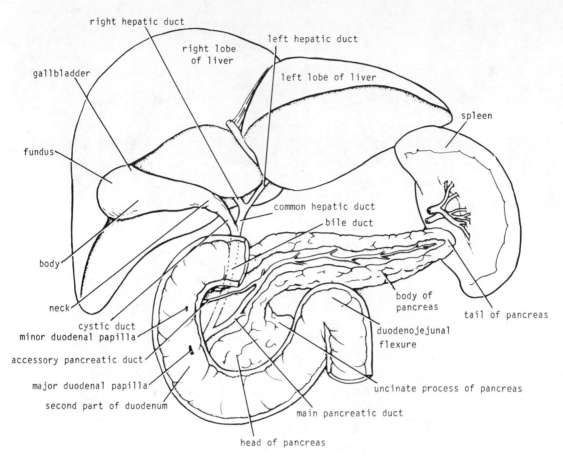

Fig. 3-10. Liver, biliary ducts, pancreas, and spleen. Note their relationship to one another and to the duodenum.

The coils of ileum occupy the lower right part of the abdominal cavity and tend to hang down into the pelvis. The ileum ends at the ileocecal junction. The coils of the jejunum and the ileum are suspended from the posterior abdominal wall by a fan-shaped fold of peritoneum called the **mesentery of the small intestine.**

Blood Supply
Arteries.   Branches of the superior mesenteric artery (Fig. 3-11) anastomose with one another to form arcades. *Veins.*   Drain into the superior mesenteric vein.

Lymph Drainage
The lymph passes to the superior mesenteric nodes via a large number of intermediate mesenteric nodes.

Nerve Supply
Sympathetic and vagus nerve fibers from the superior mesenteric plexus.

Meckel's Diverticulum
Meckel's diverticulum is a congenital anomaly representing a persistent portion of the vitello-intestinal duct. If present, it is located on the antimesenteric border of the ileum about 2 ft. (60 cm) from the ileocecal junction. It is about 2 in. (5 cm) long and occurs in about 2 percent of individuals. It is important clinically since bleeding may occur from an ulcer in its mucous membrane.

Large Intestine
The large intestine extends from the ileum to the anus (see Fig. 3-8). It is divided into the cecum, the appendix, the ascending colon, the transverse colon, the descending colon, the sigmoid colon, the rectum, and the anal canal. (The rectum and anal canal are considered in Chapter 4.)

The main functions of the large intestine include absorption of water, production of certain vitamins, storage of undigested food materials and formation of feces, and excretion of feces from the body.

Cecum
The cecum is a blind-ended pouch that lies within the right iliac fossa and is completely covered with peritoneum (see Fig. 3-8). At the junction of the cecum with the ascending colon, it is joined on the left side by the terminal part of the ileum. Attached to its posteromedial surface is the appendix.

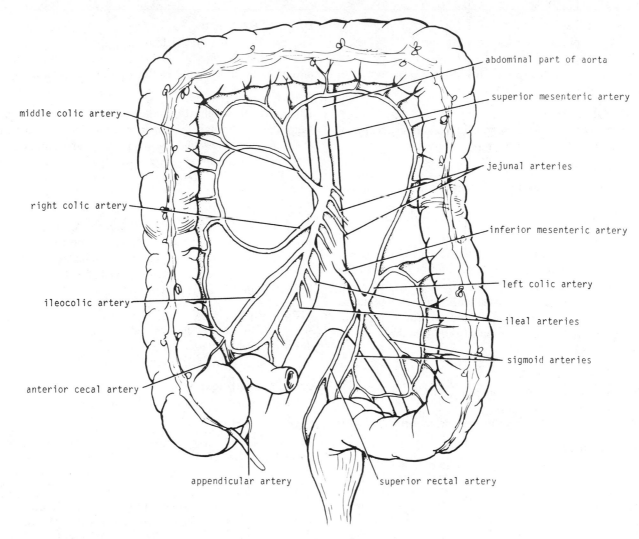

middle colic artery

right colic artery

ileocolic artery

anterior cecal artery

abdominal part of aorta

superior mesenteric artery

jejunal arteries

inferior mesenteric artery

left colic artery

ileal arteries

sigmoid arteries

appendicular artery

superior rectal artery

Fig. 3-11. Superior and inferior mesenteric arteries and their branches.

Blood Supply
Arteries.   Anterior and posterior cecal arteries from the ileocolic artery (see Fig. 3-11), a branch of the superior mesenteric artery.
Veins.   Drain into superior mesenteric vein.

Lymph Drainage
Mesenteric nodes and superior mesenteric nodes.

Nerve Supply
Sympathetic and vagus nerves via the superior mesenteric plexus.

Ileocecal Valve
A rudimentary structure, the ileocecal valve consists of two horizontal folds of mucous membrane that project around the orifice of the ileum. The valve plays little or no part in the prevention of reflux of cecal contents into the ileum. The circular muscle of the lower end of the ileum (called the **ileocecal sphincter** by physiologists) serves as a sphinc-

ter and controls the flow of contents from the ileum into the colon. The smooth muscle tone is reflexively increased when the cecum is distended; the hormone gastrin, which is produced by the stomach, causes relaxation of the muscle tone.

Appendix
The appendix (see Fig. 3-8) is a narrow, muscular tube containing a large amount of lymphoid tissue in its wall. It is attached to the posteromedial surface of the cecum about 1 in. (2.5 cm) below the ileocecal junction. It has a complete peritoneal covering, which is attached to the mesentery of the small intestine by a short mesentery of its own, the **mesoappendix.** The mesoappendix contains the appendicular vessels and nerves.

The base of the appendix can be located inside the abdomen by tracing the teniae coli of the cecum and following them to the appendix, where they converge to form a continuous muscle coat.

Blood Supply

Arteries.   Appendicular artery, a branch of the posterior cecal artery (see Fig. 3-11).
Veins.   Appendicular vein drains into the posterior cecal vein.

Lymph Drainage

One or two nodes in the mesoappendix and then eventually into the superior mesenteric lymph nodes.

Nerve Supply

Sympathetic and vagus nerves from the superior mesenteric plexus.

Ascending Colon

The ascending colon is about 5 in. (13 cm) long and extends upward from the cecum to the inferior surface of the right lobe of the liver (see Fig. 3-8). Here it turns to the left, forming the **right colic flexure,** and becomes continuous with the transverse colon. The peritoneum covers the front and sides of the ascending colon binding it to the posterior abdominal wall.

Blood Supply

Arteries.   Ileocolic and right colic branches of the superior mesenteric artery (see Fig. 3-11).
Veins.   Drain into the superior mesenteric vein.

Lymph Drainage

Colic lymph nodes and superior mesenteric nodes.

Nerve Supply

Sympathetic and vagus nerves from the superior mesenteric plexus.

Transverse Colon

The transverse colon is about 15 in. (38 cm) long and passes across the abdomen, occupying the umbilical and hypogastric regions (see Fig. 3-8). It begins at the right colic flexure below the right lobe of the liver and hangs downward, suspended by the transverse mesocolon from the pancreas. It then ascends to the **left colic flexure** below the spleen. The left colic flexure is higher than the right colic flexure and is held up by the **phrenicocolic ligament.** The **transverse mesocolon,** or mesentery of the transverse colon, is attached to the superior border of the transverse colon; the posterior layers of the greater omentum are attached to the inferior border.

Blood Supply

Arteries.   The proximal two-thirds is supplied by the middle colic artery (see Fig. 3-11), a branch of the superior mesenteric artery. The distal third is supplied by the left colic artery, a branch of the inferior mesenteric artery.
Veins.   Drain into the superior and inferior mesenteric veins.

Lymph Drainage

The proximal two-thirds drain into the colic nodes and then the superior mesenteric nodes; the distal third drains into the colic nodes and then the inferior mesenteric nodes.

Nerve Supply

The proximal two-thirds is innervated by sympathetic and vagal nerves through the superior mesenteric plexus; the distal third is innervated by sympathetic and parasympathetic pelvic splanchnic nerves through the inferior mesenteric plexus.

Descending Colon

The descending colon is about 10 in. (25 cm) long and extends downward from the left colic flexure to the pelvic brim, where it becomes continuous with the sigmoid colon (see Fig. 3-8). The peritoneum covers the front and the sides and binds it to the posterior abdominal wall.

Blood Supply

Arteries.   Left colic branch and sigmoid branches of the inferior mesenteric artery (see Fig. 3-11).
Veins.   Inferior mesenteric vein.

Lymph Drainage

Colic nodes and inferior mesenteric nodes.

Nerve Supply

Sympathetic and parasympathetic pelvic splanchnic nerves through the inferior mesenteric plexus.

Sigmoid Colon

The sigmoid colon is about 10 to 15 in. (25–38 cm) long and begins as a continuation of the descending colon in front of the pelvic brim (see Fig. 3-8). Below, it becomes continuous with the rectum in front of the third sacral vertebra. It hangs down into the pelvic cavity in the form of a loop. The sigmoid colon is attached to the posterior pelvic wall by the fan-shaped **sigmoid mesocolon.**

Blood Supply

Arteries.   Sigmoid branches of the inferior mesenteric artery (see Fig. 3-11).
Veins.   Inferior mesenteric vein.

Lymph Drainage

Colic nodes and inferior mesenteric nodes.

Nerve Supply

Sympathetic and parasympathetic nerves through the inferior hypogastric plexuses.

Differences Between the Small and Large Intestines
External Differences

1. Small intestine is more mobile (exception is the duodenum), whereas the ascending and descending parts of the colon are fixed.

2. Small intestine has a mesentery (except duodenum), whereas large intestine is retroperitoneal (except transverse colon and sigmoid colon).
3. Diameter of full small intestine is smaller than that of full large intestine.
4. Longitudinal muscle of small intestine forms a continuous layer around the gut, whereas in the large intestine (with the exception of the appendix, rectum, and anal canal), the longitudinal muscle forms three bands, the **teniae coli.**
5. The small intestine has no fatty tags attached to its wall, whereas the large intestine has the **appendices epiploicae.**
6. The wall of the small intestine is smooth, whereas that of the large intestine is sacculated.

Internal Differences

1. The mucous membrane of the small intestine has permanent folds, **plicae circulares,** which are absent from the large intestine.
2. The mucous membrane of the small intestine has **Peyer's patches,** whereas the large intestine has **solitary lymph follicles.**
3. The mucous membrane of the small intestine has villi, which are absent from the large intestine.

Accessory Organs of Gastrointestinal System

Liver

The liver is the largest organ in the body (Fig. 3-12). It occupies the upper part of the abdominal cavity just beneath the diaphragm. The liver may be divided into a large **right lobe** and a small **left lobe** by the attachment of the peritoneum of the falciform ligament (see Fig. 3-12). The right lobe is further subdivided into a **quadrate lobe** and a **caudate lobe** by the presence of the gallbladder, the fissure for the ligamentum teres, the inferior vena cava, and the fissure for the ligamentum venosum.

The liver is completely surrounded by a fibrous capsule but only partially covered with peritoneum.

Porta Hepatis, Fissures, Grooves, and Fossae
Porta Hepatis

The porta hepatis, or hilus, of the liver is found on the posteroinferior surface (see Fig. 3-12). It lies between the caudate and the quadrate lobes, and the upper part of the lesser omentum is attached to its margins. Within the porta hepatis lie the right and left hepatic ducts, the right and left branches of the hepatic artery and portal vein, nerves, and lymph vessels.

Fissure for the Ligamentum Teres

The fissure for the ligamentum teres lies between the left lobe and the quadrate lobe (see Fig. 3-12). It contains the **ligamentum teres,** which is the fibrous remains of the **umbilical vein.**

Fissure for the Ligamentum Venosum

The fissure for the ligamentum venosum lies between the left lobe and the caudate lobe (see Fig. 3-12). It contains the **ligamentum venosum,** which is the fibrous remains of the ductus venosus, and the upper part of the lesser omentum is attached to its margins.

Groove for the Inferior Vena Cava

The groove for the inferior vena cava lies between the right lobe and the caudate lobe and lodges the inferior vena cava (see Fig. 3-12); it is here that the hepatic veins join the inferior vena cava.

Fossa for the Gallbladder

The fossa for the gallbladder lies between the right lobe and the quadrate lobe (see Fig. 3-12). There is no peritoneum between the gallbladder and the right lobe of the liver.

Peritoneal Ligaments
Falciform Ligament

The falciform ligament is a two-layered fold of peritoneum that attaches the liver to the diaphragm above and to the anterior abdominal wall below (see Fig. 3-12). It has a sickle-shaped free margin that contains the **ligamentum teres,** the remains of the umbilical vein.

Coronary Ligament

The coronary ligament attaches the liver to the diaphragm (see Fig. 3-12). The peritoneal layers forming the ligament are widely separated, leaving a "bare area" of liver devoid of a peritoneal covering.

Right Triangular Ligament

The right triangular ligament is a V-shaped fold of peritoneum formed from the right extremity of the coronary ligament (see Fig. 3-12). It connects the posterior surface of the right lobe of the liver to the diaphragm.

Left Triangular Ligament

The left triangular ligament is formed by the reflection of the peritoneum from the upper surface of the left lobe of the liver to the diaphragm (see Fig. 3-12). It lies anteriorly to the abdominal part of the esophagus.

Lesser Omentum

The upper end of the lesser omentum is attached to the margins of the porta hepatis and the fissure for the ligamentum venosum. It is attached below to the lesser curvature of the stomach (see Fig. 3-7).

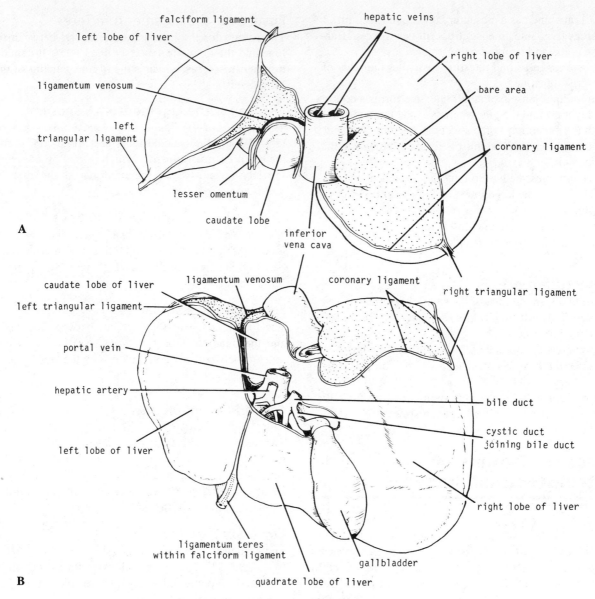

Fig. 3-12. **A.** Liver as seen from above. **B.** Liver as seen from behind. Note the position of the peritoneal reflections, the bare areas, and the peritoneal ligaments.

Blood Supply

The **hepatic artery,** a branch of the celiac artery, divides into right and left terminal branches that enter the porta hepatis. The **portal vein** divides into right and left terminal branches that enter the porta hepatis behind the arteries. The **hepatic veins** (three or more) emerge from the posterior surface of the liver and drain into the inferior vena cava.

Lymph Drainage

The lymph enters lymph nodes in the porta hepatis and then drains to the celiac nodes. Some lymph passes through the diaphragm to enter the posterior mediastinal nodes.

Nerve Supply

Sympathetic and parasympathetic (vagal) fibers from the celiac plexus. The left vagus gives rise to a large hepatic branch, which travels directly to the liver.

Gallbladder

The gallbladder is a pear-shaped sac lying on the undersurface of the liver (see Fig. 3-12). It is divided into a **fundus, body,** and **neck.** It has a capacity of about 30 ml and stores bile and concentrates it by absorbing water. The neck is continuous with the cystic duct.

Blood Supply
Arteries

Cystic artery, a branch of the right hepatic artery.

Veins

Cystic vein drains into the portal vein.

Lymph Drainage

Cystic lymph node near neck of gallbladder, then hepatic nodes, and finally celiac nodes.

Nerve Supply

Sympathetic and parasympathetic vagal fibers from the celiac plexus. The gallbladder contracts in response to the hormone **cholecystokinin,** which is produced by the mucous membrane of the duodenum on the arrival of food from the stomach.

Bile Ducts
Hepatic Ducts

The **right** and **left hepatic ducts** emerge from the right and left lobes of the liver in the porta hepatis. Each hepatic duct has been formed by the union of small bile ducts (bile canaliculi) within the liver. The **common hepatic duct** is formed by the union of the right and left hepatic ducts. It is joined on the right side by the cystic duct from the gallbladder to form the bile duct (see Fig. 3-12).

Cystic Duct

The cystic duct is an S-shaped duct that connects the neck of the gallbladder to the common hepatic duct to form the bile duct (see Fig. 3-12). The mucous membrane is raised to form a spiral fold (spiral valve) whose function is to keep the lumen constantly open.

Bile Duct (Common Bile Duct)

The bile duct is formed by the union of the cystic with the common hepatic duct (see Fig. 3-10). It runs in the right free margin of the lesser omentum, having the portal vein behind and the hepatic artery on the left. It descends in front of the opening into the lesser sac and passes behind the first part of the duodenum and then the head of the pancreas. The bile duct ends below by piercing the medial wall of the duodenum about halfway down its length (see Fig. 3-10). It is usually joined by the main pancreatic duct, and together they open into a small ampulla in the duodenal wall, the **ampulla of Vater.** The ampulla opens into the lumen of the duodenum by means of a small papilla, the **major duodenal papilla** (see Fig. 3-10). The terminal parts of both ducts and the ampulla are surrounded by circular smooth muscle called the **sphincter of Oddi.** Occasionally, the bile and pancreatic ducts open separately into the duodenum.

Pancreas

The pancreas is both an exocrine and an endocrine gland. It is an elongated structure that lies on the posterior abdominal wall behind the stomach and behind the peritoneum. It may be divided into a head, neck, body, and tail (see Fig. 3-10). The **head** is disc-shaped and lies within the concavity of the C-shaped duodenum. The **uncinate process** is a projection to the left from the lower part of the head behind the superior mesenteric vessels. The **neck** is narrow and connects the head to the body. It lies in front of the beginning of the portal vein. The **body** passes upward and to the left across the midline. The **tail** extends to the hilus of the spleen in the lienorenal ligament.

Pancreatic Ducts

The **main pancreatic duct** runs the length of the gland and opens into the second part of the duodenum with the bile duct on the major duodenal papilla (see Fig. 3-10). Sometimes the main duct drains separately into the duodenum. The **accessory duct,** when present, drains the upper part of the head and opens into the duodenum on the **minor duodenal papilla.**

Blood Supply
Arteries

Splenic artery and superior and inferior pancreaticoduodenal arteries.

Veins

The pancreatic veins drain into the portal vein.

Lymph Drainage

Lymph nodes situated along the arteries and then into celiac and superior mesenteric nodes.

Nerve Supply

Sympathetic and parasympathetic vagal nerve fibers from the celiac plexus.

Spleen

The spleen is the largest single mass of lymphoid tissue in the body (see Fig. 3-10). It lies just beneath the left half of the diaphragm close to the ninth, tenth, and eleventh ribs. The spleen is ovoid in shape, with a notched anterior border. It is surrounded by peritoneum that passes from the hilus to the stomach as the **gastrosplenic omentum** (ligament) and to the left kidney as the **lienorenal ligament** (see Fig. 3-7). The gastrosplenic omentum contains the short gastric and left gastroepiploic vessels, and the lienorenal ligament contains the splenic vessels and the tail of the pancreas.

Blood Supply
Artery
Large splenic artery, a branch of the celiac artery (see Fig. 3-9).

Vein
The splenic vein joins the superior mesenteric vein to form the portal vein.

Blood Supply of Gastrointestinal Viscera

The **celiac artery** is the artery of the foregut and supplies the gastrointestinal tract from the lower one-third of the esophagus down as far as the middle of the second part of the duodenum (see Fig. 3-9). The **superior mesenteric artery** is the artery of the midgut and supplies the gastrointestinal tract from the middle of the second part of the duodenum as far as the distal one-third of the transverse colon (see Fig. 3-11). The **inferior mesenteric artery** is the artery of the hindgut and supplies the large intestine from the distal one-third of the transverse colon to halfway down the anal canal.

Celiac Artery (Trunk)
The celiac artery is a short large artery that arises from the front of the abdominal aorta as it emerges through the diaphragm (see Fig. 3-9). It has three terminal branches: the left gastric, the splenic, and the hepatic arteries.

Left Gastric Artery
The left gastric artery is a small artery that runs to the cardiac end of the stomach, gives off a few esophageal branches, then turns to the right along the lesser curvature of the stomach. It anastomoses with the right gastric artery.

Splenic Artery
The splenic artery is the largest branch of the celiac trunk and runs to the left in a wavy course along the upper border of the pancreas and behind the stomach. On reaching the left kidney, the artery enters the lienorenal ligament and runs to the hilum of the spleen.

Branches
1. **Pancreatic branches.**
2. **Left gastroepiploic artery** arises near the hilum of the spleen and reaches the greater curvature of the stomach in the gastrosplenic omentum. It passes to the right along the greater curvature of the stomach in the greater omentum. It anastomoses with the right gastroepiploic artery.
3. **Short gastric arteries (5 or 6)** pass to the fundus of the stomach in the gastrosplenic omentum. They anastomose with the left gastric artery and the left gastroepiploic artery.

Hepatic Artery*
The hepatic artery runs forward and to the right and ascends within the lesser omentum (see Fig. 3-9). It lies in front of the opening into the lesser sac and is placed to the left of the bile duct and in front of the portal vein. At the porta hepatis it divides into right and left branches to supply the corresponding lobes of the liver.

Branches
1. **Right gastric artery** runs to the pylorus and then to the left in the lesser omentum along the lesser curvature of the stomach. It anastomoses with the left gastric artery.
2. **Gastroduodenal artery** descends behind the first part of the duodenum. It divides into the **right gastroepiploic artery,** which runs along the greater curvature of the stomach in the greater omentum, and the **superior pancreaticoduodenal artery,** which descends between the second part of the duodenum and the head of the pancreas.
3. **Right and left hepatic arteries** to the right and left lobes of the liver. The right hepatic artery usually gives off the **cystic artery,** which runs to the neck of the gallbladder.

Superior Mesenteric Artery
The superior mesenteric artery arises from the front of the abdominal aorta behind the neck of the pancreas (see Fig. 3-11). It runs downward in front of the uncinate process of the pancreas and in front of the third part of the duodenum. It continues downward to the right in the root of the mesentery of the small intestine.

Branches
1. **Inferior pancreaticoduodenal artery** passes to the right as a single or double branch along the upper border of the third part of the duodenum and below the head of the pancreas.
2. **Middle colic artery** runs into the transverse mesocolon to supply the transverse colon (see Fig. 3-11). It divides into a right branch that anastomoses with the right colic artery and a left branch that anastomoses with the left colic artery.
3. **Right colic artery** is often a branch of the ileocolic artery (see Fig. 3-11). It passes to the right to supply the ascending colon.
4. **Ileocolic artery** passes downward and to the right (see Fig. 3-11). It gives rise to a **superior branch** that anastomoses with the right colic artery and an **inferior branch** that anastomoses with the end of the superior mesenteric artery. The inferior branch gives rise to the **anterior and posterior cecal arteries; the appendicular artery** is a branch of the posterior cecal artery.

---

*Sometimes divided into the **common hepatic artery**, which extends from its origin to the gastroduodenal branch, and the **hepatic artery proper**, which is the remainder of the artery.

5. **Jejunal and ileal branches.** These are 12 to 15 in number and arise from the left side of the superior mesenteric artery (see Fig. 3-11). Each artery divides into branches that unite with adjacent branches to form arcades. Small straight branches supply the intestine.

Inferior Mesenteric Artery

The inferior mesenteric artery arises from the abdominal aorta about 1½ in. (3.8 cm) above its bifurcation (see Fig. 3-11). The artery runs downward and to the left and crosses the left common iliac artery. Here, it changes its name and becomes the superior rectal artery.

Branches

1. **Left colic artery** divides into ascending and descending branches that supply the distal third part of the transverse colon, the left colic flexure, and the upper part of the descending colon (see Fig. 3-11).
2. **Sigmoid arteries** are two or three in number and supply the descending and sigmoid colon (see Fig. 3-11).
3. **Superior rectal artery** is the continuation of the inferior mesenteric artery and descends into the pelvis behind the rectum (see Fig. 3-11). This artery supplies the rectum and the upper half of the anal canal and anastomoses with the middle and inferior rectal arteries that arise from the internal iliac and internal pudendal arteries respectively.

Marginal Artery

The colic arteries anastomose around the concave margin of the large intestine and form a single arterial trunk called the **marginal artery.** It begins at the ileocolic junction and ends at the junction of the sigmoid colon with the rectum.

Portal Venous System

The proximal tributaries drain directly into the portal vein, but the veins forming the distal tributaries correspond to the branches of the celiac artery and the superior and inferior mesenteric arteries.

Portal Vein

The portal vein is about 2 in. (5 cm) long and is formed behind the neck of the pancreas by the union of the superior mesenteric and the splenic veins (Fig. 3-13). It ascends to the porta hepatis behind the first part of the duodenum and in the free margin of the lesser omentum. In the porta hepatis it divides into right and left terminal branches.

The portal vein drains blood from the gastrointestinal tract from the lower end of the esophagus to halfway down the anal canal, as well as from the pancreas, the gallbladder, the bile ducts, and the spleen.

Tributaries

1. **Splenic vein.** This leaves the hilum of the spleen and unites with the superior mesenteric vein behind the

neck of the pancreas to form the portal vein (see Fig. 3-13). It receives the short gastric, left gastroepiploic, inferior mesenteric, and pancreatic veins.
2. **Inferior mesenteric vein.** This vein ascends on the posterior abdominal wall and joins the splenic vein behind the body of the pancreas (see Fig. 3-13). It receives the superior rectal veins, the sigmoid veins, and the left colic vein.
3. **Superior mesenteric vein.** This vein ascends in the root of the mesentery of the small intestine on the right side of the artery. It passes in front of the third part of the duodenum and joins the splenic vein behind the neck of the pancreas (see Fig. 3-13). It receives jejunal, ileal, ileocolic, right colic, middle colic, inferior pancreaticoduodenal, and right gastroepiploic veins.
4. **Left gastric vein.** This vein drains the left portion of the lesser curvature of the stomach and the distal part of the esophagus. It opens directly into the portal vein (see Fig. 3-13).
5. **Right gastric vein.** This vein drains the right portion of the lesser curvature of the stomach and drains directly into the portal vein (see Fig. 3-13).
6. **Cystic veins.** These veins drain the gallbladder either directly into the liver or join the portal vein (see Fig. 3-13).

Portal-Systemic Anastomoses

Portal-systemic anastomoses are important in patients with cirrhosis of the liver in which the portal vein may be obstructed.

1. At the lower third of the esophagus, the esophageal branches of the left gastric vein (portal tributary) anastomose with the esophageal veins draining the middle third of the esophagus into the azygos veins (systemic tributaries).
2. Halfway down the anal canal, the superior rectal veins (portal tributaries) draining the upper half of the anal canal anastomose with the middle and inferior rectal veins (systemic tributaries).
3. The **paraumbilical veins** connect the left branch of the portal vein with the superficial veins of the anterior abdominal wall (systemic tributaries). The paraumbilical veins travel in the falciform ligament and accompany the ligamentum teres.
4. The veins of the ascending colon, descending colon, duodenum, pancreas, and liver (portal tributaries) anastomose with the renal, lumbar, and phrenic veins (systemic tributaries).

Kidneys and Ureters

Kidneys

The kidneys are paired organs that lie behind the peritoneum high up on the posterior abdominal wall on either side of the

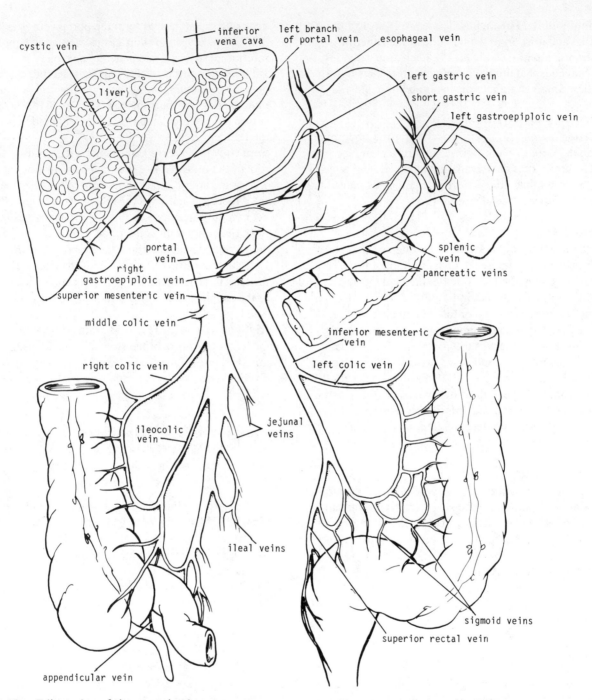

Fig. 3-13. Tributaries of the portal vein.

vertebral column (Fig. 3-14). The right kidney lies slightly lower than the left kidney due to the large size of the right lobe of the liver. With contraction of the diaphragm during respiration, both kidneys move downward in a vertical direction by as much as 1 in. (2.5 cm). On the medial concave border of each kidney is the **hilus,** which extends into a large cavity, the **renal sinus.** The hilus transmits the renal pelvis, the renal artery, the renal vein, and sympathetic nerve fibers. The kidneys have the following coverings:

1. **Fibrous capsule.** This is closely applied to its outer surface.
2. **Perirenal fat.** This is fat that covers the fibrous capsule.
3. **Renal fascia.** This is a condensation of areolar tissue and lies outside the perirenal fat. It encloses the kidneys and the suprarenal glands.
4. **Pararenal fat.** This lies external to the renal fascia and is often large in amount.

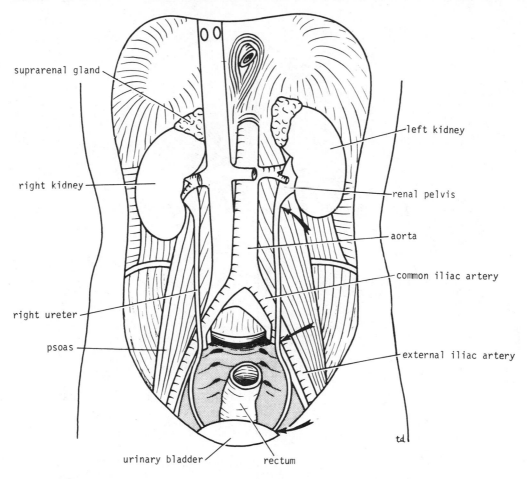

Fig. 3-14. Posterior abdominal wall, showing kidneys and ureters in situ. Arrows indicate three sites where ureter is narrowed.

The perirenal fat, renal fascia, and pararenal fat support the kidneys and hold them in position on the posterior abdominal wall.

Renal Structure

The outer **cortex** is dark brown and the inner **medulla** is light brown in color. The medulla is composed of about a dozen **renal pyramids,** each having its base oriented toward the cortex and its apex, the **renal papilla,** projecting medially (Fig. 3-15). The cortex extends into the medulla between adjacent pyramids as the **renal columns.** Extending from the bases of the renal pyramids into the cortex are striations known as **medullary rays.**

Within the renal sinus, the upper expanded end of the ureter, the **renal pelvis,** divides into two or three **major calyces,** each of which divides into two or three **minor calyces** (see Fig. 3-15). Each minor calyx is indented by the apex of the renal pyramid, the **renal papilla.**

Blood Supply

Arteries

Renal artery, a branch of the aorta.

Veins

Renal vein drains into the inferior vena cava.

Lymph Drainage

Lateral aortic lymph nodes around the origin of the renal artery.

Nerve Supply

Renal sympathetic plexus.

Ureters

The two ureters are muscular tubes that extend from the kidneys to the posterior surface of the urinary bladder (see Fig. 3-14). Each ureter measures about 10 in. (25 cm) long and has an upper expanded end called the **renal pelvis.** The renal pelvis lies within the hilus of the kidney where it receives the major calyces.

Ureteric Constrictions

1. Where the renal pelvis joins the ureter
2. Where it is kinked as it crosses the pelvic brim to enter the pelvis
3. Where it pierces the bladder wall

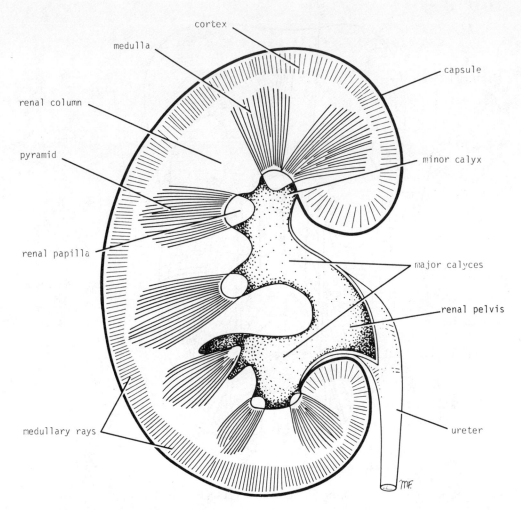

Fig. 3-15. Longitudinal section through kidney, showing cortex, medulla, pyramids, renal papillae, and calyces.

Blood Supply
Arteries
Upper end, the renal artery; middle portion, the testicular or ovarian artery; inferior end, the superior vesical artery.

Veins
Veins that correspond to the arteries.

Lymph Drainage
Lateral aortic and iliac nodes.

Nerve Supply
Renal, testicular (or ovarian), and hypogastric plexuses.

Suprarenal (Adrenal) Glands

The suprarenal glands are two in number and are located close to the upper poles of the kidneys on the posterior ab-

dominal wall (see Fig. 3-14). They are retroperitoneal and surrounded by renal fascia (but are separated from the kidneys by the perirenal fat). Each gland has a yellow-colored **cortex** and a dark brown **medulla.**

Blood Supply
Arteries
Branches from the inferior phrenic, aorta, and renal arteries.

Veins
A single vein on each side. The right suprarenal vein drains into the inferior vena cava; the left suprarenal vein drains into the left renal vein.

Lymph Drainage
Lateral aortic nodes.

Nerve Supply
Numerous preganglionic sympathetic nerves from splanchnic nerves. The majority of the fibers end on the cells in the suprarenal medulla.

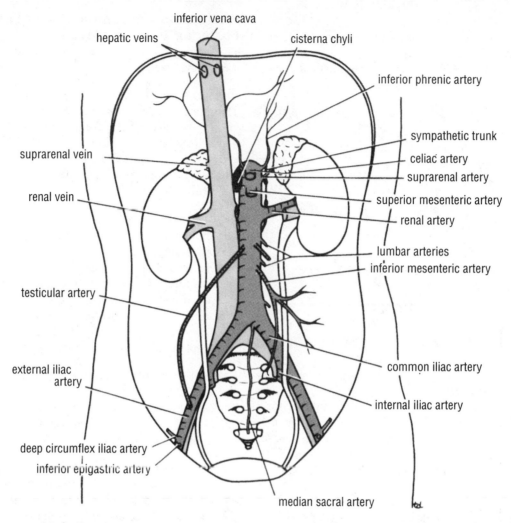

Fig. 3-16. Aorta and inferior vena cava.

Aorta and Inferior Vena Cava

Abdominal Aorta
The aorta enters the abdomen through the aortic opening of the diaphragm in front of the twelfth thoracic vertebra (Fig. 3-16). It descends on the anterior surfaces of the bodies of the lumbar vertebrae, and in front of the fourth lumbar vertebra it divides into the two common iliac arteries.

Branches
1. Three anterior visceral branches
   a. Celiac artery
   b. Superior mesenteric artery
   c. Inferior mesenteric artery
2. Three lateral visceral branches
   a. Suprarenal artery
   b. Renal artery
   c. Testicular or ovarian artery
3. Five lateral abdominal branches
   a. Inferior phrenic artery
   b. Four lumbar arteries
4. Three terminal arteries
   a. Two common iliac arteries
   b. Median sacral artery

Common Iliac Arteries
The right and left common iliac arteries are the terminal branches of the abdominal aorta (see Fig. 3-16). They run downward and laterally to end opposite the sacroiliac joint by dividing into external and internal iliac arteries. At the bifurcation, the common iliac artery on each side is crossed anteriorly by the ureter.

Inferior Vena Cava
The inferior vena cava is formed by the union of the common iliac veins at the level of the fifth lumbar vertebra (see Fig. 3-16). It ascends on the right side of the aorta, pierces the central tendon of the diaphragm at the level of the eighth thoracic vertebra, and drains into the right atrium of the heart.

Tributaries

1. Two anterior visceral tributaries, the hepatic veins
2. Three lateral visceral tributaries
   a. Right suprarenal (the left vein drains into the left renal vein)
   b. Renal veins
   c. Right testicular or ovarian vein (the left vein drains into the left renal vein)
3. Five lateral abdominal wall tributaries
   a. Inferior phrenic vein
   b. Four lumbar veins
4. Three veins of origin
   a. Two common iliac veins
   b. Median sacral vein

Lymphatics on the Posterior Abdominal Wall

Lymph Nodes

The lymph nodes form a preaortic and right and left lateral aortic chain.

Preaortic Lymph Nodes

The preaortic lymph nodes are located on the anterior surface of the abdominal aorta. Their efferent vessels form the **intestinal trunk** that drains into the cisterna chyli. These nodes may be divided into the **celiac, superior mesenteric, and inferior mesenteric groups** that lie close to the origins of these arteries.

Lateral Aortic (Para-Aortic, Lumbar) Nodes

The lateral aortic nodes are right and left groups that lie alongside the abdominal aorta. Their efferent vessels form the **right and left lumbar trunks** that drain into the cisterna chyli.

Cisterna Chyli

The **thoracic duct** commences in the abdomen as an elongated sac, the cisterna chyli. This lies on the right side of the aorta in front of the first two lumbar vertebrae. The cisterna chyli receives the intestinal trunk, the right and left lumbar trunks, and lymph vessels that descend from the lower part of the thorax.

Nerves

Lumbar Plexus

The lumbar plexus is formed from the anterior rami of the upper four lumbar nerves (Fig. 3-17). It is situated within the psoas muscle and its branches emerge from the lateral and medial borders of the muscle and from its anterior surface.

Branches of the Lumbar Plexus Seen on the Posterior Abdominal Wall

The branches of the lumbar plexus and their distribution are summarized in Table 3-2.

Iliohypogastric Nerve (L1)

The iliohypogastric nerve emerges from the lateral border of the psoas and runs in front of the quadratus lumborum muscle (see Fig. 3-17). It pierces the transversus abdominis and runs forward to supply the transversus abdominis, the internal and external oblique muscles, and the skin above the inguinal ligament.

Ilioinguinal Nerve (L1)

The ilioinguinal nerve emerges from the lateral border of the psoas and runs in front of the quadratus lumborum muscle (see Fig. 3-17). It pierces the transversus abdominis and runs forward through the inguinal canal to exit through the superficial inguinal ring. It supplies the transversus abdominis, the internal oblique, and the external oblique muscles; and it supplies the skin just above the symphysis pubis and the scrotum or labium majus.

Lateral Cutaneous Nerve of the Thigh (L2 and 3)

The lateral cutaneous nerve of the thigh emerges from the lateral border of the psoas, crosses the iliacus muscle, and enters the thigh behind the inguinal ligament (see Fig. 3-17).

Femoral Nerve (L2, 3, and 4)

The femoral nerve is the largest branch of the lumbar plexus (see Fig. 3-17). It emerges from the lateral border of the psoas and descends between the psoas and the iliacus muscles to enter the thigh lateral to the femoral vessels.

Genitofemoral Nerve (L1 and 2)

The genitofemoral nerve emerges on the anterior surface of the psoas muscle (see Fig. 3-17). It divides into a **genital branch,** which supplies the cremaster muscle, and a **femoral branch,** which supplies a small area of skin in the thigh.

Obturator Nerve (L2, 3, and 4)

The obturator nerve emerges from the medial border of the psoas muscle and runs forward on the lateral wall of the pelvis; it enters the thigh through the obturator foramen (see Fig. 3-17).

Abdominal Part of the Sympathetic Trunk and Autonomic Plexuses
Sympathetic Trunk

The abdominal part of the sympathetic trunk is continuous with the thoracic part above and with the pelvic part of the sympathetic trunk below. It runs downward along the medial border of the psoas muscle on the bodies of the lumbar vertebrae. It enters the abdomen from behind the medial

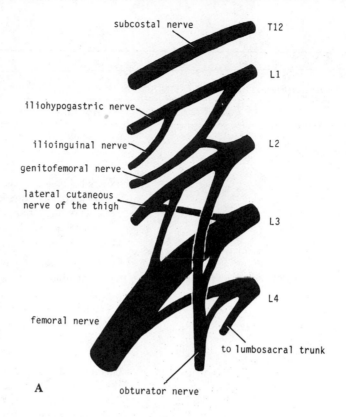

subcostal nerve — T12

L1

iliohypogastric nerve

ilioinguinal nerve — L2

genitofemoral nerve

lateral cutaneous
nerve of the thigh

L3

L4

femoral nerve

to lumbosacral trunk

A

obturator nerve

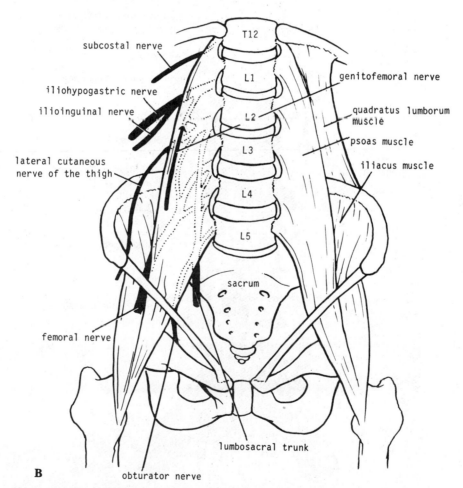

subcostal nerve

T12

L1

iliohypogastric nerve

genitofemoral nerve

ilioinguinal nerve

L2

quadratus lumborum
muscle

psoas muscle

L3

iliacus muscle

lateral cutaneous
nerve of the thigh

L4

L5

sacrum

femoral nerve

lumbosacral trunk

B

obturator nerve

Fig. 3-17. **A.** Lumbar plexus and its main branches. **B.** Lumbar plexus and its branches on the posterior abdominal wall.

Table 3-2 Branches of the Lumbar Plexus and Their Distribution

Branches	Distribution
Iliohypogastric nerve (L1)	External oblique, internal oblique, transversus abdominis muscles of anterior abdominal wall; skin over lower anterior abdominal wall and buttock
Ilioinguinal nerve (L1)	External oblique, internal oblique, transversus abdominis muscles of anterior abdominal wall; skin of upper medial aspect of thigh, root of penis and scrotum in the male, mons pubis and labia majora in the female
Lateral cutaneous nerve of thigh (L2, 3)	Skin of anterior and lateral surfaces of the thigh
Genitofemoral nerve (L1, 2)	Cremaster muscle in scrotum in male; skin over anterior surface of thigh; nervous pathway for cremasteric reflex
Femoral nerve (L2, 3, 4)	Iliacus, pectineus, sartorius, quadriceps femoris muscles, and intermediate cutaneous branches to the skin of the anterior surface of the thigh and by saphenous branch to the skin of the medial side of the leg and foot; articular branches to hip and knee joints
Obturator nerve (L2, 3, 4)	Gracilis, adductor brevis, adductor longus, obturator externus, ? pectineus, adductor magnus (adductor portion), and skin on medial surface of thigh; articular branches to hip and knee joints
Segmental branches	Quadratus lumborum and psoas muscles

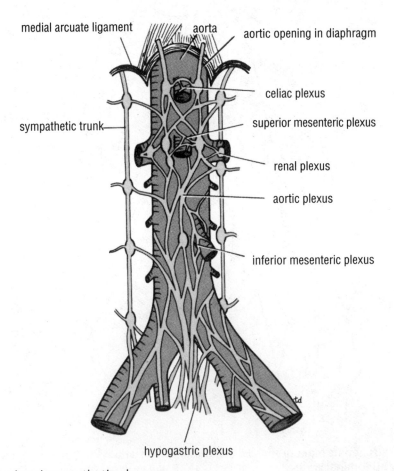

Fig. 3-18. Aorta and related sympathetic plexuses.

arcuate ligament and gains entrance to the pelvis below by passing behind the common iliac vessels. The sympathetic trunk possesses four or five segmentally arranged ganglia (Fig. 3-18).

Branches

1. **White rami communicantes** join the first two ganglia to the first two lumbar spinal nerves. A white ramus contains preganglionic nerve fibers and afferent sensory nerve fibers.
2. **Gray rami communicantes** join each ganglion to a corresponding lumbar spinal nerve. A gray ramus contains postganglionic nerve fibers.
3. Fibers pass medially to the sympathetic plexuses on the abdominal aorta and its branches (these plexuses also receive fibers from the splanchnic nerves and the vagus).
4. Fibers pass downward and medially to enter the pelvis, where, together with branches from sympathetic nerves in front of the aorta, they form a large bundle of nerve fibers called the **superior hypogastric plexus.**

Celiac Plexus

The celiac plexus is situated around the root of the celiac artery (see Fig. 3-18). It receives sympathetic fibers from the greater and lesser splanchnic nerves and parasympathetic fibers from the vagus. Branches from the plexus are distributed along branches of the celiac artery.

Superior Mesenteric Plexus

The superior mesenteric plexus is situated around the root of the superior mesenteric artery (see Fig. 3-18). It is continuous above with the celiac plexus. It receives a branch from the right vagus nerve. Branches from the plexus are distributed along the branches of the superior mesenteric artery.

Inferior Mesenteric Plexus

The inferior mesenteric plexus is situated around the root of the inferior mesenteric artery (see Fig. 3-18). It is continuous with the aortic plexus and receives branches from the lumbar part of the sympathetic trunk and parasympathetic fibers from the pelvic splanchnic nerve. Branches from the plexus are distributed along the branches of the inferior mesenteric artery.

Aortic Plexus

The aortic plexus is a continuous plexus around the abdominal aorta (see Fig. 3-18). Regional concentrations are known as the celiac, renal, superior mesenteric, and inferior mesenteric plexuses.

Muscles of the Posterior Abdominal Wall

The muscles of the posterior abdominal wall are summarized in Table 3-3.

Table 3-3 Muscles of the Posterior Abdominal Wall

Name of muscle	Origin	Insertion	Nerve supply	Action
Psoas	Transverse processes, bodies, and intervertebral discs of twelfth thoracic and five lumbar vertebrae	With iliacus into lesser trochanter of femur	Lumbar plexus	Flexes thigh on trunk; if thigh is fixed, it flexes the trunk on the thigh as in sitting up from lying down
Quadratus lumborum	Iliolumbar ligament, iliac crest, transverse processes of lower lumbar vertebrae	Twelfth rib	Lumbar plexus	Depresses twelfth rib during respiration; laterally flexes vertebral column to same side
Iliacus	Iliac fossa	With psoas into lesser trochanter of femur	Femoral nerve	Flexes thigh on trunk; if thigh is fixed, it flexes the trunk on the thigh as in sitting up from lying down

National Board Type Questions

Match the numbered structures shown in the anteroposterior radiograph of the abdomen with the appropriate lettered structures on the right.

Match the numbered structures shown in the anteroposterior radiograph of the stomach and small intestines following the ingestion of a barium meal with the appropriate lettered structures on the right.

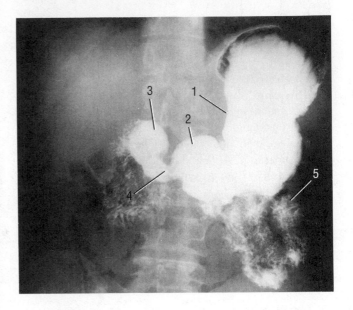

6. Structure 1.   A. Jejunum.
7. Structure 2.   B. Pylorus.
8. Structure 3.   C. Lesser curvature of stomach.
9. Structure 4.   D. First part of duodenum.
10. Structure 5.   E. Antrum of stomach.
                   F. None of the above.

1. Structure 1.   A. Sacroiliac joint.
2. Structure 2.   B. Pedicle of fourth lumbar vertebra.
3. Structure 3.   C. Iliopectineal line.
4. Structure 4.   D. Transverse process of fifth lumbar vertebra.
5. Structure 5.   E. Psoas muscle.

Match the numbered structures shown in the computed to-mographic (CT) scan of the abdomen at the level of the second lumbar vertebra following intravenous pyelography with the appropriate lettered structures on the right.

Match the numbered structures shown in the anteroposterior radiograph of the abdomen following a contrast enema with the appropriate lettered structures on the right.

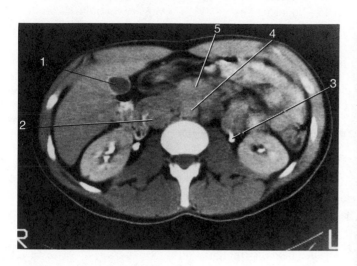

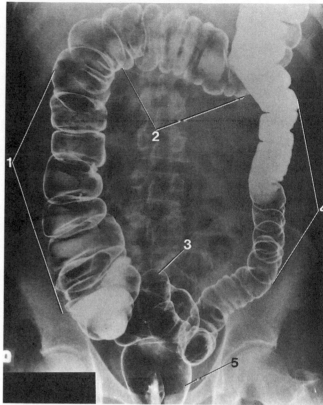

11. Structure 1.   A. Aorta.
12. Structure 2.   B. Vertebral body.
13. Structure 3.   C. Gallbladder.
14. Structure 4.   D. Pancreas.
15. Structure 5.   E. Left ureter.
                     F. Inferior vena cava.

16. Structure 1.   A. Sigmoid colon.
17. Structure 2.   B. Ascending colon.
18. Structure 3.   C. Rectum.
19. Structure 4.   D. Descending colon.
20. Structure 5.   E. Transverse colon.

Study the previous radiograph. Match the appropriate areas of the bowel numbered on the radiograph with the lettered arterial supply on the right.

21. Area 1.   A. Sigmoid arteries from inferior
22. Area 2.   mesenteric artery.
23. Area 3.   B. Middle colic artery.
24. Area 4.   C. Superior rectal artery.
25. Area 5.   D. Left colic artery.
                E. Right colic and colic branch of
                   ileocolic artery.

Match the numbered structures shown in the anteroposterior radiograph of both kidneys after intravenous injection of an iodine-containing compound with the appropriate lettered structure on the right.

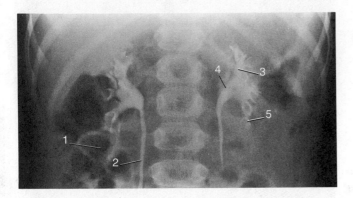

26. Structure 1.   A. Pelvis of left kidney.
27. Structure 2.   B. Pedicle of lumbar vertebra.
28. Structure 3.   C. Minor calyx.
29. Structure 4.   D. Gas in intestine.
30. Structure 5.   E. Major calyx.
                       F. Right ureter.

Match the viscera on the left with the areas of skin on the right where pain may **first** be referred.

31. Stomach.   A. Umbilical region.
32. Appendix.   B. Right iliac region.
33. Small intestine.   C. Epigastric region.
34. Left ureter   D. Penis or clitoris.
    (lower end).   E. None of the above.

Match the arterial branches on the left with their origin on the right.

35. Short gastric.   A. Superior mesenteric.
36. Right gastric.   B. Hepatic.
37. Left gastric.   C. Splenic.
38. Gastroduodenal.   D. None of the above.
39. Left gastroepiploic.

Match the autonomic innervation of the viscus on the left with the nerves on the right.

40. Parasympathetic nerves   A. Pelvic splanchnic
    to descending colon.   nerves.
41. Sympathetic nerves   B. Spinal cord segments
    to jejunum.   L1 and 2.
42. Sympathetic nerves to   C. Greater and lesser
    sigmoid colon.   splanchnic nerves.
43. Parasympathetic nerves   D. Vagus nerves.
    to ileum.   E. None of the above

Match the structure on the left with the opening of the diaphragm on the right.

44. Azygos vein.   A. Aortic hiatus.
45. Thoracic duct.   B. Esophageal hiatus.
46. Vagus nerves.   C. Vena caval hiatus.
                       D. Perforations in crura.
                       E. Between slips of origin
                          from ribs.

Matching. Indicate the position of the emerging branches of the lumbar plexus on the posterior abdominal wall relative to the psoas.

47. Iliohypogastric.   A. Medial to psoas.
48. Obturator.   B. Anterior to psoas.
49. Genitofemoral.   C. Lateral to psoas.
50. Femoral.   D. Superior to psoas.
51. Ilioinguinal.   E. None of the above.

Match the structure on the left with the appropriate lymph drainage on the right.

52. Fundus of stomach.   A. Superior mesenteric nodes.
53. Appendix.   B. Celiac nodes.
54. Right side of greater   C. Right gastroepiploic
    curvature of stomach.   nodes.
55. Sigmoid colon.   D. Inferior mesenteric
56. Proximal two-thirds   nodes.
    of transverse colon. E. Internal iliac nodes.

In each of the following questions, answer:

A. If only (1) is correct
B. If only (2) is correct
C. If both (1) and (2) are correct
D. If neither (1) nor (2) is correct

57. (1) The aorta lies in the midline of the abdomen and bifurcates into the right and left common iliac arteries in front of the body of the fifth lumbar vertebra.
    (2) The right renal vein crosses in front of the aorta to drain into the inferior vena cava.
58. (1) The superior epigastric artery descends in the rectus sheath posterior to the rectus abdominis muscle.
    (2) The superior mesenteric vessels descend posterior to the third part of the duodenum.
59. (1) The ligamentum venosum runs in the lower free margin of the falciform ligament.
    (2) The inferior mesenteric vein is usually a direct tributary of the splenic vein.
60. (1) The renal papillae open directly into the major calyces.
    (2) The right kidney lies slightly lower than the left kidney.

61. (1) The spleen is situated in the left hypochondrium and lies under cover of the ninth, tenth, and eleventh ribs.
    (2) The intestinal, right, and left lumbar trunks drain into the cisterna chyli.

62. (1) The arteries supplying the suprarenal glands include a direct branch from the aorta.
    (2) The sympathetic innervation to the suprarenal glands is from the lumbar outflow.

63. The gastrosplenic omentum contains the following important structures:
    (1) The splenic artery and vein.
    (2) The short gastric arteries.

64. The lienorenal ligament contains the following important structures:
    (1) The left renal vein.
    (2) The tail of the pancreas.

65. The sensory innervation of the peritoneum on the central part of the inferior surface of the diaphragm is derived from:
    (1) Phrenic nerves.
    (2) Greater splanchnic nerves.

66. The anterior wall of the rectus sheath at the level of the umbilicus includes:
    (1) The external oblique aponeurosis.
    (2) The anterior portion of the internal oblique aponeurosis.

67. In the angle formed by the proximal part of the superior mesenteric artery and the abdominal aorta, which of the following structure(s) is (are) found?
    (1) Body of pancreas.
    (2) Third part of duodenum.

68. Lymphatic drainage of the kidney is directly to:
    (1) Superior mesenteric nodes.
    (2) Inferior mesenteric nodes.

69. The following statements concern the thoracic duct:
    (1) It begins in the abdomen at the cisterna chyli.
    (2) It ascends through the aortic opening in the diaphragm.

70. The lymph vessels from the testes drain into the following lymph nodes:
    (1) The superficial inguinal nodes.
    (2) The para-aortic nodes.

71. The pancreas has:
    (1) A venous drainage restricted to the portal system.
    (2) A body that lies in front of the splenic artery.

Select the **best** response.

72. The following statements concerning the liver are correct **except:**
    A. Its lymph drainage is to the celiac nodes.
    B. The quadrate and caudate lobes are functionally part of the left lobe.
    C. Its parasympathetic innervation is from the vagus nerve.
    D. It receives highly oxygenated blood from the portal vein.
    E. The triangular ligaments connect the liver to the diaphragm.

73. The following differences exist between the ileum and the ascending colon **except:**
    A. The ascending colon has appendices epiploicae whereas the ileum does not.
    B. The arterial supply to the wall of the ileum is arranged in such a way that it produces areas of weakness through which mucosal herniations may take place.
    C. The ascending colon may have a well-developed marginal artery, but the ileum does not.
    D. The ileum has longitudinal muscle that forms a continuous layer around the wall, whereas the ascending colon has teniae coli.
    E. The ascending colon is retroperitoneal whereas the ileum is intraperitoneal.

74. Which of the following statements is (are) correct regarding the duodenum?
    A. The duodenum receives its entire blood supply from the superior mesenteric artery.
    B. The first part of the duodenum is completely retroperitoneal.
    C. The bile duct enters the second (vertical) portion of the duodenum.
    D. The lymph drainage of the fourth part is into the celiac nodes.
    E. The third part of the duodenum passes horizontally to the left in front of the root of the mesentery of the small intestine.

75. The following statements concerning the appendix are correct **except:**
    A. The appendix is situated in the right iliac region.
    B. At its base, the teniae coli of the cecum fuse to form a complete longitudinal muscle layer in the wall of the appendix.
    C. Afferent pain nerve fibers accompany the sympathetic nerves and enter the spinal cord at the level of the tenth thoracic segment.
    D. It receives its blood supply from a branch of the posterior cecal artery.
    E. The tip of the appendix cannot reach down into the pelvic cavity.

76. The lesser omentum contains the following important structures **except:**
    A. Left gastric artery.
    B. Portal vein.

C. Inferior vena cava.

D. Bile duct.

E. Common hepatic artery.

77. The following statements concerning the pyloric sphincter are correct **except:**

A. It receives its motor innervation from the sympathetic autonomic nerves.

B. It lies on the level of the third lumbar vertebra.

C. It is inhibited by the impulses passing down the vagus nerves.

D. Is formed by a thickening of the circular layer of smooth muscle in the stomach wall.

E. The cavity of the pylorus is called the pyloric canal.

78. The following structures form the boundaries of the entrance into the lesser sac (epiploic foramen) **except:**

A. Inferior vena cava.

B. Bile duct.

C. Portal vein.

D. Quadrate lobe of liver.

E. First part of the duodenum.

79. After complete occlusion of the origin of the inferior mesenteric artery with a blood clot, the blood supply of the left portion of the colon is maintained by the following arteries **except:**

A. The marginal artery (of Drummond).

B. The middle colic artery.

C. The left lumbar arteries.

D. Anastomoses between the superior, middle, and inferior hemorrhoidal arteries.

E. The sigmoid arteries.

80. The following statements concerning the lesser sac are true **except:**

A. It projects superiorly behind the quadrate lobe of the liver as far as the diaphragm.

B. It lies posterior to the stomach.

C. Its left margin is formed by the spleen and the lienorenal ligament and the gastrosplenic omentum (ligament).

D. It descends between the layers of the greater omentum.

E. It communicates with the greater sac through the epiploic foramen.

81. The transverse colon is characterized by all of the following **except:**

A. Appendices epiploicae.

B. Haustra.

C. Three bands of longitudinal muscle.

D. Plicae circulares.

E. Attached to the greater omentum.

82. The following statements concerning the gallbladder are correct **except:**

A. The arterial supply is from the cystic artery, a branch of the right hepatic artery.

B. The fundus of the gallbladder is located just beneath the tip of the right ninth costal cartilage.

C. The peritoneum completely surrounds the fundus, body, and neck.

D. The nerves of the gallbladder are derived from the celiac plexus.

E. Pain sensation from gallbladder disease may be referred along the phrenic and supraclavicular nerves to the skin over the shoulder.

83. In patients with an obstruction of the superior vena cava, blood may return to the right atrium using the following anastomotic channels **except:**

A. Lateral thoracic veins, lumbar veins, and superficial epigastric veins.

B. Superior and inferior epigastric veins.

C. Lateral thoracic veins, paraumbilical veins, and portal vein.

D. Posterior intercostal veins, lumbar veins.

E. Lateral thoracic veins and superior epigastric veins.

84. Which of the following structures may be eroded by a perforating gastric ulcer on the posterior wall of the stomach?

A. Inferior mesenteric artery.

B. Splenic artery.

C. Right kidney.

D. Second part of the duodenum.

E. Quadrate lobe of the liver.

85. Which of the following facts concerning the left ureter is **correct?**

A. It has the inferior mesenteric vein lying on its lateral side.

B. Its lumen is constricted at the point where it crosses the brim of the pelvis.

C. Its entire arterial supply is derived from the left renal artery.

D. It lies in direct contact with the tips of the transverse processes of the lumbar vertebrae.

E. It lies within the peritoneal cavity.

86. The following structures are connected to the liver. Which structure provides the greatest support for the liver?

A. The falciform ligament.

B. The coronary ligament.

C. The ligamentum teres.

D. The hepatic veins joining the inferior vena cava.

E. The ligamentum venosum.

87. A gallstone eroded through the posterior wall of the gallbladder and entered the intestinal tract. Which part of the intestinal tract is likely to initially contain the stone?

A. Sigmoid colon.

B. Descending colon.

C. Transverse colon.

D. Ascending colon.

E. Jejunum.

88. Which of the following statements concerning the superficial inguinal ring is **incorrect?**

    A. It is a perforation in the aponeurosis of the external oblique muscle.

    B. Its greatest width lies above and medial to the pubic tubercle.

    C. It is strengthened posteriorly by the conjoint tendon.

    D. The internal spermatic fascia is attached to its margins.

    E. In the male it permits the passage of the spermatic cord and the ilioinguinal nerve.

89. A surgeon decides to divide the anterior vagal trunk (vagotomy) as it lies on the anterior surface of the abdominal part of the esophagus as treatment for a chronic gastric ulcer that is not responding to medical treatment. Which of the following is likely to result from this procedure?

    A. A loss of secretomotor nerve supply to the mucosal glands of the stomach.

    B. The patient would become hoarse because of paralysis of the intrinsic muscles of the larynx on the left side.

    C. The heart rate would increase because of the decreased parasympathetic input to the cardiac plexus.

    D. The patient would become incontinent because of an absence of parasympathetic input to the bladder.

    E. Input into the greater splanchnic nerves would be compromised.

90. The skin of the umbilicus receives its sensory innervation from:

    A. T7.

    B. T12.

    C. L1.

    D. T10.

    E. L2.

91. Concerning the renal fascia of the kidney, all of the following statements are true **except:**

    A. It is continuous with the transversalis fascia.

    B. It surrounds the perirenal fat.

    C. It does not enclose the suprarenal gland.

    D. It separates the kidney from the pararenal fat.

    E. It covers the ureter.

92. The inguinal canal in both sexes is formed by:

    A. The descent of the gonad.

    B. The contraction of the gubernaculum.

    C. The processus vaginalis.

    D. The ilioinguinal nerve.

    E. The contraction of the cremaster muscle.

93. Which of the following statements concerning the abdominopelvic cavity is **not** correct?

    A. All the organs in the abdomen and pelvis are not located in the peritoneal cavity.

    B. Mesenteries suspend the intestines from the posterior abdominal wall.

    C. Few, if any, blood vessels or nerves are located between the two peritoneal folds of a mesentery.

    D. The suspensory ligament of the duodenum (ligament of Treitz) is attached at the site of the duodenojejunal junction.

    E. The male peritoneal cavity does not communicate with the exterior.

94. The lacunar ligament is:

    A. Formed from the conjoint tendon.

    B. Part of the posterior wall of the rectus sheath.

    C. Not continuous with the inguinal ligament.

    D. An important medial relation to the femoral ring of the femoral sheath.

    E. Attached to the inferior ramus of the pubis.

95. The following statements concerning the superficial fascia of the anterior abdominal wall are true **except:**

    A. It has a superficial fatty layer and a deep membranous layer.

    B. Scarpa's fascia fuses with the fascia lata just below the inguinal ligament.

    C. Camper's fascia is continuous with the Colles' fascia in the perineum.

    D. It is continuous with the dartos muscle in the wall of the scrotum.

    E. It does not contribute to the femoral sheath.

96. The rectus sheath contains the following structures **except:**

    A. The pyramidalis muscle (when present).

    B. The ligamentum teres.

    C. The inferior epigastric artery.

    D. T7–12 anterior rami.

    E. The rectus abdominis.

97. The following statements concerning the epididymis are correct **except:**

    A. It lies within the scrotum posterior to the testis.

    B. It has an expanded upper end called the head.

    C. The tail gives rise to the vas deferens.

    D. It is supplied by the testicular artery.

    E. Its lymph drains into the superficial inguinal nodes.

98. The jejunum and ileum can be differentiated on the basis of all the following anatomic features **except:**

    A. Numerous (4–5) arterial arcades are associated with the jejunum.

    B. The plicae circulares are much more prominent in the jejunum.

    C. Fat depositions are generally present throughout the mesentery associated with the ileum.

D. The jejunum is generally located in the upper left region of the abdominal cavity.

E. Peyer's patches are characteristic of the lower ileum and may be visible on the surface.

99. The spermatic cord contains the following structures **except:**

A. The scrotal arteries and veins.

B. The vas deferens.

C. The pampiniform plexus.

D. The testicular artery.

E. Autonomic nerves.

100. The following structures pass through the esophageal hiatus in the diaphragm **except:**

A. The left vagus nerve.

B. Branches of the left gastric artery.

C. The left phrenic nerve.

D. The right vagus nerve.

E. A tributary of the portal vein.

101. The following statements regarding the peritoneum are incorrect **except:**

A. The parietal peritoneum is sensitive to stretch only.

B. The parietal peritoneum lining the anterior abdominal wall is innervated by the lower six thoracic and first lumbar spinal nerves.

C. The visceral peritoneum is innervated by the posterior primary rami of the twelfth thoracic and lumbar spinal nerves.

D. The visceral peritoneum is sensitive to pain, temperature, and touch.

E. The peritoneum lining the lesser sac has no nerve supply.

102. The following venous anastomoses are important in uniting the portal to the systemic venous systems should the portal vein become blocked **except:**

A. Esophageal branches of the left gastric vein and esophageal branches of the azygos veins.

B. Veins of the ligamentum teres and the paraumbilical veins.

C. Superior rectal and inferior rectal veins.

D. The right colic veins and the lumbar veins.

E. Middle rectal and inferior rectal veins.

103. In a patient with cancer of the stomach that requires a total gastrectomy, the following arteries have to be ligated **except:**

A. Common hepatic artery.

B. Short gastric arteries.

C. Left and right gastroepiploic arteries.

D. Right gastric artery.

E. Left gastric artery.

104. When performing a midline abdominal paracentesis (tapping the abdominal cavity) below the umbilicus, the cannula will pass through the following anatomic structures **except:**

A. Skin and fascia.

B. Linea alba.

C. Transversalis fascia and extraperitoneal fat.

D. Rectus abdominis.

E. Parietal peritoneum.

105. The following statements are correct concerning an inguinal hernia **except:**

A. The inferior epigastric artery lies medial to the neck of an indirect inguinal hernia.

B. The occurrence of an inguinal hernia in the female is more common than the femoral hernia.

C. A direct inguinal hernia is more common in elderly men than in boys.

D. The opening of the sac of a direct inguinal hernia is wide.

E. The contents of an indirect inguinal hernia may be strangulated against the lacunar ligament.

106. The following structures normally pass through the inguinal canal in the female **except:**

A. Ilioinguinal nerve.

B. Round ligament of the ovary.

C. Lymph vessels.

D. Round ligament of the uterus.

E. Remains of the processus vaginalis.

107. The following statements concerning the portal vein are true **except:**

A. It courses through a portion of the lesser omentum.

B. It enters the liver at the porta hepatis.

C. It receives venous blood from both the large and small intestines.

D. It originates at the junction of the superior mesenteric vein and the splenic vein.

E. It passes in front of the neck of the pancreas.

108. If the common hepatic artery is unavoidably ligated during surgery, the arterial supply to the liver would be maintained by the following anastomotic connections **except:**

A. Superior pancreaticoduodenal artery anastomosing with the inferior pancreaticoduodenal artery.

B. Right gastric artery anastomosing with the left gastric artery.

C. The gastroduodenal artery anastomosing with the splenic artery.

D. The esophageal arteries anastomosing with the inferior phrenic arteries.

E. Right gastroepiploic artery anastomosing with the left gastroepiploic artery.

109. The following general statements are incorrect **except:**

A. The superior and inferior epigastric vessels anastomose on the anterior surface of the rectus abdominis muscle.

B. The inguinal ligament is attached laterally to the anterior inferior iliac spine.

C. The abdominal aorta bifurcates below in front of the fourth lumbar vertebra.

D. The foregut ends in the third (inferior) segment of the duodenum.

E. The umbilicus is located approximately at vertebral level L5.

110. The following veins drain directly into the inferior vena cava **except:**

A. Hepatic veins.

B. Renal veins.

C. Lumbar veins.

D. Inferior mesenteric vein.

E. Right testicular (ovarian) vein.

111. In order to tap a hydrocele (collection of fluid in the tunica vaginalis), the cannula would have to pierce the following structures **except:**

A. Skin.

B. Fatty subcutaneous tissue (fascia of Camper).

C. Dartos muscle.

D. Cremasteric fascia.

E. External and internal spermatic fasciae.

112. Which of the following arteries supplies the suprarenal (adrenal) gland?

A. Aorta.

B. Lumbar arteries.

C. Superior phrenic artery.

D. Testicular (ovarian) artery.

E. Subcostal artery.

113. The following structures are present in the porta hepatis **except:**

A. Lymph nodes.

B. Right and left branches of the portal vein.

C. Right and left hepatic ducts.

D. Right and left hepatic veins.

E. Right and left branches of the hepatic artery.

114. The following statements concerning the celiac plexus are true **except:**

A. The celiac plexus is not a purely sympathetic plexus.

B. The celiac plexus does not surround the celiac artery.

C. The celiac ganglia are made up of nerve cell bodies and nerve fibers.

D. Parasympathetic preganglionic fibers pass through the plexus generally synapsing with postganglionic neurons within the walls of the organs they innervate.

E. Sympathetic preganglionic fibers to the celiac plexus originate from thoracic spinal segments and travel in thoracic splanchnic nerves.

115. The right kidney has the following important relationships **except:**

A. It is related to the neck of the pancreas.

B. It is anterior to the right costodiaphragmatic recess.

C. It is related to the second part of the duodenum.

D. It is related to the right colic flexure.

E. It is anterior to the right twelfth rib.

In the next section, study each case history. Select the **best** answer to each question following it.

A 54-year-old man with a long history of duodenal ulcer was seen in the emergency department after vomiting blood-stained fluid and exhibiting all the signs and symptoms of severe hypovolemic (loss of blood) shock.

116. Assuming that the ulcer had perforated the posterior wall of the first part of the duodenum, name the artery that is most likely to have been eroded.

A. Splenic artery.

B. Right gastric artery.

C. Gastroduodenal artery.

D. Right gastroepiploic artery.

E. Right renal artery.

117. If the duodenal ulcer had perforated the anterior wall of the duodenum, where in the peritoneal cavity would the duodenal contents have entered?

A. The lesser sac.

B. The right anterior subphrenic space.

C. The left lateral paracolic gutter.

D. The right posterior subphrenic space.

E. The pouch of Douglas.

A 10-year-old girl was seen in the emergency department with a temperature of 101°F, a furred tongue, and pain in the right iliac region. On examination, the skin in the right lower quadrant was hyperesthetic and tender to touch; the abdominal muscles showed rigidity and guarding. A diagnosis of acute appendicitis was made.

118. The skin in the right lower quadrant was tender to touch because the following nerve(s) were stimulated by the inflamed appendix.

A. The eighth right intercostal nerve.

B. The tenth right intercostal nerve.

C. The subcostal nerve and the first lumbar nerve.

D. The sympathetic autonomic afferent nerves from the appendix.

E. The seventh right thoracic nerve.

119. The following muscles were contracted in the right lower quadrant producing the increased tone and guarding.

A. Right rectus abdominis muscle.

B. Right pyramidalis muscle.

C. Right psoas muscle.

D. The right external oblique, internal oblique, and transversus abdominis muscles.

E. Right quadratus lumborum muscle.

Answers and Explanations

1. D
2. B
3. E
4. A
5. C
6. C
7. E
8. D
9. B
10. A
11. C
12. F
13. E
14. A
15. D
16. B
17. E
18. A
19. D
20. C
21. E
22. B
23. A
24. D
25. C
26. D
27. F
28. E
29. A
30. C
31. C
32. A
33. A
34. D   Pain from the upper end of the ureter is referred to the back behind the kidney, pain from the middle region of the ureter is referred to the inguinal region, and pain from the lower end is referred to the penis or clitoris. The explanation is that the afferent nerves enter the spinal cord at different levels and the pain is referred along the spinal nerves originating from those spinal cord levels.
35. C
36. B
37. D
38. B
39. C
40. A

41. C
42. B
43. D
44. A
45. A
46. B
47. C
48. A
49. B
50. C
51. C
52. B
53. A
54. C

In questions 52, 53, and 54, the lymphatic vessels follow the arteries and the nodes are named according to the names of the arteries.

55. D
56. A
57. D   The aorta bifurcates at the level of L4. The right renal vein drains into the inferior vena cava, which is on the right side of the aorta.
58. A   The superior mesenteric vessels descend anterior to the third part of the duodenum.
59. B   The ligamentum teres lies in the lower margin of the falciform ligament.
60. B   The renal papillae open into the minor calyces.
61. C
62. A   The sympathetic innervation of the suprarenal glands is from the greater splanchnic nerves (T5–9).
63. B
64. B
65. A
66. C
67. B
68. D
69. C
70. B   The testes develop on the posterior abdominal wall at the level of the first lumbar vertebra. During development they descend, dragging their lymphatic drainage behind them.
71. B   Some of the veins anastomose with the lumbar veins behind the pancreas.
72. D   The liver receives highly oxygenated blood via the hepatic artery.
73. B   It is the colon that often develops mucosal herniations along its arterial supply.
74. C   The upper half of the duodenum (derived from the foregut) receives its blood supply from branches of the celiac artery (superior pancreaticoduodenal artery). The lower half of the duodenum (derived from the midgut) receives its blood supply from

branches of the superior mesenteric artery (inferior pancreaticoduodenal artery). The first inch of the duodenum is covered on its anterior and posterior surfaces with peritoneum; the remainder of the duodenum is retroperitoneal. The lymph drainage of the fourth part of the duodenum is into the superior mesenteric nodes. The third part of the duodenum passes horizontally to the left in front of the vertebral column and is crossed in front by the root of the mesentery of the small intestine and the superior mesenteric vessels.

75. E   The appendix varies in length from 3 to 5 in. (8–13 cm), or may be even longer (it tends to be longer in childhood, later atrophying). When long the tip is often found hanging down into the pelvic cavity.

76. C   The inferior vena cava is retroperitoneal behind the opening into the lesser sac.

77. B   The pylorus lies on the transpyloric plane at the level of the first lumbar vertebra.

78. D   The superior boundary of the entrance into the lesser sac is formed by the caudate process of the caudate lobe of the liver and not by the quadrate lobe.

79. C   The middle colic artery from the superior mesenteric artery, the sigmoid arteries, and the marginal artery and the anastomoses of all these arteries with the superior, middle, and inferior hemorrhoidal arteries will maintain the blood supply to the left portion of the colon.

80. A   The superior recess of the lesser sac projects superiorly as far as the diaphragm behind the caudate lobe of the liver (not the quadrate lobe).

81. D   Plicae circulares are found in the small intestine.

82. C   The fundus is the only part of the gallbladder that is completely surrounded by peritoneum.

83. E   The lateral thoracic veins and the superior epigastric veins are directly or indirectly connected with the superior vena cava only.

84. B

85. B

86. D

87. C   The transverse colon is in close posterior relation to the gallbladder.

88. D   The external spermatic fascia is attached to the margins of the superficial inguinal ring.

89. A

90. D

91. C

92. C

93. C

94. D

95. C

96. B

97. E   The lymphatic drainage of the testis and the epididymis is into the lateral aortic nodes at the level of the first lumbar vertebra.

98. A   The jejunum has usually only one or two arterial arcades with long branches passing to the intestinal wall.

99. A

100. C   The terminal branches of the left phrenic nerve pierce the diaphragm but do not pass through the hiatus.

101. B   The parietal peritoneum is sensitive to pain, temperature, and touch. The visceral peritoneum is sensitive to stretch and is innervated by autonomic afferent nerves. The peritoneum lining the lesser sac is innervated by somatic and autonomic afferent nerve fibers.

102. E   The middle and inferior rectal veins are tributaries of the systemic circulation only.

103. A

104. D   The rectus abdominis muscles lie on either side of the midline and are separated by the linea alba.

105. E   The lacunar ligament is related to the sac of a femoral hernia, **not** an inguinal hernia.

106. B   The round ligament of the ovary, which represents the remains of the upper half of the gubernaculum, extends from the medial margin of the ovary to the lateral margin of the body of the uterus. Unlike the round ligament of the uterus, the round ligament of the ovary does not pass through the inguinal canal.

107. E   The portal vein is formed behind the neck of the pancreas by the union of the superior mesenteric vein with the splenic vein.

108. C   The gastroduodenal artery does not directly anastomose with the splenic artery.

109. C   The superior and inferior epigastric arteries anastomose on the posterior surface of the rectus abdominis muscles. The inguinal ligament is attached at its lateral end to the anterior superior iliac spine. The foregut ends below halfway down the second part of the duodenum. The umbilicus is variable in position but is usually located at the level of the fourth lumbar vertebra.

110. D   The inferior mesenteric vein is a tributary of the splenic vein.

111. B   The dartos muscle takes the place of the fatty subcutaneous tissue (fascia of Camper) in the wall of the scrotum.

112. A

113. D   The hepatic veins (three or more in number) leave the posterior surface of the liver and drain directly into the inferior vena cava.

114. B   The celiac plexus completely surrounds the origin of the celiac artery.
115. A   The pancreas is not related to the right kidney.
116. C   The gastroduodenal artery is a large branch of the hepatic artery. It descends behind the first part of the duodenum and divides into the superior pancreatico-duodenal artery and the right gastroepiploic artery.
117. D   If the patient was in the erect position when the perforation occurred, the duodenal contents would then tend to gravitate downward along the right para-colic gutters.
118. C   Once the inflamed appendix comes into contact with the parietal peritoneum in the right iliac region, it stimulates the lower thoracic spinal nerves and the first lumbar spinal nerve, which supply the peritoneum. Since these nerves also supply the anterior abdominal wall muscles in this region, they are reflexly stimulated to contract. This mechanism tends to immobilize this region and assists in localizing the inflammatory process should the appendix rupture.
119. D

Pelvis and Perineum

Pelvis

The *term* pelvis is loosely used to describe the region where the trunk and lower limbs meet. The *word* pelvis means a basin and is more correctly applied to the skeleton of the region—the bony pelvis.

Bony Pelvis

The bony pelvis is composed of four bones: the two **hip bones,** the **sacrum,** and the **coccyx** (Fig. 4-1). The two hip bones articulate with each other anteriorly at the **symphysis pubis** and posteriorly with the sacrum at the **sacroiliac joints.**

The pelvis is divided into two parts by the **pelvic brim,** which is formed by the **sacral promontory** behind, the **iliopectineal lines** laterally, and the symphysis pubis anteriorly. Above the brim is the **false pelvis,** or greater pelvis, which is the expanded part of the pelvis that forms part of the abdominal cavity. Below the brim is the **true pelvis,** or lesser pelvis.

True Pelvis

The true pelvis is a bowl-shaped structure that contains and protects the lower parts of the intestinal and urinary tracts and the internal organs of reproduction. The true pelvis has an inlet, an outlet, and a cavity. The **pelvic inlet** is bounded posteriorly by the sacral promontory, laterally by the iliopectineal lines, and anteriorly by the symphysis pubis. The **pelvic outlet** is bounded posteriorly by the coccyx, laterally by the ischial tuberosities, and anteriorly by the pubic arch. Laterally, in addition to the ischial tuberosities, there are the inflexible sacrotuberous and sacrospinous ligaments (Fig. 4-2; see also Fig. 4-1). Note that these ligaments divide up the sciatic notches into the greater and lesser sciatic foramina (see Fig. 4-2). The **pelvic cavity** lies between the inlet and the outlet. It is a short, curved canal with a shallow anterior wall and a much deeper posterior wall (Fig. 4-3).

Joints of the Pelvis

1. Sacroiliac joints
2. Symphysis pubis
3. Sacrococcygeal joint

Sacroiliac Joints

The sacroiliac joints are very strong synovial joints between the auricular surfaces of the sacrum and the iliac bones (see Fig. 4-2). The very strong **posterior** and **interosseous sacroiliac ligaments** suspend the sacrum between the two iliac bones. The **anterior sacroiliac ligament** is thin and is placed in front of the joint. A small but limited amount of movement is possible at these joints. Their main function is to transmit the weight of the body from the vertebral column to the bony pelvis.

Nerve Supply

Branches of sacral plexus.

Symphysis Pubis

The symphysis pubis is a cartilaginous joint between the two pubic bones (see Fig. 4-2). The articular surfaces are covered by hyaline cartilage and are connected by a fibrocartilaginous disc. The joint is surrounded by ligaments. Almost no movement is possible at this joint.

Sacrococcygeal Joint

The sacrococcygeal joint is a cartilaginous joint between the bodies of the fifth sacral vertebra and the first coccygeal vertebra. The cornua of the sacrum and coccyx are joined by ligaments. A great deal of movement is possible.

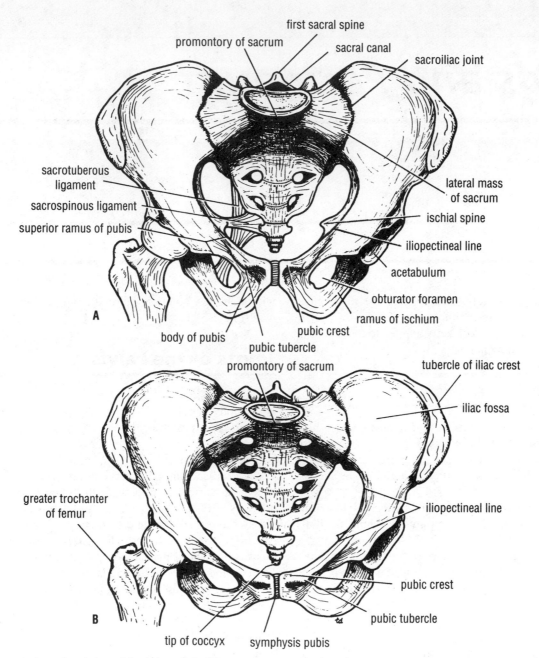

Fig. 4-1.   Male pelvis (**A**) and female pelvis (**B**), as seen on anterior view.

During pregnancy, the ligaments of the pelvic joints are relaxed by the action of the hormones estrogen, progesterone, and relaxin. This makes possible an enlargement of the pelvic cavity and thus facilitates delivery.

Sex Differences of the Pelvis

Sex differences of the pelvis exist because the female pelvis is broader than that of the male to allow easier pas-

sage of the fetal head and because female bones are more slender than those of the male (see Fig. 4-3).

1. The false pelvis is shallow in the female and deep in the male.
2. The pelvic inlet is transversely oval in the female but is heart-shaped in the male. This is due to the indentation produced by the promontory of the sacrum in the male.
3. The pelvic cavity is roomier in the female than the male, and the distance between the inlet and the outlet is much shorter.

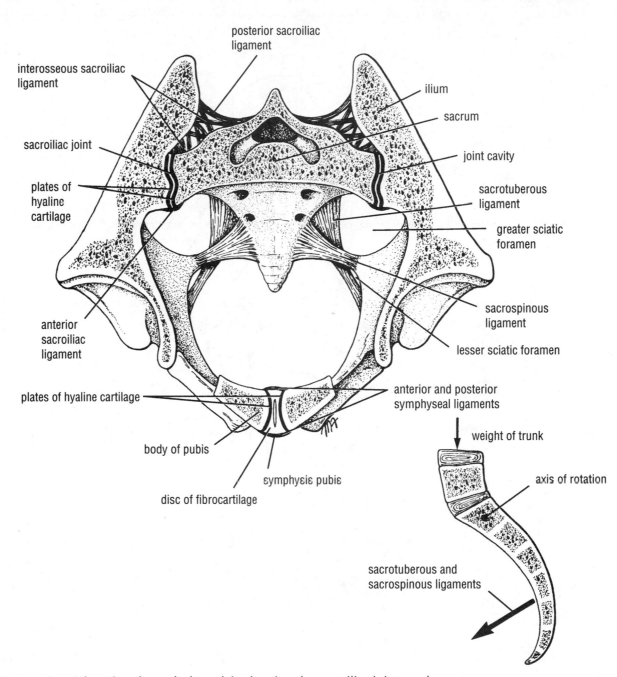

Fig. 4-2. Horizontal section through the pelvis, showing the sacroiliac joints and the symphysis pubis. The lower diagram shows the function of the sacrotuberous and sacrospinous ligaments in resisting the rotation force exerted on the sacrum by the weight of the trunk.

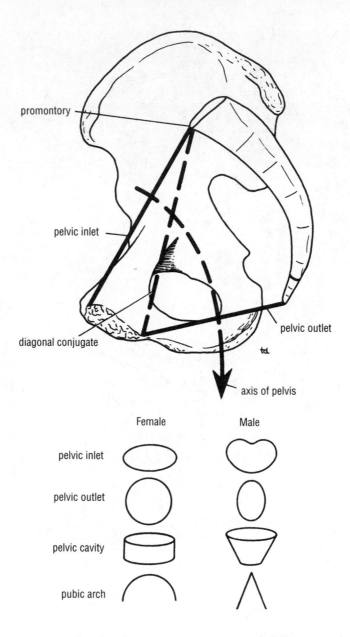

promontory

pelvic inlet

diagonal conjugate

pelvic outlet

axis of pelvis

Female   Male

pelvic inlet

pelvic outlet

pelvic cavity

pubic arch

Fig. 4-3. Pelvic inlet, pelvic outlet, diagonal conjugate, and axis of pelvis. Lower diagrams illustrate some of the main differences between female and male pelvis.

4. The pelvic outlet is larger in the female than in the male. The ischial tuberosities are everted in the female and turned in in the male.
5. The sacrum is shorter, wider, and flatter in the female than in the male.
6. The subpubic angle, or pubic arch, is more rounded and wider in the female than in the male (see Fig. 4-1).

Pelvic Walls

The walls of the pelvis are formed by bones and ligaments that are partly lined with muscles covered with fascia and parietal peritoneum. The pelvis has anterior, posterior, and lateral walls and it also has an inferior wall or floor (Fig. 4-4).

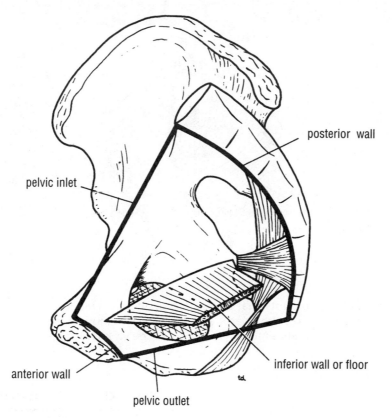

posterior wall

pelvic inlet

inferior wall or floor

anterior wall

pelvic outlet

Fig. 4-4. Right half of pelvis, showing pelvic walls.

Anterior Pelvic Wall

The anterior wall is the shallowest wall and is formed by (1) the bodies of the pubic bones, (2) the pubic rami, and (3) the symphysis pubis.

Posterior Pelvic Wall

The posterior wall is extensive and is formed by (1) the sacrum and coccyx and (2) the piriformis muscles and their covering fascia.

Lateral Pelvic Wall

The lateral wall is formed by (1) the hip bone (below the pelvic inlet), (2) the obturator membrane, (3) the sacrotuberous and sacrospinous ligaments, and (4) the obturator internus muscle and its covering fascia.

Obturator Membrane

The obturator membrane is a fibrous sheet that almost completely closes the obturator foramen in the hip bone. The small gap that is left is called the **obturator canal** and is for the passage of the obturator nerve and vessels.

Sacrotuberous Ligament

The sacrotuberous ligament is a strong ligament that extends from the lateral part of the sacrum and coccyx and

the posterior inferior iliac spine to the ischial tuberosity (see Fig. 4-2).

Sacrospinous Ligament

The sacrospinous ligament is a strong triangular-shaped ligament that is attached by its base to the lateral part of the sacrum and coccyx and by its apex to the spine of the ischium (see Fig. 4-2).

The sacrotuberous and sacrospinous ligaments prevent the lower end of the sacrum and the coccyx from being rotated at the sacroiliac joint by the weight of the body (see Fig. 4-2).

Inferior Pelvic Wall (Pelvic Floor)

The floor of the pelvis supports the pelvic viscera and is formed by the pelvic diaphragm. Below the pelvic floor lies the perineum.

The pelvic diaphragm (Fig. 4-5) is formed by (1) the important levatores ani muscles, (2) the small coccygeal muscles, and (3) the pelvic fascia covering these muscles. The pelvic diaphragm is incomplete anteriorly to allow for the passage of the urethra and also the vagina in the female.

Muscles of the Pelvic Walls

The attachments, nerve supply, and actions of the muscles of the pelvic walls are given in Table 4-1.

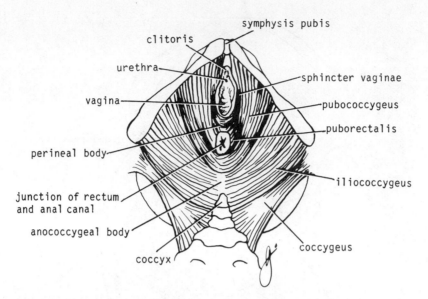

Fig. 4-5. The muscles of the pelvic floor. Note that the levator ani muscle is made up of a number of different muscle groups. The levator ani and coccygeus muscles with their fascial coverings form a continuous muscular floor to the pelvis known as the pelvic diaphragm.

Pelvic Fasciae

The **parietal pelvic fascia** lines the walls of the pelvis and is named according to the muscle it overlies. Above the pelvic inlet it is continuous with the fascia lining the abdominal walls.

The **visceral pelvic fascia** covers all the pelvic viscera. In the female, this fascia around the cervix is referred to as the **parametrium.**

Pelvic Peritoneum

The parietal peritoneum lines the pelvic walls and is reflected onto the pelvic viscera, where it becomes continuous with the visceral peritoneum.

Nerves of the Pelvis

Sacral Plexus

The sacral plexus lies on the posterior pelvic wall in front of the piriformis muscle (Fig. 4-6). It is formed from the anterior rami of the fourth and fifth lumbar nerves and the anterior rami of the first, second, third, and fourth sacral nerves (Fig. 4-7). The contribution from the fourth lumbar nerve joins the fifth lumbar nerve to form the **lumbosacral trunk.** The lumbosacral trunk passes down into the pelvis and joins the sacral nerves as they emerge from the anterior sacral foramina.

Table 4-1   Muscles of the Pelvic Walls and Floor

Name of muscle	Origin	Insertion	Nerve supply	Action
Piriformis	Front of sacrum	Greater trochanter of femur	Sacral plexus	Lateral rotator of femur at hip joint
Obturator internus	Obturator membrane and adjoining part of hip bone	Greater trochanter of femur	Sacral plexus	Lateral rotator of femur at hip joint
Levator ani	Body of pubis, fascia of obturator internus, spine of ischium	Perineal body; anococcygeal body; walls of prostate, vagina, rectum, and anal canal	Pudendal nerve, fourth sacral nerve	Supports pelvic viscera; sphincter to anorectal junction and vagina
Coccygeus	Spine of ischium	Lower end of sacrum, coccyx	Fourth and fifth sacral nerve	Assists levator ani to support pelvic viscera; flexes coccyx

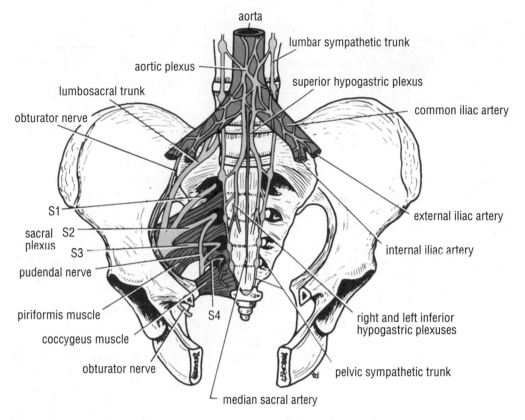

Fig. 4-6. Posterior pelvic wall, showing sacral plexus, superior hypogastric plexus, and right and left inferior hypogastric plexuses. Pelvic parts of sympathetic trunks are also shown.

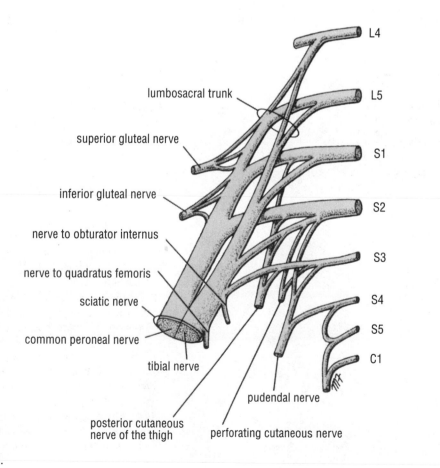

Fig. 4-7. Sacral plexus.

Branches

1. Branches to the lower limb that leave the pelvis through the greater sciatic foramen (see Fig. 4-6).
   a. **Sciatic nerve** (L4 and 5; S1, 2, and 3). This is the largest nerve in the body and the largest branch of the sacral plexus.
   b. **Superior gluteal nerve.** This supplies the gluteus medius, gluteus minimus, and the tensor fasciae latae muscles.
   c. **Inferior gluteal nerve.** This supplies the gluteus maximus muscle.
   d. **Nerve to quadratus femoris muscle.** This nerve also supplies the inferior gamellus muscle.
   e. **Nerve to obturator internus muscle.** This nerve also supplies the superior gamellus muscle.
   f. **Posterior cutaneous nerve of the thigh.** This supplies the skin of the buttock and the back of the thigh.

2. Branches to the pelvic muscles, pelvic viscera, and perineum.
   a. **Pudendal nerve** (S2, 3, and 4). This leaves the pelvis through the greater sciatic foramen and enters the perineum through the lesser sciatic foramen (see Fig. 4-6).
   b. **Nerves to the piriformis muscle.**
   c. **Pelvic splanchnic nerves.** These constitute the sacral part of the parasympathetic nervous system. They arise from the second, third, and fourth sacral nerves and are distributed to the pelvic viscera.

The branches of the sacral plexus and their distribution are summarized in Table 4-2.

Branches of the Lumbar Plexus
Lumbosacral Trunk
Part of the anterior ramus of the fourth lumbar nerve emerges from the medial border of the psoas muscle and

Table 4-2   Branches of the Sacral Plexus and Their Distribution

Branches	Distribution
Superior gluteal nerve	Gluteus medius, gluteus minimus, and tensor fasciae latae muscles
Inferior gluteal nerve	Gluteus maximus muscle
Nerve to piriformis	Piriformis muscle
Nerve to obturator internus	Obturator internus and superior gamellus muscles
Nerve to quadratus femoris	Quadratus femoris and inferior gamellus muscles
Perforating cutaneous nerve	Skin over medial aspect of buttock
Posterior cutaneous nerve of thigh	Skin over posterior surface of thigh and popliteal fossa, also over lower part of buttock, scrotum, or labium majus
Sciatic nerve (L4, 5; S1, 2, 3)	
Tibial portion	Hamstring muscles (semitendinosus, biceps femoris [long head], adductor magnus [hamstring part]), gastrocnemius, soleus, plantaris, popliteus, tibialis posterior, flexor digitorum longus, flexor hallucis longus, and via medial and lateral plantar branches to muscles of sole of foot. Sural branch supplies skin on lateral side of leg and foot
Common peroneal portion	Biceps femoris muscle (short head) and via deep peroneal branch: tibialis anterior, extensor hallucis longus, extensor digitorum longus, peroneus tertius, and extensor digitorum brevis muscles. Skin over cleft between first and second toes. The superficial peroneal branch supplies the peroneus longus and brevis muscles and skin over lower third of anterior surface of leg and dorsum of foot
Pudendal nerve	Muscles of perineum including the external anal sphincter, mucous membrane of lower half of anal canal, perianal skin, skin of penis, scrotum, clitoris, labia majora and minora

joins the anterior ramus of the fifth lumbar nerve to form the lumbosacral trunk (see Fig. 4-7). This trunk descends into the pelvis to join the sacral plexus.

Obturator Nerve (L2, 3, and 4)

The obturator nerve emerges from the medial border of the psoas muscle and accompanies the lumbosacral trunk down into the pelvis. It runs forward on the lateral pelvic wall to reach the obturator foramen (see Fig. 4-6). It splits into anterior and posterior divisions that pass through the canal to enter the adductor region of the thigh.

Branches

Sensory branches supply the parietal peritoneum on the lateral wall of the pelvis.

Autonomic Nerves
Pelvic Part of the Sympathetic Trunk

The pelvic part of the sympathetic trunk is continuous above, behind the common iliac vessels, with the lumbar part of the trunk. Below, the two trunks come together in front of the coccyx to form the **ganglion impar.** Each trunk descends behind the rectum in front of the sacrum, medial to the anterior sacral foramina. The sympathetic trunk has four or five segmentally arranged ganglia (see Fig. 4-6).

Branches
1. Gray rami communicantes to the sacral and coccygeal spinal nerves.
2. Fibers that join the hypogastric plexuses.

Pelvic Splanchnic Nerves

The pelvic splanchnic nerves constitute the parasympathetic part of the autonomic nervous system in the pelvis. They arise from S2, 3, and 4 as described with the sacral plexus. The preganglionic fibers synapse in ganglia in the hypogastric plexuses.

Superior Hypogastric Plexus

The superior hypogastric plexus is situated in front of the promontory of the sacrum (see Fig. 4-6). It divides inferiorly into right and left hypogastric nerves. The plexus is formed as a continuation of the aortic plexus and from branches of the third and fourth lumbar sympathetic ganglia. It contains sympathetic and sacral parasympathetic nerve fibers and visceral afferent nerve fibers.

Inferior Hypogastric Plexus

The inferior hypogastric plexus lies on each side of the rectum, the base of the bladder, and the vagina (see Fig. 4-6). It is formed from a hypogastric nerve (part of the superior hypogastric plexus) and from the pelvic splanchnic nerve. It contains postganglionic sympathetic fibers, pre- and postganglionic parasympathetic fibers, and visceral affer-

ent fibers. Branches pass to the pelvic viscera via small subsidiary plexuses. Note that parasympathetic fibers from the pelvic splanchnic nerve ascend through the hypogastric nerve to the superior hypogastric plexus and the inferior mesenteric plexus to supply the large bowel from the left colic flexure to the upper half of the anal canal.

Arteries of the Pelvis

Common Iliac Artery

Each common iliac artery ends at the pelvic inlet in front of the sacroiliac joint by dividing into the external and internal iliac arteries (Fig. 4-8; see also Fig. 4-6).

External Iliac Artery

The external iliac artery runs along the medial border of the psoas muscle, following the pelvic brim (see Fig. 4-8), and gives off the **inferior epigastric** and **deep circumflex iliac** branches. It leaves the false pelvis by passing under the inguinal ligament, to become the **femoral artery.**

The arteries entering the true pelvis are the

1. Internal iliac artery
2. Superior rectal artery
3. Ovarian artery
4. Median sacral artery

Internal Iliac Artery

The internal iliac artery passes down into the pelvis to the upper margin of the greater sciatic foramen, where it divides into anterior and posterior divisions (see Fig. 4-8). The branches of these divisions supply the pelvic viscera, the perineum, the pelvic walls, and the buttocks.

Branches of the Anterior Division of the Internal Iliac Artery

1. **Umbilical artery.** From the proximal patent part of the umbilical artery arises the **superior vesical artery,** which supplies the upper portion of the bladder (see Fig. 4-8).
2. **Inferior vesical artery.** This supplies the base of the bladder and the prostate and seminal vesicles in the male; it also gives off the **artery to the vas deferens.**
3. **Middle rectal artery.** Commonly, this artery arises with the inferior vesical artery (see Fig. 4-8). It supplies the muscle of the lower rectum and anastomoses with the superior rectal and inferior rectal arteries.
4. **Uterine artery.** This runs medially on the floor of the pelvis and **crosses the ureter superiorly** (see Fig. 4-13). It passes above the lateral fornix of the vagina to reach the uterus. Here, it ascends between the layers of the broad ligament along the lateral margin of the uterus. It ends by following the uterine tube laterally

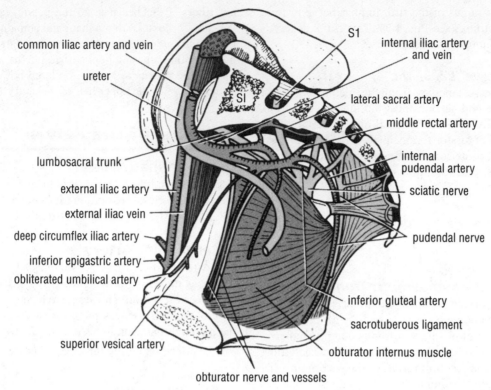

Fig. 4-8.   Lateral wall of pelvis.

where it anastomoses with the ovarian artery. The uterine artery gives a vaginal branch.

5. **Vaginal artery.** This artery usually takes the place of the inferior vesical artery present in the male. It supplies the vagina and the base of the bladder.

6. **Inferior gluteal artery.** This artery leaves the pelvis through the greater sciatic foramen below the piriformis muscle (see Fig. 4-8). It passes between the first and second or second and third sacral nerves.

7. **Obturator artery.** This artery runs forward along the lateral wall of the pelvis with the obturator nerve. It leaves the pelvis through the obturator canal.

8. **Internal pudendal artery.** This artery leaves the pelvis through the greater sciatic foramen and enters the gluteal region below the piriformis muscle (see Fig. 4-8). It enters the perineum by passing through the lesser sciatic foramen. The artery then passes forward in the pudendal canal with the pudendal nerve, and by means of its branches supplies the musculature of the anal canal and the skin and muscles of the perineum.

Branches of the Posterior Division of the Internal Iliac Artery

1. **Iliolumbar artery.** This artery ascends across the pelvic inlet posterior to the external iliac vessels, psoas, and iliacus muscles.

2. **Lateral sacral arteries.** These arteries descend in front of the sacral plexus giving off branches to neighboring structures (see Fig. 4-8).

3. **Superior gluteal artery.** This artery leaves the pelvis through the greater sciatic foramen above the piriformis muscle. It supplies the gluteal region.

Superior Rectal Artery

The superior rectal artery is a direct continuation of the inferior mesenteric artery. The name changes as the latter artery crosses the common iliac artery. It supplies the mucous membrane of the rectum and upper half of the anal canal.

Ovarian Artery

The ovarian artery arises from the abdominal part of the aorta at the level of the first lumbar vertebra. The artery is long and slender and passes downward and laterally behind the peritoneum. It crosses the external iliac artery at the pelvic inlet and enters the suspensory ligament of the ovary. It then passes into the broad ligament and enters the ovary by way of the mesovarium. (The testicular artery enters the inguinal canal and does not enter the pelvis.)

Median Sacral Artery

The median sacral artery is a small artery that arises at the bifurcation of the aorta (see Fig. 4-6). It descends over the anterior surface of the sacrum and the coccyx.

Veins of the Pelvis

External Iliac Vein

The external iliac vein begins behind the inguinal ligament as a continuation of the femoral vein. It runs along the medial side of the corresponding artery and joins the internal iliac vein to form the **common iliac vein** (see Fig. 4-8). It receives the **inferior epigastric** and **deep circumflex iliac veins.**

The veins of the true pelvis are thin-walled vessels. They include the

1. Internal iliac vein
2. Superior rectal vein
3. Ovarian vein
4. Median sacral vein

Internal Iliac Vein

The internal iliac vein is formed by the union of tributaries that correspond to the branches of the internal iliac artery (see Fig. 4-8). It ascends in front of the sacroiliac joint and joins the external iliac vein to form the common iliac vein.

Superior Rectal Vein

The superior rectal vein drains the rectal mucous membrane and the mucous membrane of the upper half of the anal canal. It anastomoses with the middle and inferior rectal veins forming an important portal-systemic anastomosis. It becomes continuous with the inferior mesenteric vein as it crosses the common iliac artery.

Ovarian Vein

The right ovarian vein ascends on the posterior abdominal wall and drains into the inferior vena cava. The left ovarian vein drains into the left renal vein.

Median Sacral Vein

The median sacral vein is a small vein that drains into the inferior vena cava or the left common iliac vein.

Lymphatics of the Pelvis

The lymph nodes and vessels are arranged as chains along the main blood vessels. The nodes are named after the blood vessels with which they are associated. Thus, there are the **external iliac nodes, internal iliac nodes, and common iliac nodes.**

Intestinal Viscera

Sigmoid Colon

The sigmoid colon is about 10 to 15 in. (25–38 cm) long and begins as a continuation of the descending colon in front of the pelvic brim. It becomes continuous below with the rectum in front of the third sacral vertebra (see Fig. 4-11). It hangs down into the pelvic cavity in the form of a loop. The sigmoid colon is attached to the posterior pelvic wall by the fan-shaped **sigmoid mesocolon.**

Blood Supply
Arteries

Sigmoid branches of the inferior mesenteric artery.

Veins

The veins are tributaries of the portal venous system.

Rectum

The rectum is about 5 in. (13 cm) long and begins in front of the third sacral vertebra as a continuation of the sigmoid colon (Fig. 4-9; see also Fig. 4-11). It passes downward, following the curve of the sacrum and coccyx, and ends in front of the tip of the coccyx by piercing the pelvic floor and becoming continuous with the anal canal. The lower part of the rectum is dilated to form the **rectal ampulla** (see Fig. 4-9). The peritoneum covers only the upper two-thirds of the rectum. The mucous membrane of the rectum, together with the circular muscle layer, form three semicircular folds: Two are placed on the left rectal wall and one on the right wall. They are called the **transverse folds of the rectum.**

Blood Supply
Arteries

Superior rectal artery, a direct continuation of the inferior mesenteric artery; middle rectal artery, a branch of the internal iliac artery; inferior rectal artery, a branch of the internal pudendal artery.

Veins

The venous blood drains into the portal vein by the superior rectal vein and into the systemic system by the middle and inferior rectal veins. The anastomosis between the rectal veins is an important portal-systemic anastomosis.

Lymph Drainage

The lymph passes to **pararectal nodes** and then upward to the **inferior mesenteric nodes.** Some lymph vessels pass to the **internal iliac nodes.**

Nerve Supply

Sympathetic and parasympathetic nerves through the inferior hypogastric plexuses.

Ureters and Urinary Bladder
Ureters

Each ureter is a muscular tube that extends from the kidney to the posterior surface of the bladder. Its abdominal course is described on page 73.

Pelvis and Perineum

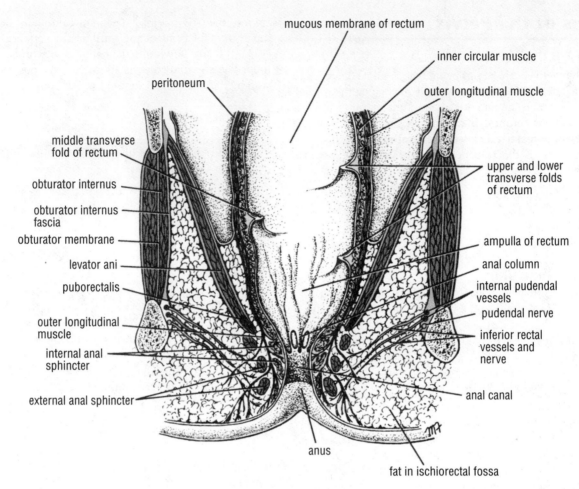

Fig. 4-9.   Coronal section through pelvis, showing rectum and pelvic floor.

The ureter enters the pelvis by crossing the bifurcation of the common iliac artery in front of the sacroiliac joint. Each ureter then runs down the lateral wall of the pelvis to the region of the ischial spine and turns forward to enter the lateral angle of the bladder (Fig. 4-10).

In the male, the ureter is crossed near its termination by the vas deferens. In the female, the ureter leaves the region of the ischial spine by turning forward and medially beneath the base of the broad ligament; here it is crossed by the uterine artery (see Fig. 4-13).

Ureteric Constrictions

The ureter possesses three constrictions: (1) where the renal pelvis joins the ureter, (2) where it is kinked as it crosses the pelvic brim, and (3) where it pierces the bladder wall.

The blood supply, lymphatic drainage, and nerve supply of the ureter are described on page 74.

Urinary Bladder

The urinary bladder is located immediately behind the pubic bones within the pelvis (Fig. 4-11; see also Fig. 4-14). The bladder has a maximum capacity of about 500 ml.

The empty bladder is pyramidal in shape and has an apex, a base, and a superior and two inferolateral surfaces (see Fig. 4-10); it also has a neck. When the bladder fills, it becomes ovoid in shape and the superior surface rises into the abdomen. In the young child, the empty bladder projects upward into the abdomen; later, when the pelvis enlarges, the bladder sinks to become a pelvic organ.

The **apex** of the bladder points anteriorly and is connected to the umbilicus by the **median umbilical ligament** (remains of urachus). The **base** of the bladder faces posteriorly and is triangular in shape. The ureters enter the supralateral angles and the urethra leaves the inferior angle. The **superior surface** of the bladder is covered with peritoneum, which is reflected laterally onto the lateral pelvic walls. As the bladder fills, the superior surface bulges upward into the abdominal cavity, peeling the peritoneum off from the lower part of the anterior abdominal wall. The **neck** of the bladder points inferiorly.

Interior of Bladder

The internal surface of the base of the bladder is called the **trigone** (see Fig. 4-10). Here the mucous membrane is

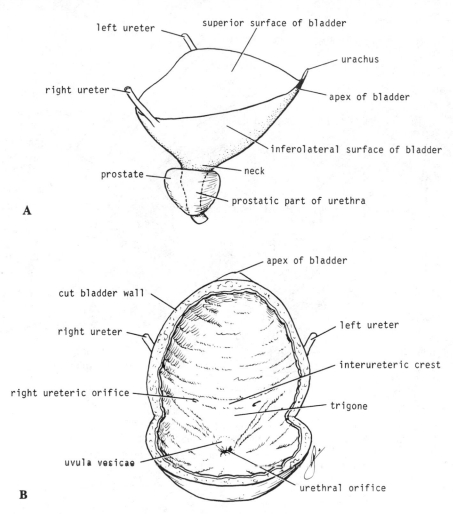

Fig. 4-10. **A.** Urinary bladder (partly filled), showing the general shape and surfaces. Note the position of the apex anteriorly and the neck inferiorly. **B.** Interior of urinary bladder in the male, as seen from in front.

firmly adhered to the underlying muscle and is always smooth. The trigone has small, slitlike openings of the ureters at its lateral angles and below the crescentic opening of the urethra. The **interureteric ridge** runs from one ureteric orifice to the other. It is caused by the underlying muscle and forms the upper limit of the trigone. In the male, the median lobe of the prostate bulges upward into the bladder slightly, behind the urethral orifice, to form a swelling, the **uvula vesicae.**

Muscular Coat of the Bladder Wall

The muscle coat, the **detrusor muscle,** consists of three interlacing layers of smooth muscle fibers. At the neck of the bladder, the circular muscle forms the **sphincter vesicae.**

Ligaments

The neck of the bladder is held in position by the **puboprostatic ligaments** in the male and the **pubovesical ligaments** in the female. These ligaments are formed of pelvic fascia.

Blood Supply

Arteries. Superior and inferior vesical arteries, branches of the internal iliac artery.
Veins. Vesical veins drain into the internal iliac veins.

Lymph Drainage

Internal and external iliac nodes.

Nerve Supply

Sympathetic and parasympathetic fibers from the inferior hypogastric plexuses.

Micturition

In the adult, when the volume of urine reaches about 300 ml, stretch receptors in the wall of the bladder transmit im-

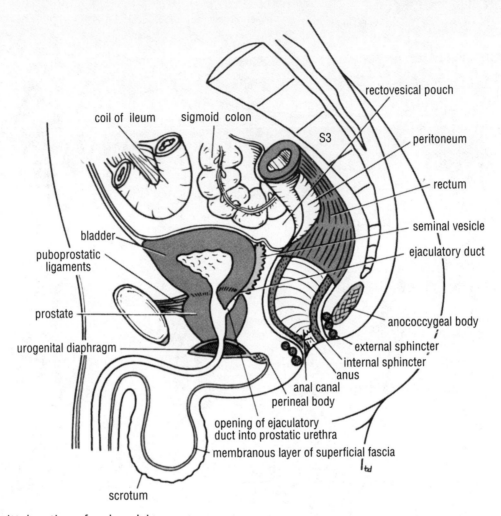

coil of ileum   sigmoid colon

rectovesical pouch

S3

peritoneum

rectum

seminal vesicle

ejaculatory duct

bladder

puboprostatic
ligaments

prostate

urogenital diaphragm

anococcygeal body

external sphincter
internal sphincter
anus

anal canal
perineal body

opening of ejaculatory
duct into prostatic urethra

membranous layer of superficial fascia

scrotum

Fig. 4-11. Sagittal section of male pelvis.

pulses to the central nervous system, and the individual has a conscious desire to micturate.

The afferent impulses enter the second, third, and fourth sacral segments of the spinal cord. Efferent impulses leave the cord from the same segments and pass via the parasympathetic preganglionic nerve fibers in the hypogastric plexuses to the bladder wall, where they synapse with postganglionic neurons. The detrusor muscle then contracts and the sphincter vesicae relaxes. Efferent impulses also pass to the urethral sphincter via the pudendal nerve and this relaxes. Micturition can be assisted by contracting the abdominal muscles so as to raise the intra-abdominal and pelvic pressures and exert external pressure on the bladder.

Voluntary control of micturition is accomplished by contracting the sphincter urethrae; this is assisted by the sphincter vesicae.

Male Genital Organs

The **testes** and **epididymides** are described on page 59.

Vas Deferens

The vas deferens is a thick-walled tube about 18 in. (45 cm) long. It emerges from the lower end or tail of the epididymis and passes through the inguinal canal into the abdomen. It then descends into the pelvis and crosses the ureter to reach the posterior surface of the bladder. Here it expands to form the **ampulla** and then joins the duct of the seminal vesicle to form the **ejaculatory duct.**

Ejaculatory Duct

There are two ejaculatory ducts. Each is formed by the union of the vas deferens and the duct of the seminal vesicle. The two ejaculatory ducts open into the prostatic part of the urethra (Fig. 4-12).

Seminal Vesicles

The seminal vesicles are paired organs that lie on the posterior surface of the bladder (see Fig. 4-11) and lateral to the terminal part of the vas deferens. Each seminal vesicle consists of a much coiled tube embedded in connective tis-

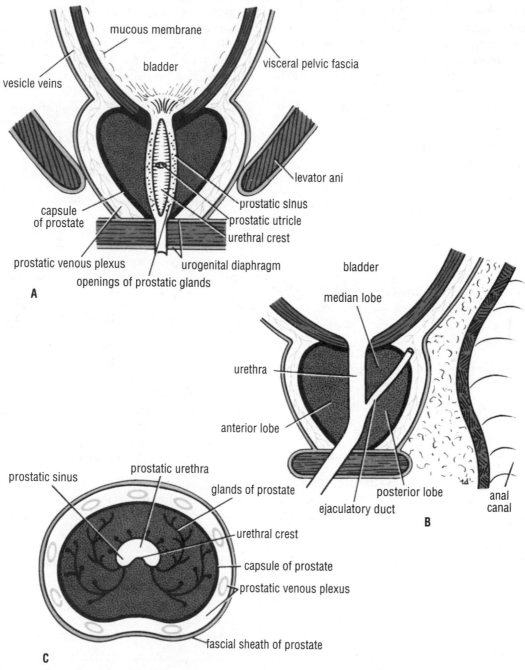

Fig. 4-12. Prostate in coronal section (**A**), sagittal section (**B**), and horizontal section (**C**).

sue. Inferiorly, the seminal vesicle narrows and joins the vas deferens of the same side to form the ejaculatory duct. The function of the seminal vesicles is to contribute fluid, fructose, ascorbic acid, amino acids, and prostaglandins to the seminal fluid.

Prostate

The prostate is a glandular structure that surrounds the prostatic urethra. It lies below the neck of the bladder

and above the urogenital diaphragm (see Fig. 4-11). The prostate has a fibrous capsule. Outside the capsule is a fibrous sheath that is part of the visceral layer of pelvic fascia (see Fig. 4-12). The prostate has a **base,** which lies superiorly against the bladder neck, and an **apex,** which lies inferiorly against the urogenital diaphragm.

The two ejaculatory ducts pierce the upper part of the posterior surface of the prostate to open into the prostatic urethra at the lateral margins of the opening of the **prostatic utricle.**

The numerous glands of the prostate are embedded in a mixture of smooth muscle and connective tissue, and their ducts open into the prostatic urethra. The prostate may be divided into a number of lobes: The **anterior lobe** lies in front of the urethra; the **middle or median lobe** lies behind the urethra and above the ejaculatory ducts; the **right and left lateral lobes** lie on either side of the urethra.

Blood Supply
Arteries. Branches of the inferior vesical and middle rectal arteries.
Veins. Prostatic venous plexus that drains into the internal iliac veins.

Lymph Drainage
Internal iliac nodes.

Nerve Supply
Inferior hypogastric plexuses.

For details concerning the **prostatic sinus,** the **prostatic utricle,** and the **urethral crest,** see the prostatic urethra, page 111. For developmental homologues of male and female genital systems, see Table 4-3.

Female Genital Organs
Ovaries
Each ovary is attached to the back of the broad ligament by the **mesovarium** (Fig. 4-13). The ovarian vessels and nerves enter the ovary at the **hilum.**

The ovary is covered by a modified area of peritoneum called the **germinal epithelium.** Beneath the germinal epithelium is a thin fibrous capsule, the **tunica albuginea.** The ovary has an outer **cortex** and an inner **medulla.** Embedded in the connective tissue of the cortex are the **ovarian follicles.**

The ovary usually lies near the lateral wall of the pelvis in a depression called the **ovarian fossa,** bounded by the external and internal iliac arteries.

Ligaments of the Ovaries
Suspensory Ligament. The suspensory ligament is the lateral part of the broad ligament connecting the mesovarium to the lateral pelvic wall (see Fig. 4-13). It contains the blood and the lymphatic vessels and nerves supplying the ovary.
Round Ligament. The round ligament of the ovary is the remains of the upper part of the gubernaculum and extends from the medial margin of the ovary to the lateral wall of the uterus (see Fig. 4-13). Note that the round ligament of the uterus is the remains of the lower part of the gubernaculum.

Blood Supply
Arteries. The **ovarian artery,** a branch of the abdominal aorta.
Veins. The **ovarian vein** drains into the inferior vena cava on the right side and into the left renal vein on the left.

Lymph Drainage
Para-aortic nodes at the level of the first lumbar vertebra.

Nerve Supply
Aortic plexus; the branches accompany the ovarian artery.

Uterine Tubes
The uterine tubes are two in number (see Fig. 4-13). Each tube lies in the upper border of the broad ligament. It connects the peritoneal cavity in the region of the ovary with the cavity of the uterus. It may be divided into four parts.

1. The **infundibulum** is the funnel-shaped lateral end that has fingerlike processes, known as **fimbrae,** which are draped over the ovary.
2. The **ampulla** is the widest part of the tube.
3. The **isthmus** is the narrowest part of the tube and lies just lateral to the uterus.

Table 4-3 Developmental Homologues of Male and Female Genital Systems

Origin	Male	Female
Developing gonad	Testis	Ovary
Mesonephric duct	Duct of epididymis, vas deferens, seminal vesicle, ejaculatory duct, trigone of bladder, and part of prostatic urethra	Ducts of epoophoron, paroophoron, trigone of bladder, and part of urethra
Paramesonephric duct	Prostatic utricle, appendix of testis	Uterine tubes, uterus
Urogenital sinus	Lower part of prostatic urethra, membranous urethra, most of penile urethra, prostate glands, and bulbourethral glands	Vagina and lower part of urethra, paraurethral glands
Glans of phallus	Glans of penis	Glans of clitoris
Genital (urethral) folds	Floor of penile urethra	Labia minora
Genital swellings	Scrotum	Labia majora
Gubernaculum	Gubernaculum testis	Round ligament of ovary, round ligament of uterus

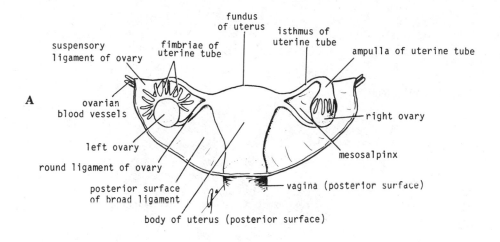

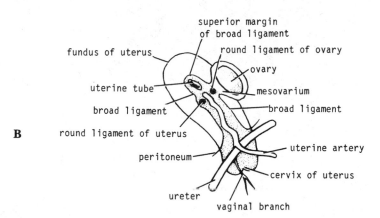

Fig. 4-13.   **A.** Posterior surface of the uterus and the broad ligaments, showing the position of the ovaries. **B.** Lateral view of the uterus, showing the attachment of the broad ligament and the relationship between the left uterine artery and the left ureter.

4. The **intramural part** is the segment that pierces the uterine wall.

The uterine tube provides a site where fertilization of the ovum can take place (usually in the ampulla). It provides nourishment for the fertilized ovum and transports the ovum to the cavity of the uterus. The tube serves as a conduit along which the spermatozoa travel to reach the ovum.

Blood Supply
Arteries.   Uterine and ovarian arteries.
Veins.   Uterine and ovarian veins.

Lymph Drainage
The lymph vessels follow the arteries and drain into the internal iliac nodes and the para-aortic nodes.

Nerve Supply
Sympathetic and parasympathetic nerves from the superior and inferior hypogastric plexuses.

Uterus
The uterus is divided into the fundus, the body, and the cervix (see Fig. 4-13). The **fundus** is the part of the uterus that lies above the entrance of the uterine tubes. The **body** is the part of the uterus that lies below the entrance of the uterine tubes. It narrows below, where it becomes continuous with the **cervix.** The cervix pierces the anterior wall of the vagina (Fig. 4-14). The cavity of the cervix, the **cervical canal,** communicates with the cavity of the body through the **internal os** and with that of the vagina through the **external os.**

Positions of Uterus
Anteversion is the term used to describe the forward bending of the uterus on the long axis of the vagina. **Anteflexion** is the term used to describe the forward bending of the body of the uterus on the cervix.

Supports of the Uterus
The main supports of the uterus are

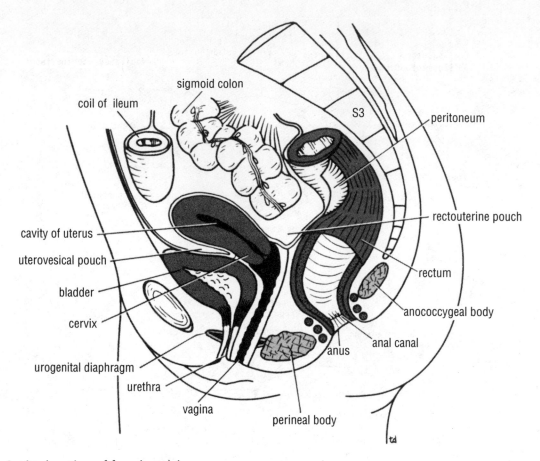

Fig. 4-14. Sagittal section of female pelvis.

1. The **pelvic diaphragm** (levatores ani and the coccygeus muscles and their fascia)
2. The **perineal body** (a fibromuscular structure in the perineum supported by the levatores ani muscles)
3. The **transverse cervical (cardinal) ligaments** that attach the cervix and upper end of the vagina to the lateral pelvic walls
4. The **pubocervical ligaments** that attach the cervix to the pubic bones
5. The **sacrocervical ligaments** that attach the cervix and the upper end of the vagina to the lower end of the sacrum

Broad Ligaments

The broad ligaments are two-layered folds of peritoneum that extend across the pelvic cavity from the lateral margins of the uterus to the lateral pelvic walls (see Fig. 4-13). Each broad ligament contains the following:

1. The uterine tube in its upper free border
2. The round ligament of the ovary and the round ligament of the uterus
3. The uterine and ovarian blood vessels, lymph vessels, and nerves

The broad ligaments give little support to the uterus.

Round Ligament of the Uterus

The round ligament extends from the superolateral angle of the uterus through the inguinal canal to the subcutaneous tissue of the labium majus. It assists in keeping the uterus anteverted and anteflexed.

Blood Supply

Arteries.   **Uterine artery** from the internal iliac artery and also the **ovarian artery.**
Veins.   The **veins** correspond to the arteries.

Lymph Drainage

From the fundus, the lymph vessels follow the ovarian artery to the para-aortic nodes at the level of the first lumbar vertebra. From the body and cervix, they drain into the internal and external iliac nodes. A few lymph vessels pass through the inguinal canal to the superficial inguinal nodes.

Nerve Supply

Sympathetic and parasympathetic nerves from the inferior hypogastric plexuses.

Vagina

The vagina is the female genital canal (see Fig. 4-14). It serves as the excretory duct for the menstrual flow from the uterus and forms part of the birth canal. The vagina is a muscular tube that extends between the vulva and the uterus. The cervix of the uterus pierces its anterior wall. The vaginal orifice in the virgin possesses a thin mucosal fold called the **hymen** that is perforated at its center. The area of the vaginal lumen that surrounds the cervix of the uterus is divided into four regions or **fornices: anterior, posterior, right lateral, and left lateral.** The upper half of the vagina lies within the pelvis between the bladder anteriorly and the rectum posteriorly; the lower half lies within the perineum between the urethra anteriorly and the anal canal posteriorly (see Fig. 4-14).

Supports of the Vagina

Upper third: levatores ani muscles; transcervical, pubocervical, and sacrocervical ligaments

Middle third: urogenital diaphragm

Lower third: perineal body

Blood Supply

Arteries.   **Vaginal artery,** a branch of the internal iliac artery, and **vaginal branch of uterine artery.**
Veins.   **Vaginal veins** drain into the internal iliac veins.

Lymph Drainage

The **upper third** drains into the internal and external iliac nodes; the **middle third** drains into the internal iliac nodes; the **lower third** drains into superficial inguinal nodes.

Nerve Supply

Inferior hypogastric plexuses.

For developmental homologues of male and female genital systems, see Table 4-3.

Visceral Pelvic Fascia in the Female Pelvis

The visceral pelvic fascia, as in the male, covers and supports the pelvic viscera. It is condensed to form the pubocervical, transverse cervical, and sacrocervical ligaments of the uterus. Clinically, the pelvic fascia in the region of the uterine cervix is often referred to as the **parametrium.**

Peritoneum in the Female Pelvis

The peritoneum passes from the anterior abdominal wall onto the upper surface of the urinary bladder (see Fig. 4-14). It then runs onto the anterior surface of the uterus upward over the fundus of the uterus and downward on the posterior surface of the uterus. It continues downward over the upper part of the posterior surface of the vagina, where it forms the anterior wall of the **rectouterine pouch (pouch of Douglas).** The peritoneum is then reflected onto the front of the rectum. The most inferior part of the peritoneal cavity is the rectouterine pouch.

Perineum

The perineum lies below the pelvic diaphragm. It is diamond-shaped and is bounded anteriorly by the **symphysis pubis,** posteriorly by the tip of the **coccyx,** and laterally by the **ischial tuberosities** (Fig. 4-15). The perineum may be divided into two triangles by joining the ischial tuberosities by an imaginary line. The anterior triangle, which contains the urogenital orifices, is called the **urogenital triangle;** the posterior triangle, which contains the anus, is called the **anal triangle** (see Fig. 4-15).

Urogenital Triangle

Urogenital Diaphragm

The urogenital diaphragm is a musculofascial diaphragm that fills in the gap of the pubic arch (Fig. 4-16). It is formed by the sphincter urethrae and the deep transverse perineal muscles, which are enclosed between a superior and an inferior layer of fascia of the urogenital diaphragm. The inferior layer of fascia is called the **perineal membrane.**

Perineal Body

The perineal body is a small mass of fibrous tissue that is attached to the center of the posterior margin of the urogenital diaphragm. It is a larger structure in the female and serves to support the posterior vaginal wall. In both sexes, it provides attachment for muscles in the perineum (see Fig. 4-16).

Perineal Pouches

Superficial Perineal Pouch

The superficial perineal pouch is a potential space that lies beneath the skin of the perineum. It is bounded below by the membranous layer of superficial fascia (Colles' fascia) and above by the urogenital diaphragm (Fig. 4-17). It is closed behind by the attachment of this membranous layer of fascia to the posterior border of the urogenital diaphragm. Laterally, it is closed by the attachment of the membranous layer of superficial fascia and the urogenital diaphragm to the pubic arch. Anteriorly, the space commu-

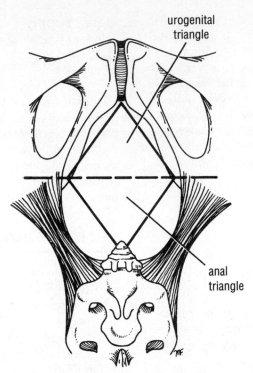

urogenital
triangle

anal
triangle

Fig. 4-15.   Diamond-shaped perineum divided by broken line into urogenital triangle and anal triangle.

nicates freely with the potential space lying between the superficial fascia of the anterior abdominal wall (Scarpa's fascia) and the anterior abdominal muscles.

Contents of the Superficial Perineal Pouch

In the **male,** the superficial perineal pouch contains the root of the penis and its associated muscles. In the **female,** the superficial perineal pouch contains the root of the clitoris and its associated muscles.

Deep Perineal Pouch

The deep perineal pouch is a closed potential space that lies within the urogenital diaphragm (see Fig. 4-17).

Contents of the Deep Perineal Pouch

In the **male,** the deep perineal pouch contains

1. The membranous part of the urethra
2. The sphincter urethrae
3. The bulbourethral glands
4. The deep transverse perineal muscles
5. The internal pudendal vessels
6. The dorsal nerves of the penis

In the **female,** the deep perineal pouch contains

1. Part of the urethra
2. Part of the vagina

3. The sphincter urethrae
4. The deep transverse perineal muscles
5. The internal pudendal vessels
6. The dorsal nerves of the clitoris

Male External Genitalia

Penis

The penis has a cylindrical **body** that hangs free and a fixed **root** (Fig. 4-18). The body of the penis has an expanded distal end called the **glans penis. The prepuce, or foreskin,** is a hoodlike fold of skin that covers the glans. The interior of the body of the penis is composed of three cylinders of erectile tissue enclosed in a tubular sheath of fascia. The erectile tissue is made up of two dorsally placed **corpora cavernosa** and a single **corpus spongiosum** applied to their ventral surface (see Fig. 4-18). At its distal end, the corpus spongiosum expands to form the glans penis, which covers the distal ends of the corpora cavernosa. The penile part of the urethra runs through the center of the corpus spongiosum and opens onto the surface of the glans at the **external urethral orifice.**

The **root of the penis** is located in the superficial perineal pouch and is made up of three masses of erectile tissue: the **bulb of the penis** and the **right and left crura of the penis** (see Fig. 4-18). The bulb is located in the midline and is traversed by the urethra. The bulb is covered on its

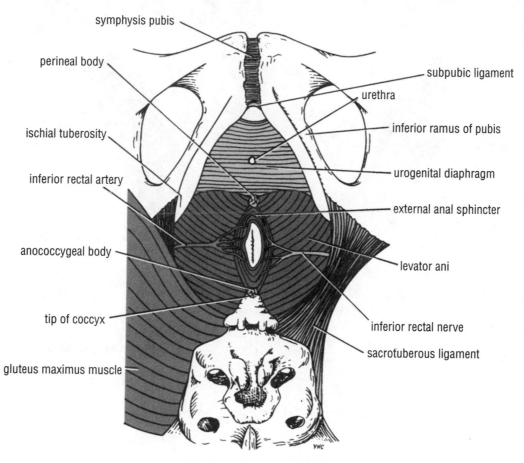

symphysis pubis

perineal body

ischial tuberosity

inferior rectal artery

anococcygeal body

tip of coccyx

gluteus maximus muscle

subpubic ligament

urethra

inferior ramus of pubis

urogenital diaphragm

external anal sphincter

levator ani

inferior rectal nerve

sacrotuberous ligament

Fig. 4-16. Anal triangle and urogenital triangle in the male as seen from below.

outer surface by the **bulbospongiosus muscles.** Each crus is attached to the side of the pubic arch and is covered on its outer surface by the **ischiocavernosus muscle.**

The bulb of the penis is continuous anteriorly with the corpus spongiosum and the two crura are continuous anteriorly with the corpora cavernosa in the body of the penis.

Blood Supply
Arteries
Deep arteries of the penis and branches of the dorsal arteries of the penis.

Veins
Deep dorsal vein.

Lymph Drainage
The glans penis drains into the deep inguinal and external iliac nodes. The skin of the remainder of the organ drains into the superficial inguinal nodes. The erectile tissue drains into the internal iliac nodes.

Nerve Supply
Pudendal nerve.

Scrotum, Testes, and Epididymides
The scrotum, testes, and epididymides are described on page 59.

Bulbourethral (Cowper's) Glands
The bulbourethral glands are two small glands situated in the deep perineal pouch among the fibers of the sphincter urethrae muscle. The ducts of the glands open into the penile urethra.

Male Urethra
The male urethra is divided into three parts: prostatic, membranous, and penile.

Prostatic Urethra
The prostatic urethra passes through the prostate from the base to the apex (see Fig. 4-17). It is the widest and most dilatable portion of the entire urethra. On the posterior wall, there is a longitudinal ridge called the **urethral crest** (see Fig. 4-12). On each side of this ridge is a groove called the **prostatic sinus,** into which open the prostatic glands. On the summit of the urethral crest is a depression, the **prostatic utricle,** which is an analog of

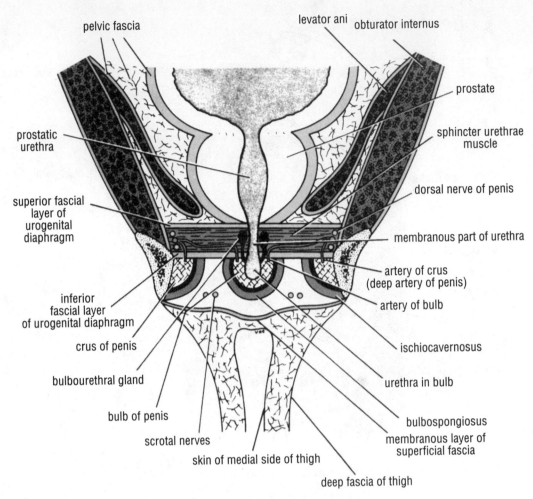

Fig. 4-17.   Coronal section of male pelvis, showing prostate, urogenital diaphragm, and contents of superficial perineal pouch.

the uterus and vagina in the female. On the edge of the mouth of the utricle are the openings of the two ejaculatory ducts.

Membranous Urethra

The membranous urethra passes through the urogenital diaphragm and is surrounded by the sphincter urethrae (see Fig. 4-17). It is the shortest and least dilatable part of the urethra.

Penile Urethra

The penile urethra passes through the bulb and the corpus spongiosum of the penis (see Fig. 4-17). The **external meatus** is the narrowest part of the entire urethra. The part of the urethra that lies within the glans penis is dilated to form the **fossa terminalis (navicular fossa).** The bulbourethral glands open into the penile urethra below the urogenital diaphragm.

For developmental homologues of male and female genital systems, see Table 4-3.

Female External Genitalia (Vulva)

The vulva includes the mons pubis (hair-bearing skin in front of pubis), the labia majora, the labia minora, the clitoris, and the greater vestibular glands.

Labia Majora

The labia major are prominent folds of skin extending from the mons pubis to unite in the midline posteriorly. They contain fat and are covered with hair on their outer surfaces. (They are equivalent to the scrotum in the male.)

Labia Minora

The labia minora are two smaller folds of skin devoid of hair that lie between the labia majora. Their posterior ends are united to form a sharp fold, the **fourchette.** Anteriorly, they split to enclose the clitoris, forming an anterior **prepuce** and a posterior **frenulum.**

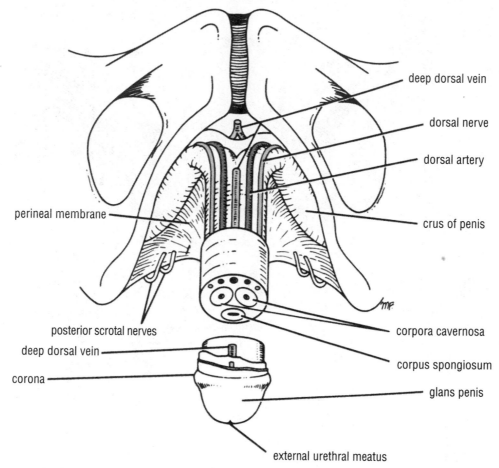

Fig. 4-18. Root and body of penis.

Vestibule of Vagina

The vestibule of the vagina is the space between the labia minora. It has the clitoris at its apex and has the openings of the urethra, the vagina, and the ducts of the greater vestibular glands in its floor.

Clitoris

The clitoris corresponds to the penis in the male. The **glans** of the clitoris is partly hidden by the **prepuce.** The **root of the clitoris** is made up of three masses of erectile tissue, which are called the bulb of the vestibule and the right and left crura of the clitoris. The **bulb of the vestibule** corresponds to the bulb of the penis, but because of the presence of the vagina, it is divided into two halves. It is attached to the undersurface of the urogenital diaphragm and is covered by the **bulbospongiosus muscles.** Anteriorly, the two halves unite to form the glans clitoris. The **crura of the clitoris** correspond to the crura of the penis. They are covered by the **ischiocavernosus muscles.**

Greater Vestibular Glands

The greater vestibular glands are a pair of mucus-secreting glands that lie under cover of the posterior parts of the bulb of the vestibule and the labia majora (Fig. 4-19). The duct of each gland opens into the groove between the hymen and the posterior part of the labium minus.

Female Urethra

The female urethra is only 1½ in. (3.8 cm) long and extends from the bladder neck to the external meatus. It passes through the urogenital diaphragm where it traverses the sphincter urethrae and opens onto the surface below the clitoris and in front of the vagina.

Muscles of the Urogenital Triangle

The attachments, nerve supply, and action of the muscles of the urogenital triangle are given in Table 4-4.

For developmental homologues of male and female genital systems, see Table 4-3.

Anal Triangle

Anal Canal

The anal canal is about 1½ in. (4 cm) long and lies below the pelvic diaphragm (see Figs. 4-11 and 4-14). It passes

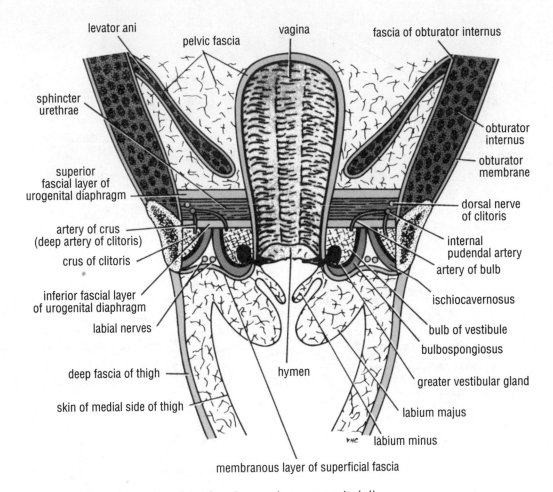

Fig. 4-19.   Coronal section of female pelvis, showing vagina, urogenital diaphragm, and contents of superficial perineal pouch.

downward and backward from the rectal ampulla to open onto the surface at the **anus.** Its lateral walls are kept in apposition by the levatores ani muscles and the anal sphincters except during defecation.

The **mucous membrane** of the upper half of the anal canal shows vertical folds called **anal columns.** These are connected together at their lower ends by small semilunar folds called **anal valves.** The mucous membrane of the lower half of the canal is smooth and merges with the skin at the anus.

The **muscle coat** is divided into outer longitudinal layers and an inner circular layer of smooth muscle. The circular layer is thickened at the upper end of the anal canal to form the **involuntary internal sphincter.** Surrounding the internal sphincter of smooth muscle is a collar of striped muscle called the **voluntary external sphincter.** The external sphincter is divided into three parts: **subcutaneous, superficial,** and **deep.** The attachments of these parts are given in Table 4-5.

The **puborectalis** fibers of the two levator ani muscles form a sling, which is attached anteriorly to the pubic bones. The sling passes around the junction of the rectum and the anal canal pulling them forward so that the rectum joins the anal canal at an acute angle.

At the anorectal junction, the internal sphincter, the deep part of the external sphincter, and the puborectalis form a distinct ring called the **anorectal ring.**

Blood Supply
Arteries
The superior rectal artery supplies the upper half, and the inferior rectal artery the lower half.

Veins
The upper half is drained by the superior rectal vein into the inferior mesenteric vein, and the lower half is drained by the inferior rectal vein into the internal pudendal vein. The anastomosis between the rectal veins forms an important portal-systemic anastomosis.

Lymph Drainage
The upper half drains into the pararectal nodes and then the inferior mesenteric nodes. The lower half drains into the medial group of superficial inguinal nodes.

Table 4-4   Muscles of the Urogenital Triangle

Name of muscle	Origin	Insertion	Nerve supply	Action
Male				
Superficial transverse perineal muscle	Ischial tuberosity	Perineal body	Perineal branch of pudendal nerve	Fixes perineal body
Bulbospongiosus	Perineal body and median raphe	Fascia of bulb of penis and corpus spongiosum and cavernosum	Perineal branch of pudendal nerve	Empties urethra after micturition and ejaculation, assists in erection
Ischiocavernosus	Ischial tuberosity and ischial ramus	Fascia covering corpus cavernosum	Perineal branch of pudendal nerve	Assists in erection
Deep transverse perineal muscle	Ramus of ischium	Perineal body	Perineal branch of pudendal nerve	Fixes perineal body
Sphincter urethrae	Pubic arch	Surrounds urethra	Perineal branch of pudendal nerve	Voluntary sphincter of urethra
Female				
Superficial transverse perineal muscle	Ischial tuberosity	Perineal body	Perineal branch of pudendal nerve	Fixes perineal body
Bulbospongiosus	Perineal body	Fascia of corpus cavernosum	Perineal branch of pudendal nerve	Sphincter of vagina, assists in erection of clitoris
Ischiocavernosus	Ischial tuberosity	Fascia covering corpus cavernosum	Perineal branch of pudendal nerve	Causes erection of clitoris
Deep transverse perineal muscle	Ramus of ischium	Perineal body	Perineal branch of pudendal nerve	Fixes perineal body
Sphincter urethrae	Pubic arch	Surrounds urethra	Perineal branch of pudendal nerve	Voluntary sphincter of urethra

Table 4-5   Muscles of the Anal Triangle

Name of muscle	Origin	Insertion	Nerve supply	Action
External Anal Sphincter				
Subcutaneous part	Encircles anal canal, no bony attachments			
Superficial part	Perineal body	Coccyx	Inferior rectal nerve and perineal branch of fourth sacral nerve	Voluntary sphincter of anal canal and closes anus
Deep part	Encircles anal canal, no bony attachments			
Puborectalis (part of levator ani)	Pubic bones	Sling around junction of rectum and anal canal	Perineal branch of fourth sacral nerve and from perineal branch of pudendal nerve	Together with external anal sphincter, forms voluntary sphincter for anal canal

Nerve Supply

The mucous membrane of the upper half is sensitive to stretch and is innervated by sensory fibers that ascend through the hypogastric plexuses. The lower half is sensitive to pain, temperature, touch, and pressure and is innervated by the inferior rectal nerves.

The internal anal sphincter is supplied by sympathetic nerves from the hypogastric plexuses. The voluntary external anal sphincter is supplied by the inferior rectal nerves.

Ischiorectal Fossa

The ischiorectal fossa is a wedge-shaped space on each side of the anal canal (see Fig. 4-9). The base of the wedge is superficial and formed by the skin. The edge of the wedge is formed by the junction of the medial and lateral walls. The medial wall is formed by the sloping levator ani muscle and the anal canal. The lateral wall is formed by the lower part of the obturator internus muscle, which is covered with pelvic fascia.

The fossa is filled with fat that supports the anal canal. Its function is to allow the anal canal to distend during the process of defecation. The pudendal nerve and the internal pudendal vessels lie in a fascial canal, the **pudendal canal,** on the medial side of the ischial tuberosity.

Muscles of the Anal Triangle

The attachments, nerve supply, and actions of the muscles of the anal triangle are given in Table 4-5.

National Board Type Questions

Match the numbered structures shown in the anteroposterior radiograph of the pelvis with the appropriate lettered structures on the right.

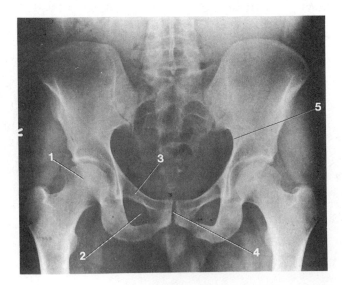

1. Structure 1.	A. Superior ramus of pubis.
2. Structure 2.	B. Symphysis pubis.
3. Structure 3.	C. Iliopectineal line.
4. Structure 4.	D. Head of femur.
5. Structure 5.	E. Obturator foramen.

Match the numbered structures shown in the anteroposterior arteriogram of the lower part of the abdominal aorta and the iliac and femoral arteries with the appropriate lettered structures on the right.

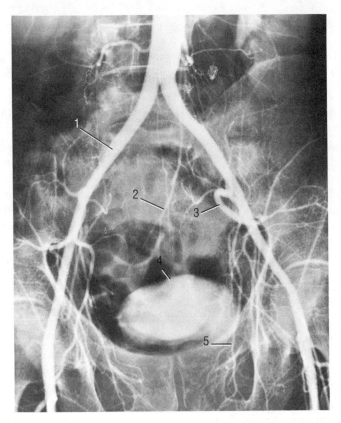

6. Structure 1.	A. Internal iliac artery.
7. Structure 2.	B. Bladder.
8. Structure 3.	C. Obturator artery.
9. Structure 4.	D. Superior rectal artery.
10. Structure 5.	E. Common iliac artery.

Match the numbered structures shown in the computed tomographic (CT) scan of the pelvis following a barium meal and intravenous pyelography with the appropriate lettered structures on the right.

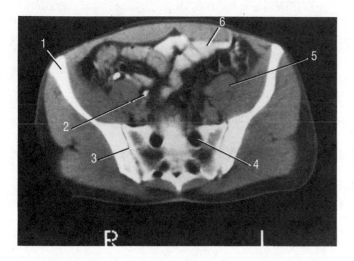

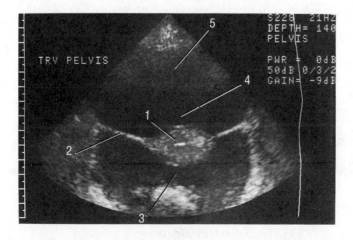

11. Structure 1.	A. Sacral foramen.
12. Structure 2.	B. Ilium.
13. Structure 3.	C. Sacroiliac joint.
14. Structure 4.	D. Small intestine.
15. Structure 5.	E. Psoas muscle.
16. Structure 6.	F. Right ureter.

Match the numbered structures shown in the transverse sonogram of the pelvis in a woman with the appropriate lettered structures on the left. The patient was involved in a serious automobile accident in which the liver was lacerated and blood escaped into the peritoneal cavity.

17. Structure 1.	A. Pouch of Douglas.
18. Structure 2.	B. Bladder.
19. Structure 3.	C. Broad ligament.

| 20. Structure 4. | D. Body of uterus. |
| 21. Structure 5. | E. Uterovesical pouch. |

Match the structures on the left with the appropriate lymphatic drainage on the right.

22. Labia majora.	A. Internal iliac nodes.
23. Skin of glans penis.	B. Superficial inguinal nodes.
24. Cervix of uterus.	C. Inferior mesenteric nodes.
25. Prostate.	D. Internal and external iliac
26. Vaginal orifice.	nodes.
	E. Para-aortic nodes at level of first lumbar vertebra.

Match the arteries on the left with their site of origin on the right.

27. Superior rectal artery.	A. Internal pudendal artery.
28. Left ovarian artery.	B. External iliac artery.
29. Inferior rectal artery.	C. Internal iliac artery.
30. Middle rectal artery.	D. Inferior mesenteric artery.
	E. None of the above.

Match the muscles on the left with their nerve supply on the right.

31. Obturator internus.	A. Obturator nerve.
32. Internal anal sphincter.	B. Pudendal nerve.
33. Sphincter urethrae.	C. Pudendal nerve and perineal branch of S4.
34. Levator ani.	D. Hypogastric plexuses.
	E. None of the above.

Match the veins on the left with their site of termination on the right.

35. Superior rectal vein.	A. Inferior vena cava.
36. Right ovarian vein.	B. Internal iliac vein.
37. Left testicular vein.	C. External iliac vein.
38. Prostatic veins.	D. Inferior mesenteric vein.
	E. None of the above.

In each of the following questions, answer:

> A. If only (1) is correct
> B. If only (2) is correct
> C. If both (1) and (2) are correct
> D. If neither (1) nor (2) is correct

39. (1) As the ureter passes forward on the upper surface of the pelvic diaphragm in the female, it passes inferior to the uterine artery.
(2) As the ureter approaches the bladder wall in the male, it passes superior to the vas deferens.

40. (1) In the majority of women, the uterus is anteverted and retroflexed.
(2) The uterus receives its blood supply from the uterine and ovarian arteries.

41. (1) It is judicious to examine the para-aortic lymph nodes at the level of the first lumbar vertebra in women with ovarian cancer.
(2) The ovarian fossa is bounded by the internal and external iliac arteries.

42. (1) The narrowest part of the entire male urethra is the external urethral meatus.
(2) The widest part of the male urethra is the membranous urethra.

43. (1) Erection of the penis is a sympathetic response.
(2) Ejaculation is a parasympathetic response.

44. (1) The posterior wall of the lower third of the vagina is supported by the perineal body.
(2) The rectouterine pouch (pouch of Douglas) may be entered by a surgical incision through the posterior fornix of the vagina.

45. (1) Cancer of the prostate can metastasize (spread) to the skull via the vertebral venous plexus.
(2) The middle (median) lobe of the prostate lies inferior to the ejaculatory ducts and posterior to the urethra.

46. (1) The promontory of the sacrum is formed by the anterior surface of the second sacral vertebra.
(2) The pelvic outlet is bounded posteriorly by the coccyx, laterally by the sacrotuberous ligaments and ischial tuberosities, and anteriorly by the pubic arch.

47. (1) The pelvic diaphragm is formed by the levatores ani and coccygeal muscles and their covering fasciae.
(2) The urogenital diaphragm is attached laterally to the inferior ramus of the pubis and the ischial ramus.

48. (1) When passing an instrument into the rectum, a physician must remember the presence of two crescentic transverse mucosal folds on the right rectal wall and one on the left.
(2) No part of the rectum can be felt by introducing a gloved index finger into the anal canal.

49. (1) During the act of defecation, the levator ani muscles continue to support the pelvic viscera, but the puborectalis fibers relax with the anal sphincters.
(2) During the second stage of labor, the gutter shape of the pelvic floor tends to cause the baby's head to rotate.

50. Concerning the anal canal:
(1) The mucous membrane of the upper half of the anal canal is sensitive to touch and pain.
(2) Anastomoses between the tributaries of the superior and inferior rectal veins occur in the anal columns.

51. Concerning the development of the rectum and anal canal:
(1) The rectum and the upper half of the anal canal are developed from the hindgut.
(2) The lower half of the anal canal is formed from the proctodeum.

52. Concerning the female urethra:
(1) The female urethra measures about 3 in. (7.62 cm) long.
(2) The female urethra is difficult to dilate.

53. Concerning the ischiorectal fossa:
(1) The lateral wall is formed by the lower part of the obturator internus muscle.
(2) The inferior wall or floor is formed by the membranous layer of superficial fascia.

54. Concerning the process of orgasm in the female:
(1) The nervous impulses that pass to the smooth muscle in the vaginal wall arise in spinal cord segments S2, 3, and 4.
(2) The bulbospongiosus muscles undergo rhythmic contractions as the result of sympathetic nerve fibers from L1 and 2 segments of the spinal cord.

55. Concerning the structure of the penis:
(1) The glans is formed by the expanded end of the corpus spongiosum.
(2) The urethra passes through the corpus spongiosum.

56. Concerning the erectile tissue of the penis:
(1) The bulb of the penis is supplied by the deep artery of the penis.
(2) The two corpora cavernosa are located on the ventral surface of the body of the penis.

57. Concerning the rectouterine pouch (pouch of Douglas):
(1) It commonly contains the sigmoid colon and coils of ileum.
(2) It is the most inferior extension of the female peritoneal cavity.

Select the **best** response.

58. The following structures are closely related to the rectouterine pouch (pouch of Douglas) **except:**
A. Anteriorly is situated the posterior surface of the upper part of the vagina.
B. Posteriorly is situated the upper part of the rectum.
C. The trigone of the bladder is directly related to its anterior wall.
D. Anteriorly is situated the posterior surface of the body of the uterus.
E. Laterally is situated the sacrocervical ligaments passing forward to the cervix.

59. The deep perineal pouch in the female contains the following structures **except:**
A. Part of the urethra.
B. The sphincter urethrae.
C. The internal pudendal vessels.
D. The greater vestibular glands.
E. Part of the vagina.

60. The pudendal nerve and/or its branches have the following characteristics **except:**
A. Traverse the pudendal (Alcock's) canal.

B. Enter the perineum through the lesser sciatic foramen.

C. Give origin to the posterior scrotal nerves.

D. Supply the ischiocavernosus muscle.

E. Arise from S1, 2, and 3 spinal nerves.

61. Support for the uterus, either directly or indirectly, is provided by the following structures **except:**
    A. Perineal body.
    B. Mesosalpinx.
    C. Transverse cervical (cardinal) ligaments.
    D. Levator ani muscles.
    E. Pubocervical ligaments.

62. The following statements concerning the ovary are correct **except:**
    A. It is attached to the posterior layer of the broad ligament.
    B. It ovulates an ovum into the peritoneal cavity.
    C. It is attached to the lateral pelvic wall by the round ligament of the ovary.
    D. It is normally not related to the posterior fornix of the vagina.
    E. The right ovarian vein drains into the inferior vena cava.

63. The following statements concerning the urinary bladder are true **except:**
    A. It lies in the visceral layer of pelvic fascia beneath the peritoneum.
    B. The internal surface, when the bladder is empty, is wrinkled except at the trigone, which is smooth.
    C. Parasympathetic nerve fibers innervate the detrusor muscle.
    D. The trigone is the area between the openings of the urethra and the two ureters.
    E. The bladder in a child is located entirely within the pelvis.

64. The broad ligament contains all of the following **except** the:
    A. Round ligament of the ovary.
    B. Uterine artery.
    C. Round ligament of the uterus.
    D. Uterine tubes.
    E. Ureters.

65. Malignant tumors of the trigone of the bladder spread (metastasize) to the following lymph nodes:
    A. Lumbar.
    B. Sacral.
    C. External iliac only.
    D. External and internal iliac.
    E. Superficial inguinal.

66. All the following statements concerning the hypogastric plexuses are correct **except:**
    A. The superior hypogastric plexus is situated in front of the bifurcation of the aorta and the promontory of the sacrum.

B. The plexus contains sympathetic and parasympathetic fibers and numerous small ganglia.

C. The pelvic splanchnic nerve enters the plexus from S2, 3, and 4 segments of the spinal cord.

D. The preganglionic sympathetic fibers are derived from L4 and 5 segments of the spinal cord.

E. The pelvic parts of the sympathetic trunks are distinct from the plexuses.

67. Traumatic injury to the perineum in the male may rupture the bulb of the penis or penile urethra. The resulting leakage of blood or urine may be found in all of the following **except:**
    A. Anterior abdominal wall.
    B. Ischiorectal fossa.
    C. Scrotum.
    D. Penis.
    E. Superficial perineal pouch.

68. Which statement regarding the pudendal (Alcock's) canal is **correct?**
    A. It is in the deep fascia of the levator ani muscle.
    B. It extends from the greater sciatic foramen to the posterior border of the urogenital diaphragm.
    C. It is in the deep fascia of the obturator internus muscle.
    D. It is in the medial wall of the ischiorectal fossa.
    E. It is in the deep fascia of the obturator externus muscle.

In the next section, study each case history. Select the **best** answer to each question following it.

A 21-year-old woman was seen in the emergency department complaining of severe pain in the right iliac region. Just before admission she had fainted. On physical examination, her abdominal wall was extremely tender on palpation in the right iliac region; some rigidity and guarding of the lower abdominal muscles were noticed. A vaginal examination revealed that a tender "doughlike mass" could be felt through the posterior fornix. The patient had missed her last period. A diagnosis of a ruptured ectopic pregnancy was made.

69. In this ectopic pregnancy the embryo was implanted in the right uterine tube. Rupture of the tube produced the symptoms and signs noted above. Using your knowledge of anatomy, explain the relationship of the uterine tube to the peritoneal cavity.
    A. The tube is retroperitoneal.
    B. The tube lies within the peritoneal cavity.
    C. The tube lies within the broad ligament near its base.
    D. The tube lies in the free margin of the broad ligament.
    E. The tube lies within the parametrium.

70. The tender doughlike mass felt through the posterior wall of the vagina was due to:
    A. Retroverted uterus.

B. Blood in the pouch of Douglas.

C. A full bladder.

D. Blood in the uterovesical pouch.

E. A prolapsed ovary.

71. What is responsible for the guarding (partial contraction) of the lower abdominal muscles?

   A. Blood in the peritoneal cavity causing irritation of the parietal peritoneum and reflex contraction of the lower abdominal muscles.

   B. Apprehension of the patient.

   C. Reflex contraction of the abdominal muscles due to damage to the wall of the uterine tube.

   D. Stretching of the peritoneum of the broad ligament.

   E. The high level of blood estrogen causing irritability of the abdominal muscles.

A 56-year-old man was seen by his physician because he had noticed that his "bowel" was protruding from his anus after defecation. On questioning, the patient stated that for the past 2 years he had frequently passed blood-stained stools. Digital rectal examination revealed nothing abnormal. Proctoscopic examination showed that the mucous membrane above the level of the anal valves tended to bulge downward in three areas when the patient strained. The swollen mucous membrane contained large congested veins beneath the surface.

72. Based on your knowledge of anatomy, make a reasoned diagnosis.

   A. External hemorrhoids.

   B. Perianal abscess.

   C. Complete rectal prolapse.

   D. Internal hemorrhoids.

   E. Chronic anal fissures.

73. What is the venous drainage of the mucous membrane of the anal canal?

   A. Middle rectal veins only.

   B. Inferior rectal veins only.

   C. Superior and inferior rectal veins.

   D. Middle and inferior rectal veins.

   E. Internal pudendal veins.

Answers and Explanations

1. D
2. E
3. A
4. B
5. C
6. E
7. D
8. A
9. B
10. C
11. B
12. F
13. C
14. A
15. E
16. D
17. D
18. C
19. A
20. E
21. B
22. B
23. B
24. D
25. A
26. B
27. D
28. E
29. A
30. C
31. E
32. D
33. B
34. C
35. D
36. A
37. E
38. B
39. A   In the male, the vas deferens crosses the ureter superiorly.
40. B   In the majority of women, the uterus is anteverted and anteflexed.
41. C
42. A   The widest and most easily dilatable part of the male urethra is the prostatic part.
43. D   Erection of the penis is a parasympathetic response; ejaculation is a sympathetic response.
44. C
45. A   The middle or median lobe of the prostate lies above the ejaculatory ducts and posterior to the urethra.
46. B   The promontory of the sacrum is formed by the anterior superior border of the first sacral vertebra.
47. C
48. D   The rectal mucosal folds are placed two on the left and one on the right. The rectum can be entered by the index finger since the anal canal is only 1½ in. (3.8 cm) long.
49. C
50. B   The mucous membrane of the upper half of the anal canal is only sensitive to stretch.
51. C
52. D   The female urethra measures about 1½ in. (3.8 cm) long and it is moderately easy to dilate.

53. C
54. D   The smooth muscle in the vaginal wall is made to contract by sympathetic nerve fibers (L1 and 2). The bulbospongiosus muscles are innervated by the perineal branch of the pudendal nerve.
55. C
56. D   The bulb of the penis is supplied by the artery of the bulb. The two corpora cavernosa are located on the dorsal surface of the body of the penis.
57. C
58. C   The pouch of Douglas is separated from the trigone of the bladder by the uterus and vagina.
59. D   The greater vestibular glands are situated under cover of the posterior parts of the bulb of the vestibule and the labia majora in the superficial perineal pouch.
60. E   The pudendal nerve is a branch of the sacral plexus (S2, 3, and 4).
61. B   The mesosalpinx is an area of the broad ligament located between the uterine tube and the attachment of the mesovarium. It provides no support for the uterus.
62. C   The ovary is attached (suspended) from the lateral pelvic wall by the suspensory ligament. It contains the blood and the lymphatic vessels and nerves supplying the ovary. The round ligament of the ovary is the remains of the upper part of the gubernaculum; it extends from the medial border of the ovary to the lateral wall of the body of the uterus.
63. E   In the young child, because the pelvis is small, there is insufficient room for the urinary bladder. Even the empty bladder projects upward into the abdomen; later, when the pelvis enlarges, the bladder sinks to become a pelvic organ.
64. E   The ureters pass forward inferior to the broad ligaments.
65. D
66. D   The preganglionic sympathetic nerve fibers that influence the activities of the pelvic viscera arise from L1 and 2 segments of the spinal cord.
67. B   The arrangement of the membranous layer of superficial fascia (Colles' fascia) in the perineum prevents fluid from traveling backward into the ischiorectal fossae; the fascia is attached to the posterior border of the urogenital diaphragm.

68. C
69. D
70. B   Tubal pregnancies commonly take place where the infundibulum narrows to join the isthmus. Rupture of the tube almost invariably occurs with severe intraperitoneal hemorrhage. The blood gravitates downward into the lowest part of the peritoneal cavity, namely, the pouch of Douglas. Here, it clots and forms a doughlike mass that can usually be felt through the posterior vaginal wall.
71. A   Blood is an irritant to the peritoneum. The parietal peritoneum of the lower part of the abdomen and pelvis receives its sensory nerve supply from the lumbar spinal nerves (iliohypogastric, ilioinguinal, and obturator nerves). The muscles of the lower part of the anterior abdominal wall are innervated by the iliohypogastric and ilioinguinal nerves and there is a reflex spasm of these abdominal muscles. This is essentially a protective mechanism and an attempt to keep that area of the abdomen at rest so that the irritant will remain localized.
72. D   Internal hemorrhoids are dilatations of the tributaries of the superior rectal vein. They are covered by mucous membrane in the upper half of the anal canal and lie within the anal columns. One large tributary occurs on the left of the canal and two on the right. As these dilatations enlarge, they are pushed down the anal canal during defecation and may be extruded through the anus. At first they return to the canal at the end of defecation but with further enlargement and elongation they remain outside the anus.
73. C   The venous drainage of the mucous membrane of the upper half of the anal canal is from the superior rectal vein; that from the lower half is from the inferior rectal veins. The middle rectal veins are mainly concerned with draining blood from the muscle and connective tissue of the canal. Since the superior rectal vein is a tributary of the portal vein and the inferior and the middle rectal veins are tributaries of systemic veins, the anal canal is an important site for portal-systemic anastomoses in patients with a blocked portal vein.

Pelvis and Perineum

Upper Limb

The upper limb should be regarded as a multijointed lever that is freely movable on the trunk at the shoulder joint; at its distal end is the important organ, the hand.

It is suggested that the upper limb be reviewed in the following order:

1. The mammary gland. This organ, situated in the pectoral region, is of great clinical importance.
2. A brief overview of the bones and major joints, preferably with the help of an articulated skeleton.
3. A consideration of the more important muscles, concentrating on their actions and nerve supply.
4. A brief review of the blood supply and lymphatic drainage of the upper limb.
5. A detailed overview of the nerves and their distribution, with special emphasis on the branches of the brachial plexus (frequently injured).

To assist students in the review process, tables have been used extensively.

Mammary Gland

The mammary glands are specialized accessory glands of the skin that are capable of secreting milk (Fig. 5-1). They are present in both sexes. In the male and the immature female, mammary glands are similar in structure. The **nipples** are small and surrounded by a colored area of skin called the **areola.** The breast tissue consists of little more than a system of ducts embedded in connective tissue; the ducts do not extend beyond the margin of the areola.

Under the influence of the ovarian hormones, the mammary glands gradually enlarge and assume their hemispherical shape (see Fig. 5-1) in the female at puberty. The ducts elongate, but the increased size of the glands is mainly due to the deposition of adipose tissue. The base of the breast extends from the second to the sixth rib and from the lateral margin of the sternum to the midaxillary line. The greater part of the gland lies in the superficial fascia. A small part, the **axillary tail,** pierces the deep fascia at the lower border of the pec-toralis major muscle and enters the axilla. The mammary glands are separated from the deep fascia covering the underlying muscle by loose areolar tissue known as the **retromammary space.**

Each gland consists of 15 to 20 **lobes** that radiate out from the nipple. There is no capsule. Each lobe is separated from its neighbor by connective tissue septa that extend from the skin to the deep fascia and serve as **suspensory ligaments.** The main **lactiferous duct** from each lobe opens separately on the summit of the nipple and possesses a dilated **ampulla** or **lactiferous sinus** just prior to its termination (see Fig. 5-1). The ampulla serves as a small reservoir for the secreted milk.

Blood Supply
Arteries
Lateral thoracic and thoracoacromial arteries, branches of the axillary artery, and perforating branches of the internal thoracic and intercostal arteries.

Veins
These correspond to the arteries.

Lymphatic Drainage
The lateral quadrants of the breast drain into the anterior axillary or pectoral nodes (Fig. 5-2); the medial quadrants drain into the internal thoracic nodes. A few lymph vessels drain posteriorly into the posterior intercostal nodes, whereas some communicate with the lymphatic vessels of the opposite breast and with those of the anterior abdominal wall.

The lymphatic drainage of the breast is of great clinical importance because of the high incidence of cancer of the breast and the subsequent spread to lymph nodes that drain the breast.

Bones

Bones of the Shoulder Girdle
The clavicle and scapula form the shoulder girdle.

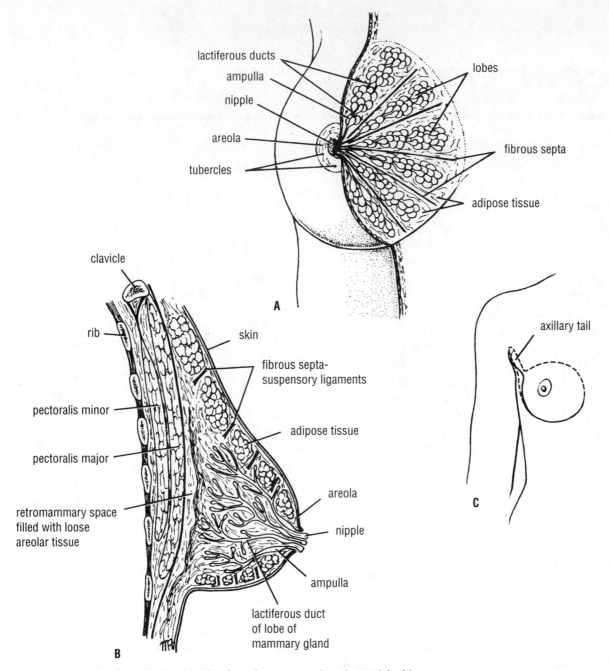

Fig. 5-1. Mature mammary gland in the female. **A.** Anterior view with skin partially removed to show internal structure. **B.** Sagittal section. **C.** The axillary tail, which pierces the deep fascia and extends into the axilla.

Clavicle

The clavicle acts as a strut that holds the upper limb away from the trunk. It articulates medially with the sternum and first costal cartilage and with the scapula. The clavicle lies just beneath the skin; its medial two-thirds is convex forward, and its lateral third is concave forward.

Scapula

The scapula is a flat triangular bone (Fig. 5-3). On its posterior surface, the **spine of the scapula** projects backward. The lateral end of the spine forms the **acromion,** which articulates with the clavicle. The superolateral angle of the scapula forms the **glenoid cavity** for articulation with the head of the humerus. The **coracoid process** projects upward and forward and provides attachment for muscles and ligaments. Medial to the base of the coracoid process is the **suprascapular notch.** The **subscapular fossa** is the name given to the concave anterior surface of the scapula. The **supraspinous fossa** lies above the spine, and the **infraspinous fossa** lies below the spine on the posterior surface of the scapula.

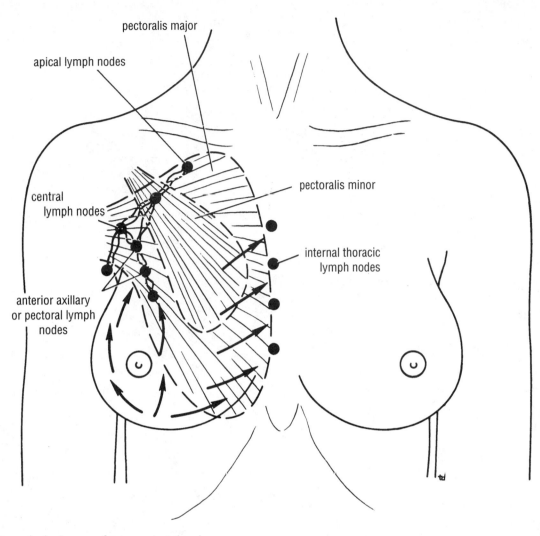

Fig. 5-2. Lymph drainage of mammary gland.

Bones of the Arm
Humerus

The **head** of the humerus lies at the upper end and forms about a third of a sphere that articulates with the glenoid cavity of the scapula (Fig. 5-4). Immediately below the head is the **anatomical neck.** Below the neck are the **greater and lesser tuberosities,** separated from each other by the **bicipital groove.** Distal to the tuberosities is a narrow region that is frequently fractured called the **surgical neck.** The **deltoid tuberosity** is a roughened area about halfway down the lateral aspect of the shaft (see Fig. 5-4); it is for the insertion of the deltoid muscle. Behind and below the tuberosity is a **spiral groove,** in which lies the radial nerve.

The **medial and lateral epicondyles** lie at the lower end of the humerus for the attachment of muscles and ligaments. The rounded **capitulum** articulates with the head of the radius, and the pulley-shaped **trochlea** articulates with the trochlear notch of the ulna. The **radial fossa** lies

above the capitulum and receives the head of the radius in full flexion of the elbow joint. Above the trochlea anteriorly is the **coronoid fossa,** which receives the coronoid process of the ulna during full flexion of the elbow joint. Above the trochlea posteriorly is the **olecranon fossa,** which receives the olecranon process of the ulna when the elbow joint is extended.

Bones of the Forearm
Radius

The radius is the lateral bone of the forearm (see Fig. 5-4). The **head** lies at the upper end and is small and circular. The upper concave surface articulates with the convex capitulum. The circumference of the head articulates with the radial notch of the ulna. The bone is constricted below the head to form the **neck.** Below the neck is the **bicipital tuberosity** for the insertion of the biceps brachii muscle.

The **shaft** has a sharp **interosseous border** medially for the attachment of the **interosseous membrane** that binds the

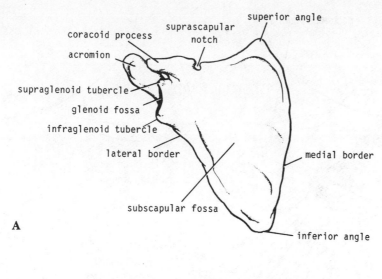

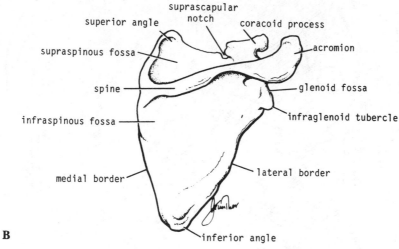

Fig. 5-3.   A. Anterior surface of right scapula. **B.** Posterior surface of right scapula.

radius and ulna together. The **pronator tubercle,** for the insertion of the pronator teres muscle, lies halfway down on its lateral side.

The **styloid process** projects distally from the lateral margin of the lower end (see Fig. 5-4). On the medial surface of the lower end is the **ulnar notch,** which articulates with the head of the ulna. The inferior surface of the lower end articulates with the scaphoid and lunate bones. On the posterior surface of the lower end is the small **dorsal tubercle,** which is grooved on its medial side by the tendon of extensor pollicis longus.

Ulna

The ulna is the medial bone of the forearm (see Fig. 5-4). The **olecranon process** is the large upper end that forms the prominence of the elbow. The **trochlear notch** lies on the

anterior surface of the olecranon process and articulates with the trochlea of the humerus. Below the trochlear notch is the triangular **coronoid process,** which has on its lateral surface the radial notch for articulation with the head of the radius.

The **shaft** has a sharp **interosseous border** laterally for the attachment of the interosseous membrane.

The small rounded **head** lies at the lower end of the ulna. The **styloid process** projects from the medial aspect of the head.

Bones of the Hand
Carpal Bones

There are eight carpal bones arranged in two rows of four (Fig. 5-5). The **proximal row** consists of (from lateral to medial) the **scaphoid** (navicular), **lunate, triquetral,** and

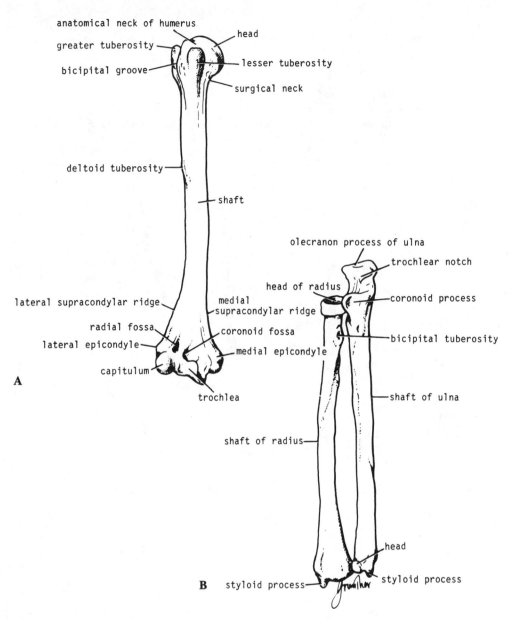

Fig. 5-4. **A.** Anterior surface of right humerus. **B.** Anterior surface of right radius and ulna.

pisiform bones. The **distal row** consists of (from lateral to medial) the **trapezium, trapezoid, capitate,** and **hamate** bones. The bones are united to one another by strong ligaments. The bones together form a concavity on their anterior surface, to the lateral and medial edges of which is attached a strong membranous band, the **flexor retinaculum,** which forms a bridge. The bridge and the bones form the **carpal tunnel** for the passage of the median nerve and the long flexor tendons of the fingers (see Fig. 5-13). Clinically, the median nerve may be compressed in the carpal tunnel, giving rise to the **carpal tunnel syndrome.**

Metacarpals

There are five metacarpal bones, each of which has a proximal **base,** a **shaft,** and a distal **head** (see Fig. 5-5). The bases of the metacarpals articulate with the distal row of the carpal bones. The heads, which form the knuckles, articulate with the proximal phalanges.

Phalanges

There are three phalanges for each of the fingers, but only two for the thumb (see Fig. 5-5). Each phalanx has a proximal **base,** a **shaft,** and a distal **head.**

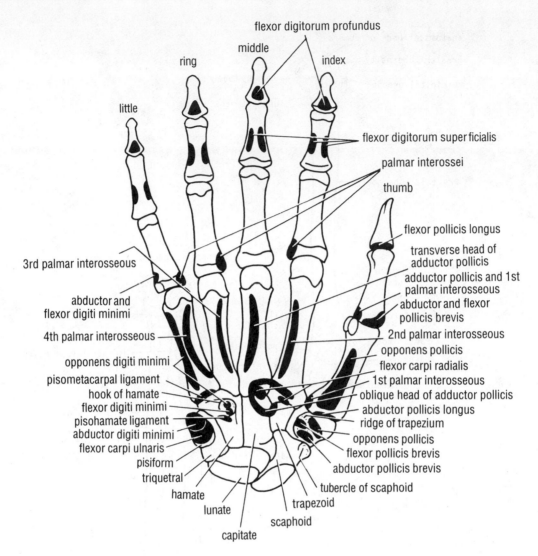

flexor digitorum profundus

middle

ring

index

little

flexor digitorum superficialis

palmar interossei

thumb

flexor pollicis longus

3rd palmar interosseous

transverse head of adductor pollicis

adductor pollicis and 1st palmar interosseous

abductor and flexor digiti minimi

abductor and flexor pollicis brevis

4th palmar interosseous

2nd palmar interosseous

opponens digiti minimi

opponens pollicis

pisometacarpal ligament

flexor carpi radialis

hook of hamate

1st palmar interosseous

flexor digiti minimi

oblique head of adductor pollicis

pisohamate ligament

abductor pollicis longus

abductor digiti minimi

ridge of trapezium

flexor carpi ulnaris

opponens pollicis

pisiform

flexor pollicis brevis

triquetral

abductor pollicis brevis

hamate

tubercle of scaphoid

lunate

trapezoid

scaphoid

capitate

Fig. 5-5. Anterior surface of bones of right hand, showing important muscular attachments.

Joints

Shoulder Joint
Articulation
Rounded head of the humerus and the shallow glenoid cavity of the scapula covered with hyaline cartilage (Fig. 5-6). The glenoid cavity is deepened by the fibrocartilaginous rim, the **glenoid labrum.**

Type
Synovial ball-and-socket joint.

Capsule
Thin and lax allowing a wide range of movement. It is attached around the outside of the glenoid labrum and to the anatomical neck of the humerus. The capsule is strength-ened by the tendons of the short muscles around the joint, namely, the subscapularis anteriorly, the supraspinatus superiorly, and the infraspinatus and teres minor posteriorly. Collectively, these muscle tendons are called the **rotator cuff.** The cuff plays a very important role in stabilizing the shoulder joint.

Ligaments
Glenohumeral Ligaments
These are three weak bands that strengthen the anterior part of the capsule.

Transverse Humeral Ligament
This strengthens the capsule and bridges the gap between the greater and lesser tuberosities of the humerus. It holds the tendon of the long head of the biceps muscle in place.

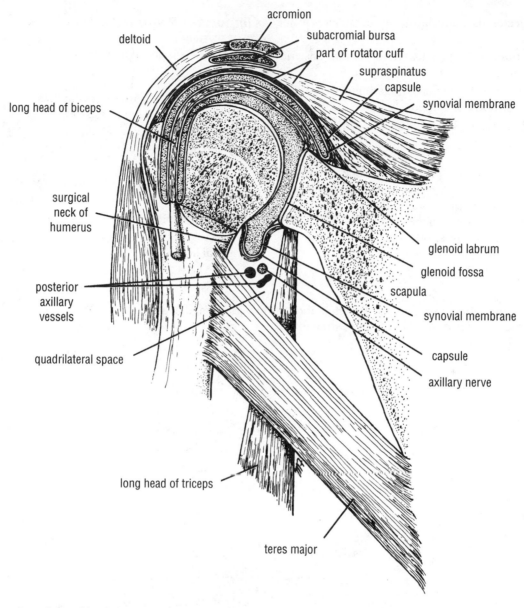

Fig. 5-6. Interior of shoulder joint.

Coracohumeral Ligament

This strengthens the capsule above and extends from the root of the coracoid process to the greater tuberosity of the humerus.

Accessory Ligament

The **coracoacromial ligament** extends from the coracoid process to the acromion. It protects the superior aspect of the joint.

Synovial Membrane

Lines the capsule (see Fig. 5-6), surrounds the tendon of the biceps, and protrudes forward through the capsule to form a bursa beneath the subscapularis muscle.

Nerve Supply

Axillary and suprascapular nerves.

Movements and the Muscles that Produce Movements

The shoulder joint has a wide range of movement.

Flexion: Anterior fibers of deltoid, pectoralis major, biceps, coracobrachialis

Extension: Posterior fibers of deltoid, latissimus dorsi, teres major

Abduction: Middle fibers of deltoid, assisted by supraspinatus

Adduction: Pectoralis major, latissimus dorsi, teres major, teres minor

Lateral rotation: Infraspinatus, teres minor, posterior fibers of deltoid

Medial rotation: Subscapularis, latissimus dorsi, teres major, anterior fibers of deltoid

Circumduction: A combination of all these movements

Stability

The strength of the joint depends on the tone of the short muscles that cross in front, above, and behind the joint, namely, the subscapularis, supraspinatus, infraspinatus, and teres minor (tendons from the rotator cuff). The weakest part of the joint lies inferiorly since there is little support and the capsule is weakest in this area.

Important Relations
Anteriorly
Brachial plexus and axillary vessels.

Inferiorly
Axillary nerve and posterior circumflex humeral vessels as they lie in the quadrangular space.

Elbow Joint
Articulation
Trochlea and capitulum of the humerus and the trochlear notch of the ulna and head of the radius (Fig. 5-7).

Type
Synovial hinge joint.

Fig. 5-7.   A. Right elbow joint, lateral aspect. **B.** Right elbow joint, medial aspect. **C.** Inferior radioulnar joint, wrist joint, and carpal joints.

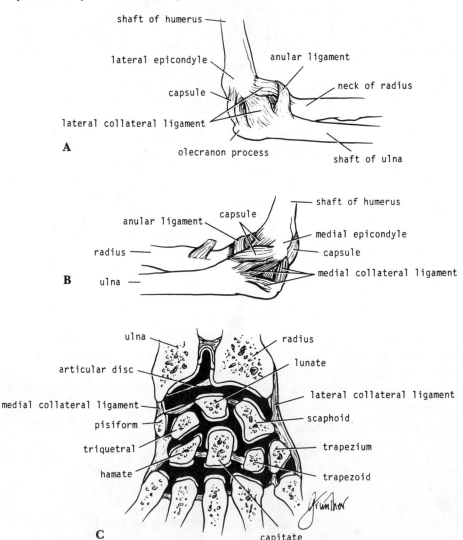

Capsule
Encloses the joint.

Ligaments
Lateral Collateral Ligament
This is triangular in shape and is attached by its apex to the lateral epicondyle of the humerus and by its base to the superior margin of the anular ligament and the ulna (see Fig. 5-7).

Medial Collateral Ligament
This is triangular in shape and is attached by its apex to the medial epicondyle of the humerus and by its base to the coronoid process and olecranon process of the ulna (see Fig. 5-7). **It is closely related to the ulnar nerve.**

Synovial Membrane
Lines the capsule and is continuous below with the synovial membrane of the superior radioulnar joint.

Nerve Supply
Median, ulnar, musculocutaneous, and radial nerves.

Movements and the Muscles that Produce Movement
Flexion: Brachialis, biceps, brachioradialis, pronator teres

Extension: Triceps, anconeus

Carrying Angle
The carrying angle, which opens laterally, is seen when the elbow joint is extended and lies between the arm and forearm. The angle is 170 degrees in the male and 167 degrees in the female. The angle disappears when the elbow joint is flexed.

Important Relations
Anteriorly: Median nerve and brachial artery

Medially: Ulnar nerve as it passes **behind** the medial epicondyle of the humerus

Superior Radioulnar Joint
Articulation
Circumference of the head of the radius and the anular ligament and radial notch of the ulna. The superior radioulnar joint forms a collar around the head of the radius.

Synovial Membrane
Lines the capsule and is continuous above with that of the elbow joint.

Movements and the Muscles that Produce Movement
Pronation and supination of the forearm.

Pronation: Pronator teres, pronator quadratus. This movement results in the hand rotating medially so that the palm faces posteriorly and the thumb lies on the medial side.

Supination: Biceps, supinator. This movement returns the pronated hand to the anatomic position; thus, the palm faces anteriorly and the thumb lies on the lateral side. Supination is more powerful than pronation.

Inferior Radioulnar Joint
Articulation
Head of ulna and ulnar notch of radius (see Fig. 5-7).

Type
Synovial pivot joint.

Capsule and Ligaments
Capsule encloses the joint and is strengthened by anterior and posterior ligaments.

Articular Disc
The articular disc is triangular in shape and composed of fibrocartilage. Its apex is attached to the base of the styloid process of the ulna and its base is attached to the lower border of the ulnar notch of the radius. It binds the distal ends of the radius and ulna together and shuts off the inferior radioulnar joint from the wrist joint.

Movements and the Muscles that Produce Movement
There are rotary movements around a vertical axis at the superior and inferior radioulnar joints (see movements for superior radioulnar joint, above).

Wrist Joint (Radiocarpal Joint)
Articulation
Distal end of radius and the triangular cartilaginous articular disc above, and the scaphoid, lunate, and triquetral bones below (see Fig. 5-7).

Type
Synovial condyloid joint.

Capsule
Encloses the joint.

Ligaments
Anterior and Posterior Ligaments
These strengthen the capsule.

Medial Ligament
This connects the styloid process of the ulna to the triquetral bone.

Lateral Ligament

This connects the styloid process of the radius to the scaphoid bone.

Synovial Membrane

Lines the capsule.

Nerve Supply

Anterior interosseous nerve from the median and deep branches of the radial and ulnar nerves.

Movements and the Muscles that Produce Movement

Rotation is **not** possible because the articular surfaces are ovoid in shape. The lack of rotation is compensated for by the movements of pronation and supination of the forearm.

Flexion: Flexor carpi radialis, flexor carpi ulnaris, palmaris longus, flexor digitorum superficialis, flexor digitorum profundus, flexor pollicis longus

Extension: Extensor carpi radialis longus, extensor carpi radialis brevis, extensor carpi ulnaris, extensor digitorum, extensor indicis, extensor digiti minimi, extensor pollicis longus

Abduction: Flexor carpi radialis, extensor carpi radialis longus and brevis, abductor pollicis longus, extensor pollicis longus and brevis

Adduction: Flexor and extensor carpi ulnaris

Important Relations
Anteriorly

Median and ulnar nerves.

Laterally

Radial artery.

Carpometacarpal Joints

The carpometacarpal joints are synovial gliding joints having **anterior, posterior,** and **interosseous ligaments** (see Fig. 5-7).

Carpometacarpal Joint of the Thumb
Articulation

Trapezium and saddle-shaped base of the first metacarpal bone (see Fig. 5-7).

Type

Synovial saddle joint (biaxial joint).

Movements and the Muscles that Produce Movement

Flexion: Flexor pollicis brevis and longus, opponens pollicis

Extension: Extensor pollicis longus and brevis

Abduction: Abductor pollicis longus and brevis

Adduction: Adductor pollicis

Rotation (as in opposition): Opponens pollicis (rotates the thumb medially)

Metacarpophalangeal Joints
Articulation

Convex heads of metacarpal bones and concave bases of the proximal phalanges (see Fig. 5-7).

Type

Synovial condyloid.

Ligaments
Palmar Ligaments

These are strong and contain fibrocartilage.

Collateral Ligaments

These are cordlike bands that join the head of the metacarpal bone to the base of the phalanx. The collateral ligaments are taut when the joint is in flexion and lax when the joint is in extension (fingers can thus be abducted and adducted in extension).

Movements and the Muscles that Produce Movement

Flexion: Lumbricals and interossei assisted by flexor digitorum superficialis and profundus

Extension: Extensor digitorum, extensor indicis, extensor digiti minimi

Abduction (movement away from the midline of the third finger): dorsal interossei, i.e., **D. AB.** (dorsal abduct)

Adduction (movement toward the midline of the third finger): palmar interossei, i.e., **P. AD.** (palmar adduct)

In the metacarpophalangeal joint of the thumb, flexion is performed by the flexor pollicis longus and brevis; extension is performed by the extensor pollicis longus and brevis. The movements of abduction and adduction are performed at the carpometacarpal joint.

Interphalangeal Joints

The interphalangeal joints are synovial hinge joints that have a structure similar to that of the metacarpophalangeal joints.

Muscles of the Upper Limb

Shoulder Region

The muscles connecting the upper limb to the vertebral column are shown in Table 5-1; the muscles connecting the upper limb to the thoracic wall are shown in Table 5-2; and the muscles connecting the scapula to the humerus are shown in Table 5-3.

Table 5-1   Muscles Connecting the Upper Limb to the Vertebral Column

Name of muscle	Origin	Insertion	Nerve supply	Action
Trapezius	Occipital bone, ligamentum nuchae, spine of seventh cervical vertebra, spines of all thoracic vertebrae	Upper fibers into lateral third of clavicle, middle fibers and lower fibers into acromion and spine of scapula	Spinal part of accessory nerve and C3 and 4	Upper fibers elevate the scapula, middle fibers pull scapula medially, lower fibers pull medial border of scapula downward
Latissimus dorsi	Iliac crest, lumbar fascia, spines of lower six thoracic vertebrae, lower three or four ribs, inferior angle of scapula	Floor of bicipital groove of humerus	Thoracodorsal nerve	Extends, adducts, and medially rotates the arm
Levator scapulae	Transverse process of first four cervical vertebrae	Medial border of scapula	C3 and 4 and dorsal scapular nerve	Raises medial border of scapula
Rhomboid minor	Ligamentum nuchae and spines of seventh cervical and first thoracic vertebrae	Medial border of scapula	Dorsal scapular nerve	Raises medial border of scapula upward and medially
Rhomboid major	Second to fifth thoracic spines	Medial border of scapula	Dorsal scapular nerve	Raises medial border of scapula upward and medially

Table 5-2   Muscles Connecting the Upper Limb to the Thoracic Wall

Name of muscle	Origin	Insertion	Nerve supply	Action
Pectoralis major	Clavicle, sternum, and upper six costal cartilages	Lateral lip of bicipital groove of humerus	Medial and lateral pectoral nerves from brachial plexus	Adducts arm and rotates it medially; clavicular fibers also flex arm
Pectoralis minor	Third, fourth, and fifth ribs	Coracoid process of scapula	Medial pectoral nerve from brachial plexus	Depresses point of shoulder; if the scapula is fixed, it elevates ribs of origin
Subclavius	First costal cartilage	Clavicle	Nerve to subclavius from upper trunk of brachial plexus	Depresses the clavicle and steadies this bone during movements of the shoulder girdle
Serratus anterior	Upper eight ribs	Medial border and inferior angle of scapula	Long thoracic nerve	Draws the scapula forward around the chest wall; rotates the scapula

Upper Limb

Table 5-3   Muscles Connecting the Scapula to the Humerus

Name of muscle	Origin	Insertion	Nerve supply	Action
Deltoid	Lateral third of clavicle, acromion, spine of scapula	Middle of lateral surface of shaft of humerus	Axillary nerve	Abducts arm; anterior fibers flex and medially rotate arm, posterior fibers extend and laterally rotate arm
Supraspinatus	Supraspinous fossa of scapula	Greater tuberosity of humerus; capsule of shoulder joint	Suprascapular nerve	Abducts arm and stabilizes shoulder joint
Infraspinatus	Infraspinous fossa of scapula	Greater tuberosity of humerus; capsule of shoulder joint	Suprascapular nerve	Laterally rotates arm and stabilizes shoulder joint
Teres major	Lower third lateral border of scapula	Medial lip of the bicipital groove of humerus	Lower subscapular nerve	Medially rotates and adducts arm and stabilizes shoulder joint
Teres minor	Upper two-thirds of lateral border of scapula	Greater tuberosity of humerus; capsule of shoulder joint	Axillary nerve	Laterally rotates arm and stabilizes shoulder joint
Subscapularis	Subscapular fossa	Lesser tuberosity of humerus	Upper and lower subscapular nerves	Medially rotates arm and stabilizes shoulder joint

Axilla

The axilla, or armpit, is a pyramid-shaped space between the upper part of the arm and the side of the chest (Fig. 5-8). The upper end, or **apex,** is directed into the root of the neck and is bounded in front by the clavicle, behind by the upper border of the scapula, and medially by the outer border of the first rib. The lower end, or **base,** is bounded in front by the anterior axillary fold (formed by the lower border of the pectoralis major muscle), behind by the posterior axillary fold (formed by the tendon of latissimus dorsi and the teres major muscle), and medially by the chest wall.

The axilla contains the principal vessels (axillary artery and vein) and nerves (brachial plexus and its branches) to the upper limb and many lymph nodes (see Fig. 5-8).

Axillary Sheath

The axillary sheath encloses the axillary vessels and the brachial plexus. It is continuous above in the neck with the prevertebral layer of deep cervical fascia. The axillary sheath is important when carrying out a nerve block of the brachial plexus since the sheath localizes the anesthetic solution to the nerve plexus.

Rotator Cuff

The rotator cuff has already been alluded to in the section on the shoulder joint. It is the name given to the tendons of the subscapularis, supraspinatus, infraspinatus, and the teres minor muscles, which are fused to the underlying capsule of the shoulder joint. The cuff is important in stabilizing the shoulder joint. Lesions of the cuff are a common cause of pain in the shoulder region.

Quadrilateral Space

The quadrilateral space is bounded above by the subscapularis in front and the teres minor behind. It is bounded below by the teres major. The space is bounded laterally by the surgical neck of the humerus and medially by the long head of the triceps.

The quadrilateral space is located immediately below the shoulder joint. The axillary nerve, which passes through the space, may be damaged in dislocation of the shoulder joint.

Upper Arm

The muscles of the upper arm (Figs. 5-9 and 5-10) are shown in Table 5-4.

Fascial Compartments of the Upper Arm

The upper arm is enclosed in a sheath of deep fascia. Two fascial septa, one on the medial and one on the lateral side, extend from this sheath and are attached to the medial and lateral borders of the humerus respectively. By this means, the upper arm is divided into an anterior and a posterior fascial compartment, each having its own muscles, nerves, and arteries.

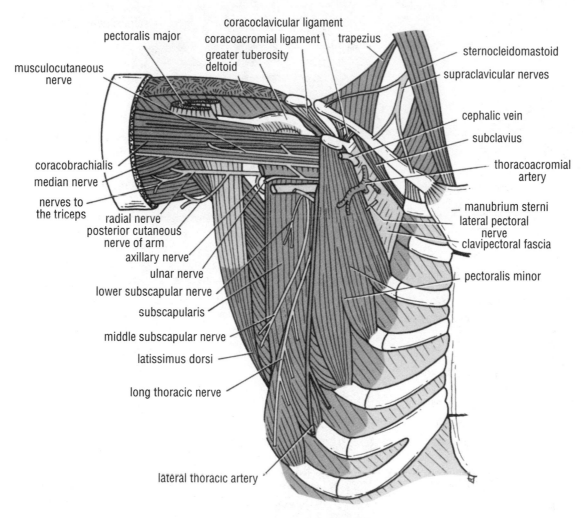

Fig. 5-8. Pectoral region and axilla; pectoralis major muscle has been removed to display underlying structures.

Cubital Fossa

The cubital fossa is a skin depression that lies in front of the elbow and is triangular in shape (Fig. 5-11). It is bounded **laterally** by the brachioradialis muscle and **medially** by the pronator teres muscle. The **base** of the triangle is formed by an imaginary line drawn between the two epicondyles of the humerus.

The cubital fossa contains the following structures (from the medial to the lateral side): the median nerve, the bifurcation of the brachial artery into ulnar and radial arteries, the tendon of the biceps muscle, and the radial nerve and its deep branch.

Lying in the superficial fascia covering the cubital fossa are the **cephalic** and **basilic veins** and their tributaries.

Forearm

The muscles of the anterior fascial compartment (Fig. 5-12) are shown in Table 5-5; the muscles of the lateral fascial compartment (see Fig. 5-12) are shown in Table 5-6; and

the muscles of the posterior fascial compartment are shown in Table 5-7.

Fascial Compartments of the Forearm

The forearm is enclosed in a sheath of deep fascia, which is attached to the periosteum of the posterior subcutaneous border of the ulna. This fascial sheath, together with the interosseous membrane and fibrous intermuscular septa, divides up the forearm into a number of compartments, each having its own muscles, nerves, and blood supply.

Interosseous Membrane

The interosseous membrane is a strong membrane that unites the shafts of the radius and the ulna. Its fibers are taut, and therefore the forearm is most stable when it is in the midprone position (position of function). The membrane provides attachment for neighboring muscles.

Upper Limb

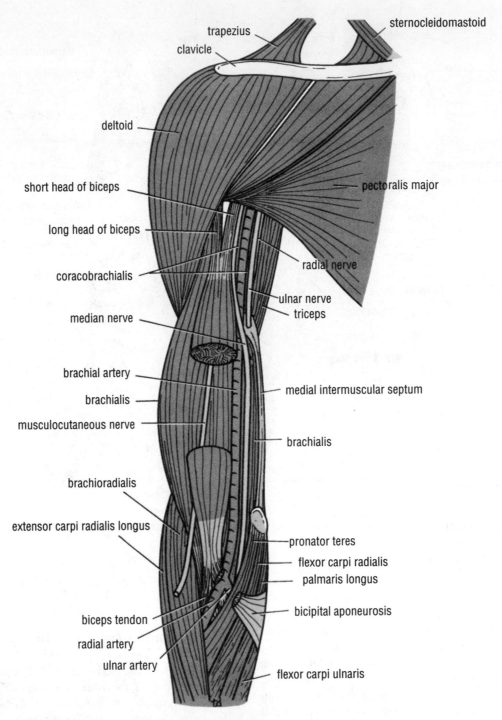

Fig. 5-9.   Anterior view of upper arm. Middle portion of biceps brachii has been removed to show musculocutaneous nerve lying in front of brachialis.

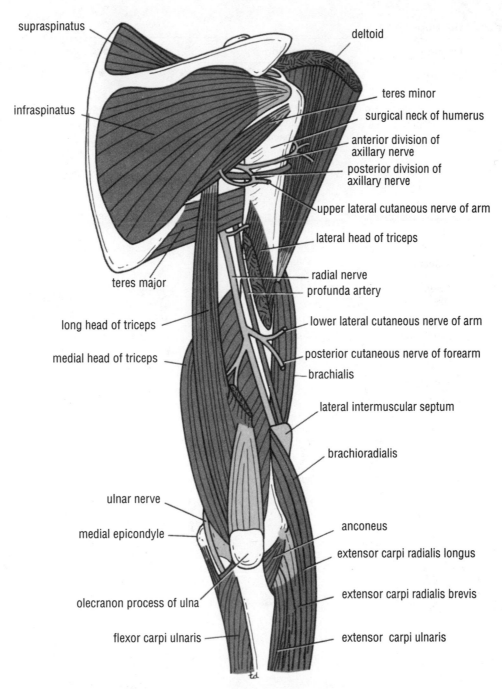

supraspinatus

deltoid

infraspinatus

teres minor

surgical neck of humerus

anterior division of
axillary nerve

posterior division of
axillary nerve

upper lateral cutaneous nerve of arm

lateral head of triceps

teres major

radial nerve
profunda artery

long head of triceps

lower lateral cutaneous nerve of arm

medial head of triceps

posterior cutaneous nerve of forearm

brachialis

lateral intermuscular septum

brachioradialis

ulnar nerve

anconeus

medial epicondyle

extensor carpi radialis longus

extensor carpi radialis brevis

olecranon process of ulna

extensor carpi ulnaris

flexor carpi ulnaris

Fig. 5-10. Posterior view of upper arm. Lateral head of triceps has been divided to
display radial nerve and profunda artery in spiral groove of humerus.

Upper Limb

Table 5-4   Muscles of the Upper Arm

Name of muscle	Origin	Insertion	Nerve supply	Action
Anterior Fascial Compartment				
Biceps brachii				
Long head	Supraglenoid tubercle of scapula	Tuberosity of radius and bicipital aponeurosis into deep fascia of forearm	Musculocutaneous nerve	Supinator of forearm, flexor of elbow joint, weak flexor of shoulder joint
Short head	Coracoid process of scapula			
Coracobrachialis	Coracoid process of scapula	Medial aspect of shaft of humerus	Musculocutaneous nerve	Flexes arm, weak adductor
Brachialis	Front of lower half of humerus	Coronoid process of ulna	Musculocutaneous nerve	Flexor of elbow joint
Posterior Fascial Compartment				
Triceps				
Long head	Infraglenoid tubercle of scapula	Olecranon process of ulna	Radial nerve	Extensor of elbow joint
Lateral head	Upper half of posterior surface of shaft of humerus			
Medial head	Lower half of posterior surface of shaft of humerus			

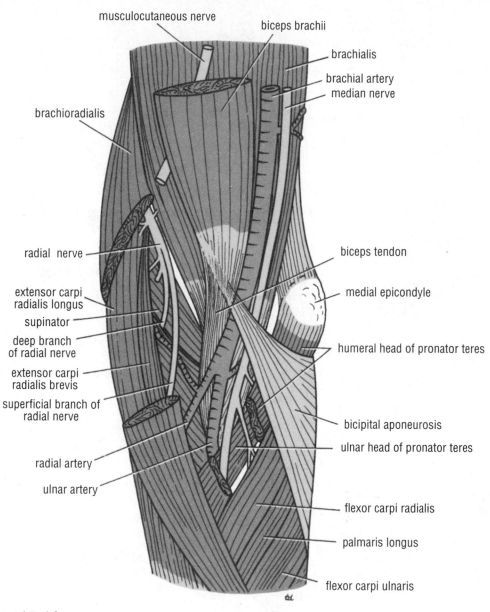

Fig. 5-11.   Right cubital fossa.

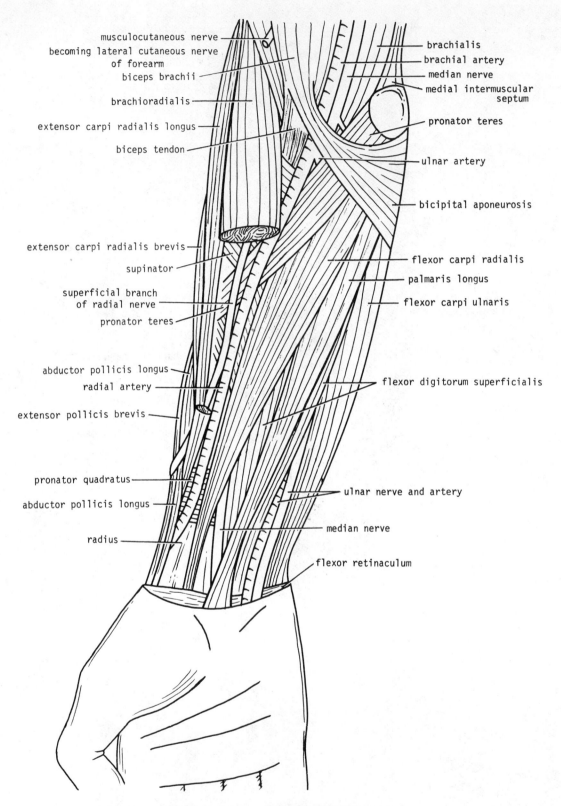

musculocutaneous nerve
becoming lateral cutaneous nerve
of forearm
biceps brachii
brachioradialis
extensor carpi radialis longus
biceps tendon

brachialis
brachial artery
median nerve
medial intermuscular
septum
pronator teres
ulnar artery

bicipital aponeurosis

extensor carpi radialis brevis
supinator
superficial branch
of radial nerve
pronator teres

flexor carpi radialis
palmaris longus
flexor carpi ulnaris

abductor pollicis longus
radial artery
extensor pollicis brevis

flexor digitorum superficialis

pronator quadratus
abductor pollicis longus
radius

ulnar nerve and artery
median nerve
flexor retinaculum

Fig. 5-12. Anterior view of forearm. Middle portion of brachioradialis muscle has been removed to display superficial branch of radial nerve and the radial artery.

Table 5-5 Muscles of the Anterior Fascial Compartment of the Forearm

Name of muscle	Origin	Insertion	Nerve supply	Action
Pronator teres				
Humeral head	Medial epicondyle of humerus	Lateral aspect of shaft of radius	Median nerve	Pronation and flexion of forearm
Ulnar head	Coronoid process of ulna			
Flexor carpi radialis	Medial epicondyle of humerus	Bases of second and third metacarpal bones	Median nerve	Flexes and abducts hand at wrist joint
Palmaris longus (often absent)	Medial epicondyle of humerus	Flexor retinaculum and palmar aponeurosis	Median nerve	Flexes hand
Flexor carpi ulnaris				
Humeral head	Medial epicondyle of humerus	Pisiform bone, hook of hamate, base of fifth metacarpal bone	Ulnar nerve	Flexes and adducts the hand at the wrist joint
Ulnar head	Olecranon process and posterior border of ulna			
Flexor digitorum superficialis				
Humeroulnar head	Medial epicondyle of humerus	Middle phalanx of medial four fingers	Median nerve	Flexes middle phalanx of fingers and assists in flexing proximal phalanx and hand
Radial head	Oblique line on anterior surface of shaft of radius			
Flexor pollicis longus	Anterior surface of shaft of radius	Distal phalanx of thumb	Anterior interosseous branch of median nerve	Flexes distal phalanx of thumb
Flexor digitorum profundus	Anterior surface of shaft of ulna, interosseous membrane	Distal phalanges of medial four fingers	Ulnar (medial half) and median (lateral half) nerves	Flexes distal phalanx of the fingers, then assists in flexion of middle and proximal phalanges and the wrist
Pronator quadratus	Anterior surface of shaft of ulna	Anterior surface of shaft of radius	Anterior interosseous branch of median nerve	Pronates forearm

Table 5-6   Muscles of the Lateral Fascial Compartment of the Forearm

Name of muscle	Origin	Insertion	Nerve supply	Action
Brachioradialis	Lateral supracondylar ridge of humerus	Styloid process of radius	Radial nerve	Flexes forearm at elbow joint, rotates forearm to midprone position
Extensor carpi radialis longus	Lateral supracondylar ridge of humerus	Base of second metacarpal bone	Radial nerve	Extends and abducts hand at wrist joint

Table 5-7   Muscles of the Posterior Fascial Compartment of the Forearm

Name of muscle	Origin	Insertion	Nerve supply	Action
Extensor carpi radialis brevis	Lateral epicondyle of humerus	Base of third metacarpal bone	Deep branch of radial nerve	Extends and abducts the hand at wrist joint
Extensor digitorum	Lateral epicondyle of humerus	Middle and distal phalanges of the medial four fingers	Deep branch of radial nerve	Extends fingers and hand
Extensor digiti minimi	Lateral epicondyle of humerus	Extensor expansion of little finger	Deep branch of radial nerve	Extends metacarpophalangeal joint of little finger
Extensor carpi ulnaris	Lateral epicondyle of humerus	Base of fifth metacarpal bone	Deep branch of radial nerve	Extends and adducts the hand at wrist joint
Anconeus	Lateral epicondyle of humerus	Olecranon process of ulna	Radial nerve	Extends elbow joint
Supinator	Lateral epicondyle of humerus, anular ligament of superior radioulnar joint and ulna	Neck and shaft of radius	Deep branch of radial nerve	Supination of forearm
Abductor pollicis longus	Shafts of radius and ulna	Base of first metacarpal bone	Deep branch of radial nerve	Abducts and extends thumb
Extensor pollicis brevis	Shaft of radius and interosseous membrane	Base of proximal phalanx of thumb	Deep branch of radial nerve	Extends metacarpophalangeal joints of thumb
Extensor pollicis longus	Shaft of ulna and interosseous membrane	Base of distal phalanx of thumb	Deep branch of radial nerve	Extends distal phalanx of thumb
Extensor indicis	Shaft of ulna and interosseous membrane	Extensor expansion of index finger	Deep branch of radial nerve	Extends metacarpophalangeal joint of index finger

Wrist
Flexor and Extensor Retinacula

The retinacula are bands of deep fascia that hold the long flexor and extensor tendons in position at the wrist (Fig. 5-13). The flexor retinaculum is attached medially to the pisiform bone and the hook of the hamate, and laterally to the tubercle of the scaphoid and the trapezium. The extensor retinaculum is attached medially to the pisiform bone and the hook of the hamate, and laterally to the distal end of the radius.

Carpal Tunnel

The bones of the hand and the flexor retinaculum form the carpal tunnel (see Fig. 5-13). The median nerve lies in a **restricted** space between the flexor digitorum superficialis and the flexor carpi radialis muscles and may become compressed at this site.

Hand

The small muscles of the hand (Figs. 5-14, 5-15, and 5-16) are shown in Table 5-8.

Fibrous Flexor Sheaths

The anterior surface of each finger, from the metacarpal head to the base of the distal phalanx, is provided with a strong fibrous sheath that is attached to the sides of the phalanges (see Fig. 5-14). It forms a blind tunnel in which the long flexor tendons of the fingers lie.

Synovial Flexor Sheaths

In the hand, the tendons of the flexor digitorum superficialis and profundus invaginate a common synovial sheath from the lateral side (Fig. 5-17). The medial part of this common sheath extends distally without interruption on the tendons of the little fingers. The lateral part of the sheath stops abruptly on the middle of the palm, and the distal ends of the long flexor tendons of the index, middle, and ring fingers acquire **digital synovial sheaths** as they enter the fingers. The flexor pollicis longus tendon has its own synovial sheath that passes into the thumb. The function of these sheaths is to allow the long tendons to move smoothly, with a minimum of friction, beneath the flexor retinaculum and the fibrous flexor sheaths.

Fig. 5-13. Cross section of hand, showing relationship of tendons, nerves, and arteries to flexor and extensor retinacula.

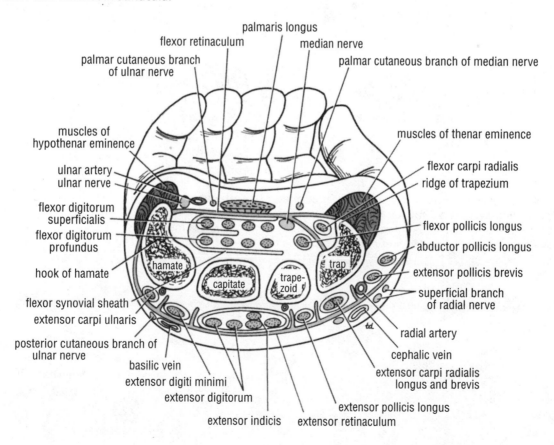

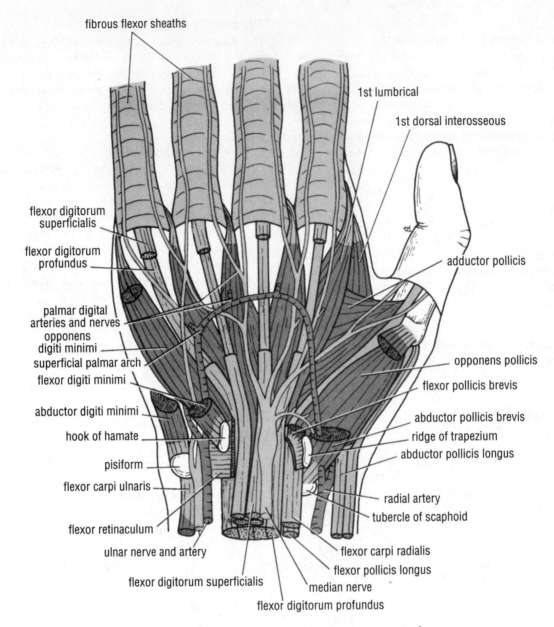

Fig. 5-14.   Anterior view of palm of hand. Palmar aponeurosis and greater part of flexor retinaculum have been removed to display superficial palmar arch, median nerve, and long flexor tendons. Segments of tendons of flexor digitorum superficialis have been removed to show underlying tendons of flexor digitorum profundus.

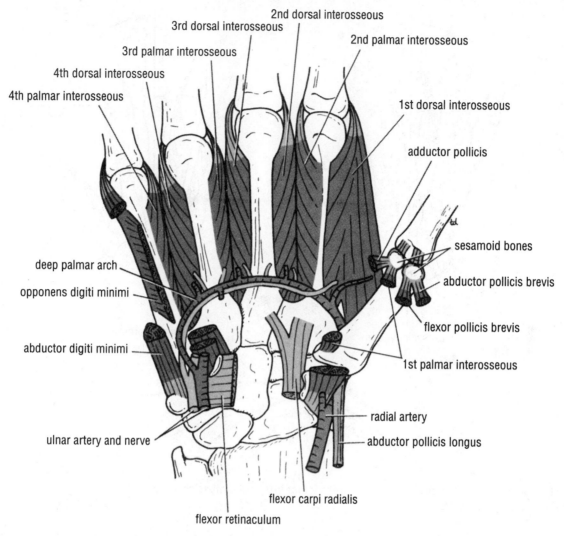

Fig. 5-15. Anterior view of palm of hand, showing deep palmar arch and deep terminal branch of ulnar nerve; interossei are also shown.

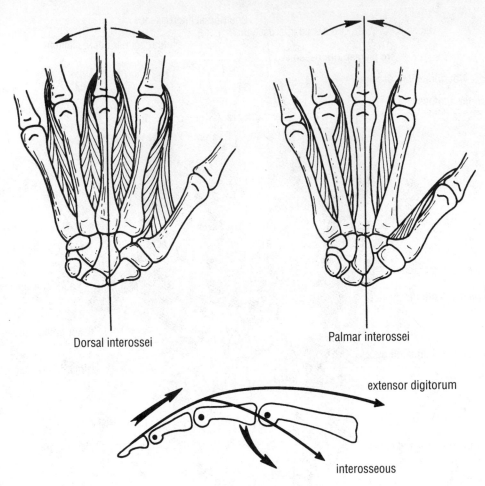

Dorsal interossei

Palmar interossei

extensor digitorum

interosseous

Fig. 5-16.   Origins and insertions of palmar and dorsal interossei muscles; actions of these muscles are also shown.

Table 5-8 Small Muscles of the Hand

Name of muscle	Origin	Insertion	Nerve supply	Action
Lumbricals (4)	Tendons of flexor digitorum profundus	Extensor expansion of medial four fingers	First and second (lateral two) median nerve, third and fourth ulnar nerve	Flex metacarpophalangeal joints and extend interphalangeal joints of fingers (except thumb)
Interossei (8)				
Palmar (4)	First, second, fourth, and fifth metacarpal bones	Base of proximal phalanges of fingers, extensor expansion	Deep branch of ulnar nerve	Adduct fingers toward center of third finger
Dorsal (4)	Contiguous sides of five metacarpal bones	Base of proximal phalanges of fingers, extensor expansion	Deep branch of ulnar nerve	Abduct fingers from center of third finger **Both** palmar and dorsal interossei flex the metacarpophalangeal joints and extend the interphalangeal joints
Palmaris brevis	Flexor retinaculum and palmar aponeurosis	Skin of palm	Superficial branch of ulnar nerve	Corrugates the skin to improve grip of palm

Short Muscles of Thumb

Name of muscle	Origin	Insertion	Nerve supply	Action
Abductor pollicis brevis	Scaphoid, trapezium, flexor retinaculum	Base of proximal phalanx of thumb	Median nerve	Abduction of thumb
Flexor pollicis brevis	Flexor retinaculum	Base of proximal phalanx of thumb	Median nerve	Flexes metacarpophalangeal joint of thumb
Opponens pollicis	Flexor retinaculum	Shaft of metacarpal bone of thumb	Median nerve	Pulls thumb medially and forward across palm
Adductor pollicis				
Oblique head	Second and third metacarpal bones	Base of proximal phalanx of thumb	Deep branch of ulnar nerve	Adducts thumb
Transverse head	Third metacarpal bone			

Short Muscles of Little Finger

Name of muscle	Origin	Insertion	Nerve supply	Action
Abductor digiti minimi	Pisiform bone	Base of proximal phalanx of little finger	Deep branch of ulnar nerve	Flexes little finger
Flexor digiti minimi	Flexor retinaculum	Base of proximal phalanx of little finger	Deep branch of ulnar nerve	Flexes little finger
Opponens digiti minimi	Flexor retinaculum	Shaft of metacarpal bone of little finger	Deep branch of ulnar nerve	Pulls fifth metacarpal bone forward as in cupping the palm

Upper Limb

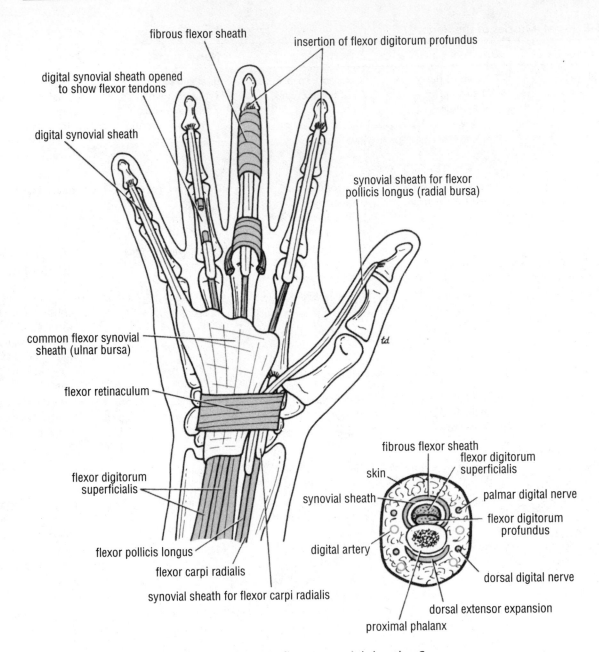

Fig. 5-17.   Anterior view of palm of hand, showing flexor synovial sheaths. Cross section of a finger is also shown.

Insertion of the Long Flexor Tendons

Each tendon of the flexor digitorum superficialis divides into two halves that pass around the profundus tendon and meet on its posterior surface, where partial decussation of the fibers takes place. The superficialis tendon, having united again, divides into two further slips that are attached to the borders of the middle phalanx. Each tendon of the flexor digitorum profundus, having passed through the superficialis tendon, is inserted into the base of the distal phalanx.

Insertion of the Long Extensor Tendons

The four tendons of the extensor digitorum fan out over the dorsum of the hand. The tendon to the index finger is joined on its **medial side** (Fig. 5-18) by the tendon of the extensor indicis. The tendon of the little finger is joined on its **medial side** by the two tendons of the extensor digiti minimi.

On the posterior surface of each finger, the extensor tendon widens out to form the **extensor expansion.** Near the proximal interphalangeal joint, the extensor expansion splits

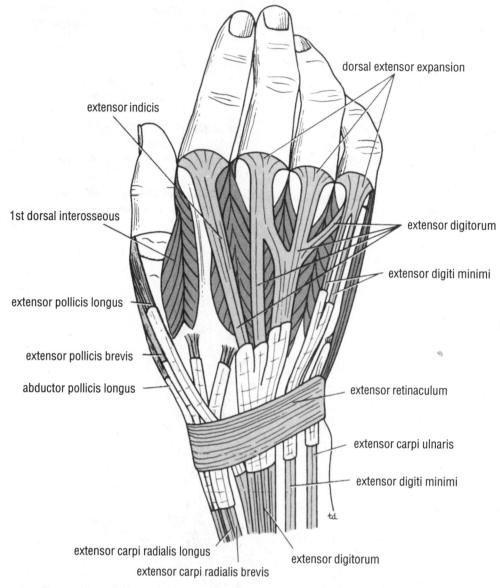

extensor indicis

dorsal extensor expansion

1st dorsal interosseous

extensor digitorum

extensor pollicis longus

extensor digiti minimi

extensor pollicis brevis

abductor pollicis longus

extensor retinaculum

extensor carpi ulnaris

extensor digiti minimi

extensor carpi radialis longus

extensor carpi radialis brevis

extensor digitorum

Fig. 5-18. Dorsal surface of hand, showing long extensor tendons and their synovial sheaths.

into three parts: a **central part,** which is inserted into the base of the middle phalanx, and **two lateral parts,** which converge to be inserted into the base of the distal phalanx.

The extensor expansion also receives the tendon of insertion of the corresponding interosseous muscle one each side and, further distally, receives the tendon of the lumbrical muscle on the lateral side.

Palmar Aponeurosis

In the palm, the deep fascia is greatly thickened to protect the underlying tendons, nerves, and blood vessels. It is known as the **palmar aponeurosis.** It is continuous proximally with the palmaris longus tendon and is attached to the flexor retinaculum. It is also continuous with the fasciae covering the thenar and hypothenar eminences.

Fascial Spaces of the Palm

The **thenar space** and the **midpalmar space** are potential spaces lying deep to the palmar aponeurosis that are filled with loose connective tissue. They are separated by an oblique fascial septum. The thenar space lies lateral to the third metacarpal bone, posterior to the long flexor tendons to the index finger, and in front of the adductor pollicis muscle. The midpalmar space lies medial to the third metacarpal bone and posterior to the long flexor tendons to the middle, ring, and little fingers.

Arteries of the Upper Limb

Axillary Artery

The axillary artery begins at the lateral border of the first rib as a continuation of the subclavian artery and, at the lower border of the teres major muscle, it becomes the brachial artery (Fig. 5-19). Throughout its course, the artery is closely related to the cords of the brachial plexus and their branches and is enclosed with them in a connective tissue sheath called the **axillary sheath.**

The pectoralis minor muscle crosses in front of the artery, dividing it (for purposes of description) into three parts: the lateral border of the first rib to the upper border of the pectoralis minor, the portion that lies posterior to the pectoralis minor, and from the lower border of the pectoralis minor to the lower border of the teres major.

Branches

The first part of the artery gives off one branch, the second part two branches, and the third part three branches.

Branch of the First Part

The **highest thoracic artery** is small and runs to the chest wall along the upper border of the pectoralis minor.

Branches of the Second Part

The **thoracoacromial artery** immediately divides into four terminal branches. The **lateral thoracic artery** runs to

Fig. 5-19.   The main arteries of the upper limb.

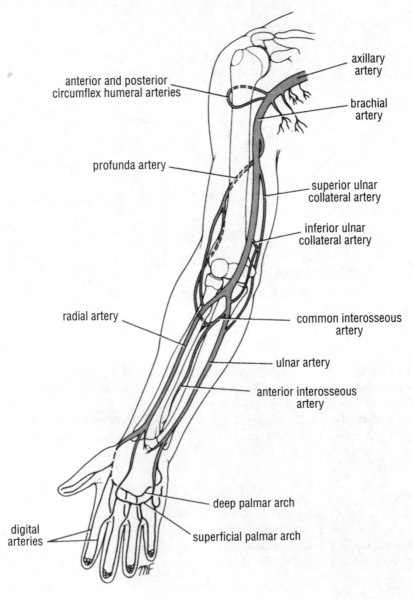

anterior and posterior
circumflex humeral arteries

axillary
artery

brachial
artery

profunda artery

superior ulnar
collateral artery

inferior ulnar
collateral artery

radial artery

common interosseous
artery

ulnar artery

anterior interosseous
artery

deep palmar arch

digital
arteries

superficial palmar arch

the chest wall along the lower border of the pectoralis minor. In the female it supplies the mammary gland.

Branches of the Third Part

The **subscapular artery** runs along the lower border of the subscapularis muscle. The **anterior** and **posterior circumflex humeral arteries** wind around the front and back of the surgical neck of the humerus, respectively.

Arterial Anastomosis Around the Shoulder Joint

The suprascapular and superficial cervical arteries (branches of the thyrocervical trunk from the first part of the subclavian artery) anastomose with the subscapular and anterior and posterior circumflex humeral arteries (branches of the third part of the axillary artery).

Brachial Artery

The brachial artery begins at the lower border of the teres major muscle as a direct continuation of the axillary artery (see Fig. 5-19). It descends through the anterior compartment of the arm on the brachialis muscle (see Fig. 5-9), enters the cubital fossa, and ends at the level of the neck of the radius by dividing into the radial and ulnar arteries (see Fig. 5-11).

The artery is superficial and is overlapped from the lateral side by the coracobrachialis and the biceps. The median nerve crosses its middle part and the bicipital aponeurosis its lower part. The lower part of the artery has the tendon of the biceps on its lateral side.

Branches

1. **Muscular branches.**
2. **Nutrient artery** to the humerus.
3. **Profunda artery** is a large branch that follows the radial nerve into the posterior compartment of the arm (in the spiral groove).
4. **Superior ulnar collateral artery** follows the ulnar nerve.
5. **Inferior ulnar collateral artery** takes part in the anastomosis around the elbow joint.

Radial Artery

The radial artery is the smaller of the two terminal branches of the brachial artery (see Fig. 5-19). It begins in the cubital fossa at the level of the neck of the radius and descends through the anterior and lateral compartments of the forearm, lying superficially throughout most of its course (see Fig. 5-12). In the middle third of its course, the radial nerve lies on its lateral side. At the wrist, the artery winds backward around the lateral side of the carpus to the proximal end of the space between the first and second metacarpal bones. Here, it passes anteriorly into the palm between the two heads of the first dorsal interosseous mus-

cle and joins the deep branch of the ulnar artery, forming the **deep palmar arch** (see Fig. 5-15).

In the lower part of the forearm, the radial artery lies on the anterior surface of the radius and is covered only by skin and fascia. Here, the artery has the tendon of brachioradialis on its lateral side and the tendon of the flexor carpi radialis on its medial side (site for taking radial pulse).

Branches

1. **Muscular branches.**
2. **Recurrent branch** takes part in the arterial anastomosis around the elbow joint.
3. **Superficial palmar branch** arises just above the wrist. It enters the palm and frequently joins the ulnar artery to form the **superficial palmar arch.**
4. **First dorsal metacarpal artery** supplies the adjacent sides of the thumb and index finger.
5. **Arteria princeps pollicis** divides into two branches that supply the sides of the thumb.
6. **Arteria radialis indicis** supplies the lateral side of the index finger.

Deep Palmar Arch

The deep palmar arch is deeply placed in the palm and extends from the proximal end of the space between the first and second metacarpal bones to the base of the fifth metacarpal bone (see Fig. 5-15). It is formed as a continuation of the radial artery and terminates by anastomosing with the deep branch of the ulnar artery.

Branches

1. **Palmar**
2. **Metacarpal**
3. **Perforating**
4. **Recurrent**

Ulnar Artery

The ulnar artery is the larger of the two terminal branches of the brachial artery (see Fig. 5-19). It begins in the cubital fossa at the level of the neck of the radius (see Fig. 5-11). It descends through the anterior compartment of the forearm and enters the palm **in front of** the flexor retinaculum in company with the ulnar nerve (see Fig. 5-13). It ends by forming the superficial palmar arch, often anastomosing with the superficial palmar branch of the radial artery (see Fig. 5-14).

In the upper part of its course, the ulnar artery lies deep to the flexor muscles (see Fig. 5-12). Below, it becomes superficial and lies between the tendons of the flexor carpi ulnaris and the tendons of the flexor digitorum superficialis. In front of the flexor retinaculum, it lies just lateral to the pisiform bone and is covered only by skin and fascia (site for taking ulnar pulse).

Branches
1. **Muscular branches.**
2. **Recurrent branches** take part in the arterial anastomosis around the elbow joint.
3. **Common interosseous artery** arises from the upper part of the ulnar artery and divides into the **anterior and posterior interosseous arteries.** These arteries descend on the anterior and posterior surfaces of the interosseous membrane, respectively.
4. **Deep palmar branch** arises in front of the flexor retinaculum and joins the radial artery to complete the deep palmar arch.

Superficial Palmar Arch
The superficial palmar arch lies just beneath the palmar aponeurosis on the long flexor tendons (see Fig. 5-14). It is a continuation of the ulnar artery and is often completed on the lateral side by the superficial palmar branch of the radial artery.

Branches
Digital arteries to the four medial fingers.

Veins of the Upper Limb

The superficial veins lie in the superficial fascia and are of **great clinical importance.** The deep veins accompany the main arteries.

Superficial Veins
Dorsal Venous Network
The dorsal venous network lies on the dorsum of the hand. It is drained on the lateral side by the cephalic vein and on the medial side by the basilic vein.

Cephalic Vein
The cephalic vein arises from the lateral side of the dorsal venous network. It ascends around the lateral border of the forearm, just lateral to the styloid process of the radius (Fig. 5-20). It ascends on the anterior aspect of the forearm and runs along the lateral border of the biceps in the arm. On reaching the interval between the deltoid and pectoralis major muscles, the cephalic vein pierces the deep fascia and joins the axillary vein.

Basilic Vein
The basilic vein arises from the medial side of the dorsal venous network and ascends on the posterior surface of the forearm. Just below the elbow, it inclines forward to reach the cubital fossa (see Fig. 5-20). The vein then ascends medial to the biceps and pierces the deep fascia at about the middle of the arm. It then joins the venae comitantes of the brachial artery to form the axillary vein.

Median Cubital Vein
The median cubital vein connects the cephalic to the basilic vein (see Fig. 5-20). It lies superficial to the bicipital aponeurosis, which separates it from the brachial artery.

Median Vein of the Forearm
The median vein is a small vein that arises in the palm and ascends on the front of the forearm (see Fig. 5-20). It drains into the basilic vein or the median cubital vein, or divides into two branches, one of which joins the basilic (**median basilic vein**) and the other joins the cephalic (**median cephalic vein**).

Deep Veins
Venae Comitantes
The deep veins accompany the respective arteries as venae comitantes. The two venae comitantes of the brachial artery join the basilic vein at the lower border of the teres major to form the axillary vein.

Axillary Vein
The axillary vein is formed by the union of the venae comitantes of the brachial artery with the basilic vein. It ascends along the medial border of the axillary artery and becomes the subclavian vein at the outer border of the first rib. It receives tributaries that correspond to the branches of the axillary artery; it also receives the cephalic vein.

Lymphatic Drainage of the Upper Limb

The **superficial lymph vessels** ascend the limb in the superficial fascia and accompany the superficial veins. The **deep lymph vessels** lie deep to the deep fascia and follow the deep arteries and veins. All the lymph vessels of the upper limb ultimately drain into lymph nodes that are situated in the axilla.

Axillary Lymph Nodes
The axillary lymph nodes drain lymph vessels from the entire upper limb. In addition, they drain vessels from the **lateral part of the breast** and superficial lymph vessels from the thoracoabdominal walls above the level of the umbilicus.

The lymph nodes are 20 to 30 in number and located as follows:

1. **Anterior (pectoral) nodes** lie along the lower border of pectoralis minor and behind pectoralis major muscles. They receive lymph from the lateral part of the breast and the superficial vessels from the thoracoabdominal wall above the level of the umbilicus.

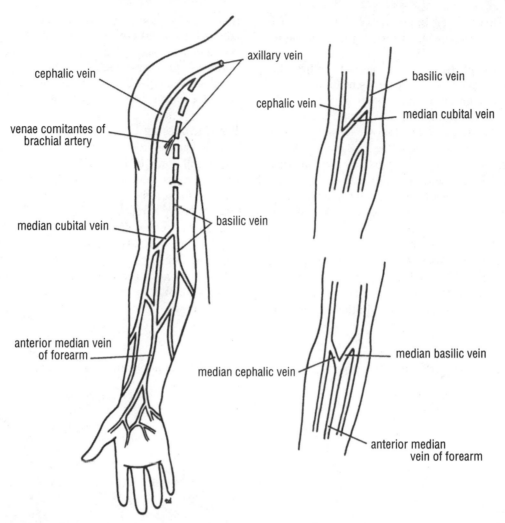

Fig. 5-20.   Superficial veins of upper limb. Note common variations seen in region of elbow.

2. **Posterior (subscapular) nodes** lie in front of the subscapularis muscle. They receive superficial lymph vessels from the back, down as far as the level of the iliac crests.
3. **Lateral nodes** lie along the axillary vein. They receive most of the lymph vessels from the upper limb (except the superficial vessels draining the lateral side, described below in infraclavicular nodes).
4. **Central nodes** lie in the center of the axilla. They receive lymph from the above three groups.
5. **Infraclavicular (deltopectoral) nodes.** These nodes are not strictly axillary nodes since they are located outside the axilla. They lie in the interval between the deltoid and pectoralis major muscles. They receive lymph from the superficial vessels from the lateral side of the hand, the forearm, and the arm; the lymph vessels accompany the cephalic vein.
6. **Apical group** lies at the apex of the axilla. They receive lymph from all the other axillary nodes. The apical nodes drain into the subclavian trunk in the neck.

Supratrochlear (Cubital) Lymph Node

The supratrochlear lymph node lies in the superficial fascia in the cubital fossa close to the trochlea of the humerus. It receives lymph from the medial fingers, the medial part of the hand, and the medial side of the forearm. The efferent lymph vessels ascend to the lateral axillary lymph nodes.

Nerves of the Upper Limb

Brachial Plexus

The brachial plexus is formed by the union of the anterior rami of the fifth, sixth, seventh, and eighth cervical spinal nerves and the first thoracic spinal nerve (Fig. 5-21). The plexus may be divided up into **roots, trunks, divisions,** and **cords.**

The roots of the plexus enter the base of the neck between the scalenus anterior and scalenus medius muscles. The trunks and divisions cross the posterior triangle of the neck, and the cords become arranged around the axillary artery in the axilla. Here, the brachial plexus and the axillary artery and vein are enclosed in the axillary sheath.

The branches of the brachial plexus and their distribution are summarized in Table 5-9. Dermatomal charts for the anterior and posterior surfaces of the body are shown in Figures 1-7 and 1-8.

Musculocutaneous Nerve

The musculocutaneous nerve (see Fig. 5-21) arises from the lateral cord of the brachial plexus (C5, 6, and 7). It pierces the coracobrachialis muscle and descends between the biceps and the brachialis muscles. In the region of the elbow, it pierces the deep fascia and is distributed to the skin as the **lateral cutaneous nerve of the forearm.** The musculocutaneous nerve supplies the coracobrachialis, both heads of biceps, and the greater part of the brachialis muscles. Figure 5-22 summarizes the main branches of the musculocutaneous nerve.

Median Nerve

The median nerve (see Fig. 5-21) arises from the medial and lateral cords of the brachial plexus (C5, 6, 7, 8, and T1). The nerve descends on the lateral side of the axillary and brachial arteries. Halfway down the arm, it crosses the brachial artery to reach its medial side. The nerve descends through the forearm between the two heads of pronator teres and runs on the posterior surface of the flexor digitorum superficialis. At the wrist, it lies behind the tendon of palmaris longus. The median nerve enters the palm by passing **behind** the flexor retinaculum and through the carpal tunnel.

Fig. 5-21. Roots, trunks, divisions, cords, and terminal branches of brachial plexus.

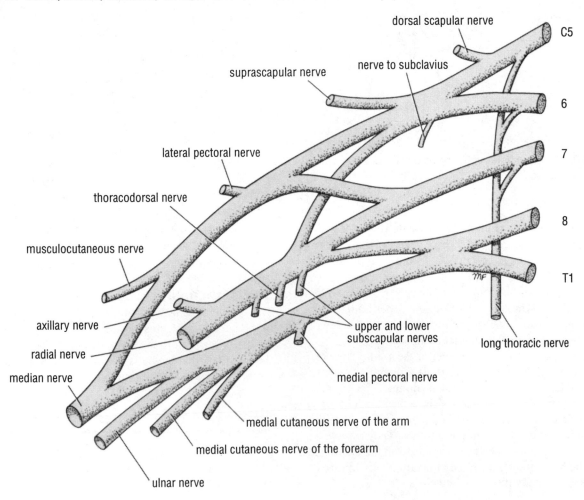

Table 5-9 Branches of the Brachial Plexus and Their Distribution

Branches	Distribution
Roots	
Dorsal scapular nerve (C5)	Rhomboid minor, rhomboid major, levator scapulae muscles
Long thoracic nerve (C5, 6, 7)	Serratus anterior muscle
Upper trunk	
Suprascapular nerve	Supraspinatus and infraspinatus muscles
Nerve to subclavius	Subclavius muscle
Lateral cord	
Lateral pectoral nerve	Pectoralis major muscle
Musculocutaneous nerve (C5, 6, 7)	Coracobrachialis, biceps brachii, brachialis muscles; supplies skin along lateral border of forearm when it becomes the lateral cutaneous nerve of forearm
Lateral root of median nerve	See Medial root of median nerve
Posterior cord	
Upper subscapular nerve	Subscapularis muscle
Thoracodorsal nerve	Latissimus dorsi muscle
Lower subscapular nerve	Subscapularis and teres major muscles
Axillary nerve (C5, 6)	Deltoid and teres minor muscles; upper lateral cutaneous nerve of arm supplies skin over lower half of deltoid muscle
Radial nerve (C5, 6, 7, 8, T1)	Triceps, anconeus, part of brachialis, extensor carpi radialis longus; via deep radial nerve branch supplies extensor muscles of forearm: supinator, extensor carpi radialis brevis, extensor carpi ulnaris, extensor digitorum, extensor digiti minimi, extensor indicis, abductor pollicis longus, extensor pollicis longus, extensor pollicis brevis; skin, lower lateral cutaneous nerve of arm, posterior cutaneous nerve of arm and posterior cutaneous nerve of forearm; skin on lateral side of dorsum of hand and dorsal surface of lateral 3½ fingers; articular branches to elbow, wrist, and hand
Medial cord	
Medial pectoral nerve	Pectoralis major and minor muscles
Medial cutaneous nerve of arm joined by intercostal brachial nerve from second intercostal nerve	Skin of medial side of arm
Medial cutaneous nerve of forearm	Skin of medial side of forearm
Ulnar nerve (C8 and T1)	Flexor carpi ulnaris and medial half of flexor digitorum profundus, flexor digiti minimi, opponens digiti minimi, abductor digiti minimi, adductor pollicis, third and fourth lumbricals, interossei, palmaris brevis, skin of medial half of dorsum of hand and palm, skin of palmar and dorsal surfaces of medial 1½ fingers
Medial root of median nerve (with lateral root) forms median nerve (C5, 6, 7, 8, T1)	Pronator teres, flexor carpi radialis, palmaris longus, flexor digitorum superficialis, abductor pollicis brevis, flexor pollicis brevis, opponens pollicis, first two lumbricals (by way of anterior interosseous branch), flexor pollicis longus, flexor digitorum profundus (lateral half), pronator quadratus; palmar cutaneous branch to lateral half of palm and digital branches to palmar surface of lateral 3½ fingers; articular branches to elbow, wrist, and carpal joints

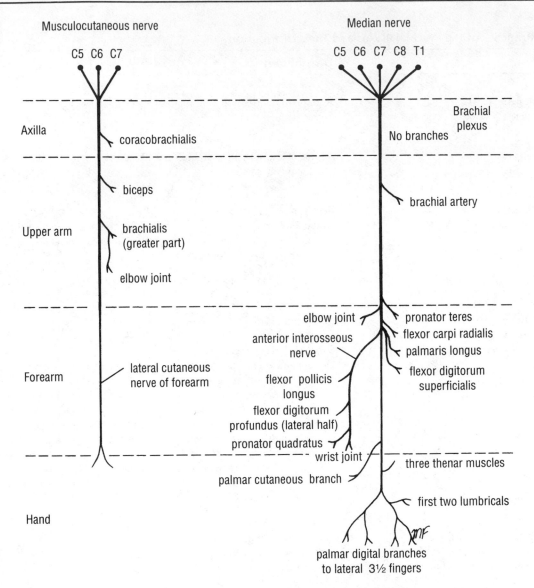

Fig. 5-22. Summary diagram of main branches of musculocutaneous and median nerves.

Branches of the Median Nerve in the Forearm

1. **Muscular branches:** Pronator teres, flexor carpi radialis, palmaris longus, flexor digitorum superficialis.
2. **Articular branches:** Elbow joint.
3. **Anterior interosseous nerve**
   a. **Muscular branches:** Flexor pollicis longus, pronator quadratus, lateral half of flexor digitorum profundus.
   b. **Articular branches:** Wrist and carpal joints.
4. **Palmar branch** supplies the skin over the lateral part of the palm.

Branches of the Median Nerve in the Palm

1. **Muscular branches:** Abductor pollicis brevis, flexor pollicis brevis, opponens pollicis, the first and second lumbrical muscles

2. **Cutaneous branches** to the palmar aspect of the lateral 3½ fingers and the distal half of the dorsal aspect of each finger as well

Figure 5-22 summarizes the main branches of the median nerve.

Ulnar Nerve

The ulnar nerve (see Fig. 5-21) arises from the medial cord of the brachial plexus (C8, T1). It descends along the medial side of the axillary and brachial arteries in the anterior compartment of the arm. At the middle of the arm, it pierces the medial intermuscular septum and passes down **behind** the medial epicondyle of the humerus. It then enters the anterior compartment of the forearm and descends behind the flexor carpi ulnaris medial to the ulnar artery. At the wrist, it passes **anterior to the flexor retinaculum** and **lateral to the pisiform bone.** It then divides into superficial and deep terminal branches.

Branches of the Ulnar Nerve in the Forearm
1. **Muscular branches:** Flexor carpi ulnaris, medial half of flexor digitorum profundus
2. **Articular branches:** Elbow joint
3. **Dorsal cutaneous branch:** Supplies skin over medial side of back of hand and back of medial 1½ fingers over the proximal phalanges

Branches of the Ulnar Nerve in the Hand
The **superficial terminal branch** descends into the palm. It gives off the following branches:

1. **Muscular branch:** Palmaris brevis
2. **Cutaneous branches:** Supply skin over palmar aspect of medial 1½ fingers including their nail beds

The **deep terminal branch** runs backward between the abductor digiti minimi and the flexor digiti minimi and pierces the opponens digiti minimi. It gives off the following branches:

1. **Muscular branches:** Abductor digiti minimi, flexor digiti minimi, opponens digiti minimi, all palmar and all dorsal interossei, third and fourth lumbricals, adductor pollicis
2. **Articular branches:** Carpal joints

Figure 5-23 summarizes the main branches of the ulnar nerve.

Radial Nerve
The radial nerve (see Fig. 5-21) arises from the posterior cord of the brachial plexus (C5, 6, 7, 8, and T1). It descends behind the axillary and brachial arteries and enters the posterior compartment of the arm. The radial nerve winds around the back of the humerus in the spiral groove

Fig. 5-23. Summary diagram of main branches of ulnar nerve.

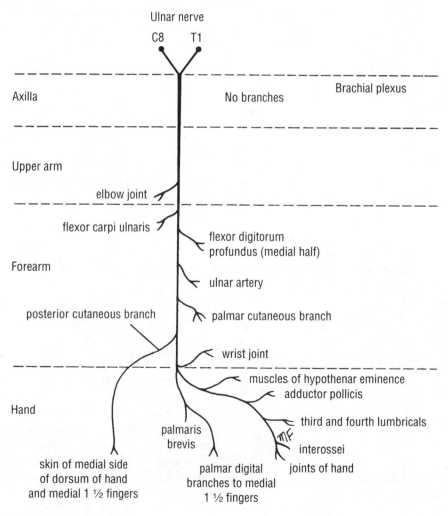

with the profunda artery. Piercing the lateral intermuscular septum just above the elbow, it descends in front of the lateral epicondyle and divides into superficial and deep terminal branches.

Branches of the Radial Nerve in the Axilla
1. **Muscular branches:** Long and medial heads of triceps
2. **Cutaneous branch:** Posterior cutaneous nerve of the arm

Branches of the Radial Nerve in the Spiral Groove Behind the Humerus
1. **Muscular branches:** Lateral and medial heads of triceps, anconeus
2. **Cutaneous branches:** Lower lateral cutaneous nerve of the arm, posterior cutaneous nerve of the forearm

Branches of the Radial Nerve in the Anterior Compartment of the Arm close to the Lateral Epicondyle
1. **Muscular branches:** Brachialis, brachioradialis, extensor carpi radialis longus
2. **Articular branches:** Elbow joint

Superficial Branch of Radial Nerve
The superficial branch descends undercover of the brachioradialis muscle on the lateral side of the radial artery. It emerges from beneath the brachioradialis tendon and descends on the back of the hand.
Cutaneous Branches.   Cutaneous branches run to the lateral ⅔ of the dorsal surface of the hand (variable) and the posterior surface of the lateral 3½ fingers over the proximal phalanges.

Deep Branch of the Radial Nerve
The deep branch winds around the lateral side of the neck of the radius within the supinator muscle. It enters the posterior compartment of the forearm and descends between the muscles to reach the interosseous membrane.

1. **Muscular branches:** Extensor carpi radialis brevis, supinator, extensor carpi ulnaris, abductor pollicis longus, extensor pollicis brevis, extensor pollicis longus, extensor indicis
2. **Articular branches:** Wrist and carpal joints

Figure 5-24 summarizes the main branches of the radial nerve.

Axillary Nerve
The axillary nerve (see Fig. 5-21) arises from the posterior cord of the brachial plexus (C5 and 6). It passes backward through the quadrilateral space below the shoulder joint in company with the posterior circumflex humeral vessels.

Branches
1. **Articular branch** to the shoulder joint.

2. **Anterior terminal branch** winds around the surgical neck of the humerus and supplies the deltoid muscle and the skin that covers its lower half (supraclavicular nerves supply skin over upper half of deltoid).
3. **Posterior terminal branch** supplies the teres minor muscle, the deltoid muscle, and then becomes the **upper lateral cutaneous nerve of the arm,** which also supplies the skin over the lower part of the deltoid muscle.

Injuries to the Brachial Plexus and its Branches
Upper Trunk Lesions of the Brachial Plexus (Erb-Duchenne Palsy)
Caused by displacement of head to opposite side and depression of shoulder on the same side as in falls on shoulder or in infants during a difficult delivery. The limb hangs limply by the side and is medially rotated, and the forearm is pronated (**waiter's tip hand**).

Lower Trunk Lesions of the Brachial Plexus (Klumpke Palsy)
Caused by traction injury as in excessive abduction of the arm. The first thoracic nerve is usually torn. All the small muscles of the hand are paralyzed and the patient develops a **claw hand.**

Long Thoracic Nerve (C5, 6, and 7) Lesions
Caused by blows or surgical injury to nerve in the axilla. Paralysis of the serratus anterior muscle allows the inferior angle of the scapula to protrude (**winged scapula**). The patient also has difficulty in raising the arm above the head.

Axillary Nerve (C5 and 6) Lesions
Caused by inferior dislocations of shoulder joint or fracture of the surgical neck of the humerus. Damages nerve in the quadrilateral space. The deltoid muscle is paralyzed and rapidly atrophies; there is loss of cutaneous sensation over the lower half of the deltoid muscle.

Radial Nerve (C5, 6, 7, 8, and T1) Lesions
Commonly caused by fracture of the midshaft of the humerus, injuring the nerve in the spiral groove. The patient is unable to extend the wrist and the fingers, and there is **wrist drop.**

Median Nerve (C5, 6, 7, 8, and T1) Lesions
Commonly caused by supracondylar fractures of the humerus and wounds just proximal to the flexor retinaculum. Among the signs present, the muscles of the thenar eminence are paralyzed and wasted so that the eminence is flattened and the thumb is laterally rotated and adducted. The hand looks **apelike.**

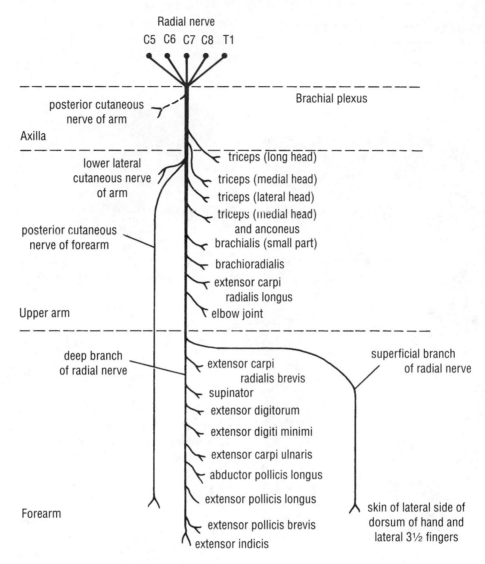

Fig. 5-24. Summary diagram of main branches of radial nerve.

Ulnar Nerve (C8 and T1) Lesions

Caused by injuries where the nerve lies behind the medial epicondyle of the humerus and at the wrist where it lies in front of the flexor retinaculum. The small muscles of the hand will be paralyzed except the muscles of the thenar eminence and the first two lumbricals (median nerve). The patient is unable to adduct and abduct the fingers. Also, the thumb cannot be adducted because the adductor pollicis is paralyzed. The metacarpophalangeal joints become hyperextended due to paralysis of the lumbrical and interosseous muscles, and this is most prominent in the joints of the fourth and fifth fingers. The interphalangeal joints are flexed, also due to paralysis of the lumbrical and interosseous muscles. In long-standing cases, the hand assumes the characteristic **claw deformity.**

National Board Type Questions

Match the numbered structures shown in the anteroposterior radiograph of the shoulder region with the appropriate lettered structures on the right.

Match the numbered structures shown in the lateral radiograph of the elbow region with the appropriate lettered structures on the right.

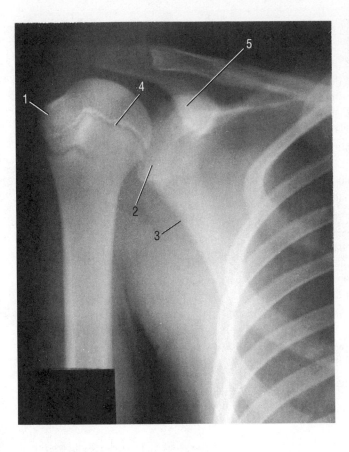

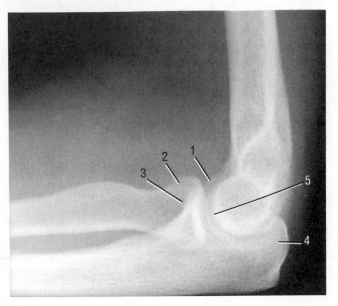

6. Structure 1.   A. Olecranon process.
7. Structure 2.   B. Capitulum.
8. Structure 3.   C. Head of radius.
9. Structure 4.   D. Coronoid process.
10. Structure 5.   E. Trochlea.

1. Structure 1.   A. Coracoid process.
2. Structure 2.   B. Epiphyseal line.
3. Structure 3.   C. Glenoid fossa.
4. Structure 4.   D. Lateral border of scapula.
5. Structure 5.   E. Greater tuberosity.

Match the numbered structures listed below with the appropriate lettered structures (A–E) shown in the anteroposterior radiograph of the hand.

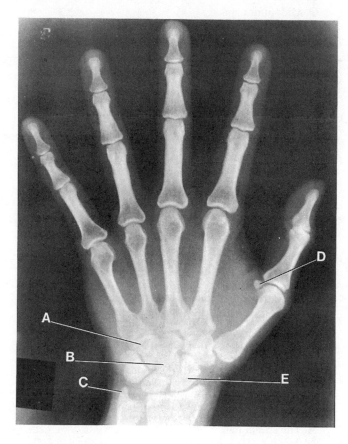

11. Styloid process of ulna.
12. Scaphoid.
13. Hook of hamate.
14. Capitate.
15. Sesamoid bone.

Match each statement on the left with the correct response on the right.

16. Receives contributions from the C8 spinal nerve.
17. Has a terminal branch that supplies the skin on the medial side of the arm.
18. Is formed by the anterior division of two trunks of the brachial plexus.
19. Has branches that supply extensor muscles of the arm.

A. Lateral cord of the brachial plexus.
B. Posterior cord of the brachial plexus.
C. Both A and B.
D. Neither A nor B.

Match each statement on the left with the correct response on the right.

20. During its course in the upper limb, the axillary nerve lies _____.
21. During its course in the upper limb, the ulnar nerve lies _____.
22. During its course in the upper limb, the median nerve lies _____.
23. During its course in the upper limb, the radial nerve lies _____.

A. In front of the lateral epicondyle of the humerus.
B. Against the spiral groove of the humerus.
C. Medial to the brachial artery in the cubital fossa.
D. Against the surgical neck of the humerus.
E. Behind the medial epicondyle of the humerus.

Match the muscles on the left with their appropriate nerve supply on the right.

24. Extensor carpi radialis brevis.
25. Dorsal interossei.
26. Extensor indicis.
27. Extensor carpi ulnaris.
28. Extensor carpi radialis longus.

A. Radial nerve.
B. Ulnar nerve.
C. Superficial radial nerve.
D. Deep branch of radial nerve.
E. Deep branch of ulnar nerve.

In each of the following questions, answer:

A. If only (1) is correct
B. If only (2) is correct
C. If both (1) and (2) are correct
D. If neither (1) nor (2) is correct

29. (1) The pulsations of the radial artery may be felt anterior to the distal third of the radius between the tendons of the brachioradialis and the flexor carpi radialis.
(2) The axillary vein is formed by the union of the venae comitantes of the brachial artery and the basilic vein.

30. (1) The subacromial bursa communicates with the shoulder joint.
(2) The inferior part of the capsule of the shoulder joint is the weakest part.

31. (1) The brachial plexus and the axillary artery lie within the axillary sheath.
(2) The axillary sheath is derived from the prevertebral layer of deep cervical fascia.

32. (1) The ulnar nerve enters the palm by passing posterior to the flexor retinaculum.
(2) The dorsal cutaneous branch of the ulnar nerve enters the back of the forearm by passing beneath the tendon of the flexor carpi ulnaris muscle.

33. Facts concerning the cephalic vein:
    (1) It arises on the dorsum of the hand.
    (2) It drains into the axillary vein.
34. In the cubital fossa, the bicipital aponeurosis covers the following structures:
    (1) The brachial artery.
    (2) The musculocutaneous nerve.
35. Muscle(s) that attach(es) to both the scapula and the humerus:
    (1) Supraspinatus.
    (2) Rhomboid major.
36. At the wrist, the flexor retinaculum is attached to the following bone(s) on the lateral side:
    (1) Trapezium.
    (2) Trapezoid.
37. **True** statements concerning the pectoralis minor muscle include:
    (1) It is inserted into the acromion of the scapula.
    (2) It crosses the cords of the brachial plexus.
38. Enlargement of the infraclavicular (deltopectoral) lymph nodes is most likely caused by:
    (1) Infection of the nail bed of the little finger.
    (2) An infected cut of the index finger.

Select the **best** response.

39. Collateral circulation around the shoulder joint would involve all of the following **except:**
    A. Subscapular artery.
    B. Superficial cervical artery.
    C. Suprascapular artery.
    D. Anterior circumflex humeral artery.
    E. Lateral thoracic artery.
40. The proximal row of carpal bones comprises all of the following carpal bones **except:**
    A. Pisiform.
    B. Capitate.
    C. Lunate.
    D. Triquetral.
    E. Scaphoid.
41. A shoulder separation (which involves the lateral end of the clavicle sliding onto the superior aspect of the acromion) would most likely result from damage to:
    A. The costoclavicular ligament.
    B. The sternoclavicular ligament.
    C. The coracoclavicular ligament.
    D. The glenohumeral ligament.
    E. The coracoacromial ligament.
42. Select the muscle that will compensate in part for the paralysis of the supinator muscle.
    A. Extensor carpi ulnaris muscle.
    B. Brachialis muscle.
    C. Triceps brachii muscle.
    D. Biceps brachii muscle.
    E. Anconeus muscle.

43. The synovial sheath of the flexor pollicis longus muscle forms the:
    A. Thenar space.
    B. Ulnar bursa of the wrist.
    C. Midpalmar space.
    D. Radial bursa of the wrist.
    E. Digital synovial sheath for the index finger.
44. To test trapezius muscle paralysis, you would ask the patient to:
    A. Flex the arm fully.
    B. Adduct the arm against resistance.
    C. Push against a wall with both hands.
    D. Shrug the shoulder.
    E. Abduct the arm fully.
45. The lymph from the medial quadrants of the breast drain mainly into:
    A. Posterior axillary (subscapular) nodes.
    B. Internal thoracic nodes.
    C. Anterior axillary (pectoral) nodes.
    D. Lateral axillary (brachial) nodes.
    E. Infraclavicular (deltopectoral) nodes.
46. Cutting the dorsal scapular nerve would most likely result in paralysis of the:
    A. Supraspinatus muscle.
    B. Deltoid muscle.
    C. Rhomboid major muscle.
    D. Trapezius muscle.
    E. Infraspinatus muscle.
47. Concerning "winged scapula," the following facts are correct **except:**
    A. The spinal part of the accessory nerve is damaged.
    B. The inferior angle of the scapula projects backward.
    C. The serratus anterior muscle may be wasted.
    D. The long thoracic nerve is damaged.
    E. The scapula can no longer be pulled anteriorly around the chest wall as in thrusting the upper limb anteriorly when reaching.
48. Following injury to a nerve at the wrist, the thumb is laterally rotated and adducted. The hand looks flattened and "apelike." Which nerve has been damaged?
    A. Anterior interosseous nerve.
    B. Ulnar nerve.
    C. Deep branch of the radial nerve.
    D. Median nerve.
    E. Superficial branch of the radial nerve.
49. The dermatome present over the lateral side of the wrist is:
    A. C8.
    B. C6.
    C. T1.
    D. T2.
    E. C5.

50. A 14-year-old boy fell off a wall and fractured his right humerus at midshaft. The wrist joint immediately assumed a flexed position that the patient was unable to correct. Extension and supination of the forearm was weakened, but not abolished. Skin sensation over the lateral side of the dorsum of the hand was diminished. Select the one peripheral nerve, which, if damaged, could account for the symptoms and signs.
    A. Ulnar nerve.
    B. Median nerve.
    C. Radial nerve.
    D. Axillary nerve.
    E. Musculocutaneous nerve.

51. A patient is able to raise (abduct) her arm to a 15-degree position but no further. This would suggest paralysis of which muscle?
    A. Rhomboid minor.
    B. Deltoid.
    C. Supraspinatus.
    D. Teres minor.
    E. Subscapularis.

52. The following statements concerning the rotator cuff are incorrect **except:**
    A. It adds to the stability of the shoulder joint.
    B. It is formed of the tendons of the long muscles found in the region of the shoulder joint.
    C. It is composed of muscle fibers that are located close to the shoulder joint.
    D. All the muscle tendons are innervated by the suprascapular nerve.
    E. The muscle tendons are not fused to the capsule of the shoulder joint.

53. The quadrilateral space in the region of the shoulder transmits the following structures **except:**
    A. Axillary nerve.
    B. Posterior circumflex humeral artery.
    C. Lymphatic vessels.
    D. Radial nerve.
    E. Posterior circumflex humeral vein.

54. A patient is seen in the emergency department with a knife wound to the front of the left wrist. Examination reveals that two superficial tendons on either side of the median nerve have been severed. Which of the following tendons are these?
    A. Flexor pollicis longus and the pronator quadratus tendons.
    B. Flexor carpi ulnaris and the flexor digitorum profundus tendons.
    C. Flexor digitorum superficialis and the flexor carpi radialis tendons.
    D. Brachioradialis and the pronator teres tendons.
    E. Abductor pollicis longus and the extensor pollicis brevis tendons.

55. The following statements concerning the lateral cord of the brachial plexus are true **except:**
    A. Contains sympathetic nerve fibers.
    B. Has a branch that supplies the pectoralis major muscle.
    C. Has a branch that supplies the skin on the lateral side of the forearm.
    D. Has a branch that supplies the skin on the lateral side of the upper arm.
    E. It lies lateral to the second part of the axillary artery.

56. The examination of a patient with the carpal tunnel syndrome may reveal all the following symptoms and signs **except:**
    A. Atrophy of the muscles of the thenar eminence.
    B. Weakness of the movement of opposition of the thumb.
    C. Loss of sensation of the skin of the medial part of the palm.
    D. Loss of sensation of the skin on the ventral surface of the index finger.
    E. Normal skin sensation on the ventral surface of the little finger.

57. The following movements would be expected to be normal following complete section of the medial cord of the brachial plexus **except:**
    A. Extension of the wrist.
    B. Flexion of the elbow.
    C. Abduction of the shoulder joint.
    D. Metacarpophalangeal flexion and interphalangeal extension of the medial four fingers.
    E. Metacarpophalangeal flexion and interphalangeal extension of the thumb.

58. The following statements concerning the sympathetic innervation of the arteries of the upper limb are correct **except:**
    A. Have preganglionic nerve fibers originating in spinal cord segments T2–8.
    B. Cause vasoconstriction of the arteries and veins of the skin.
    C. Have preganglionic nerve fibers synapsing in the middle cervical, inferior cervical, and first thoracic ganglia.
    D. Many of the postganglionic fibers are distributed within the branches of the brachial plexus.
    E. The sympathetic nerves do not innervate the sweat glands.

59. In an automobile accident, a patient fractured the neck of her right radius and damaged a closely related nerve. On examination the patient exhibited the following **except:**
    A. Weakness in extending the terminal phalanx of the thumb.
    B. A loss of skin sensation on the lateral part of the dorsum of the hand.

C. An inability to extend the metacarpophalangeal joint of the index finger.

D. A normal ability to adduct the thumb at the carpometacarpal joint.

E. Normal skin sensation down the medial border of the hand.

60. Diminished sweating and an increased warmth and vasodilation of the skin vessels over the hypothenar eminence and the ring and little fingers could be caused by the following **except:**

A. Lesion of the posterior cord of the brachial plexus.

B. Ulnar nerve damage behind the medial epicondyle of the humerus.

C. Lesion of the medial cord of the brachial plexus.

D. Eighth cervical nerve lesion.

E. Ulnar nerve damage over the front of the wrist.

In the next section, study each case history. Select the **best** answer to each question following it.

A 17-year-old girl was thrown from her horse while she was attempting a difficult jump. She landed on her right shoulder and the right side of her head. After a week of hospitalization, it was noticed that she kept her right arm medially rotated and close to her side with the forearm pronated. On examination an area of anesthesia was found along the lateral side of the upper part of the arm.

61. Which nerve(s) was (were) damaged during the accident?

A. Lower trunk of the brachial plexus.

B. Axillary nerve.

C. Radial nerve.

D. Middle trunk of the brachial plexus.

E. C5 and C6 roots of the brachial plexus.

62. The position adopted by the right arm in this patient can be explained by the paralysis of one of the following groups of muscles.

A. Supraspinatus, deltoid, biceps brachii, greater part of brachialis, infraspinatus, and teres minor.

B. Latissimus dorsi, triceps, anconeus, brachioradialis, and supinator.

C. Flexor digitorum superficialis, flexor pollicis longus, flexor carpi radialis longus, and flexor carpi ulnaris.

D. Extensor carpi ulnaris, supinator, extensor indicis, and extensor digiti minimi.

E. Pectoralis major, abductor pollicis longus, extensor pollicis brevis, and extensor pollicis longus.

A 30-year-old plasterer was finishing off a difficult ceiling in a remodeled kitchen. He was standing on the top of a stepladder with his right arm above his head. As he moved the trowel filled with plaster across the ceiling in his right hand, he suddenly felt an acute spasm of pain over the tip of the right shoulder. On examination of the patient in the emergency department, it was found that the pain in the right shoulder recurred in the middle range of abduction and there was tenderness over the greater tuberosity of the humerus.

63. Name the most likely structural damage in this case.

A. Fracture of the upper end of the humerus.

B. Tearing of the deltoid muscle.

C. Fracture of the acromion.

D. Lesion of the rotator cuff.

E. Tearing of the trapezius muscle.

Answers and Explanations

1. E
2. C
3. D
4. B
5. A
6. B
7. C
8. D
9. A
10. E
11. C
12. E
13. A
14. B
15. D
16. B
17. D
18. A
19. B
20. D
21. E
22. C
23. B
24. D
25. E
26. D
27. D
28. A
29. C
30. B   The subacromial bursa does **not** communicate with the cavity of the shoulder joint; the subscapular bursa, however, does communicate with the joint.
31. C
32. B   The ulnar nerve enters the palm by passing anterior to the flexor retinaculum at the wrist. This is a common site for the nerve to be damaged.
33. C   The cephalic vein is one of the important superficial veins in the upper limb.

34. A   The musculocutaneous nerve lies on the lateral side of the biceps tendon and is therefore not covered by the bicipital aponeurosis, which is situated on the medial side of the tendon.
35. A   The rhomboid major muscle arises from the second to the sixth thoracic vertebral spines and is inserted into the medial border of the scapula.
36. A   The flexor retinaculum is attached medially to the pisiform bone and the hook of the hamate and laterally to the tubercle of the scaphoid and the ridge on the trapezium.
37. B   The pectoralis minor is inserted into the tip of the coracoid process of the scapula.
38. B   Lymph from the little finger drains upward through the supratrochlear (cubital) lymph node to the lateral group of axillary nodes.
39. E
40. B
41. C
42. D   The biceps brachii is a powerful supinator of the superior and inferior radioulnar joints.
43. D
44. D   The upper fibers of the trapezius muscle elevate the scapula and the shoulder, as in shrugging the shoulder.
45. B   The internal thoracic nodes lie within the thoracic cavity along the internal thoracic artery.
46. C
47. A   The spinal part of the accessory nerve supplies the sternocleidomastoid and the trapezius muscles.
48. D   The muscles of the thenar eminence are innervated by the recurrent branch of the median nerve. A lesion of the median nerve causes these muscles to atrophy and the eminence becomes flattened. The adductor pollicis is supplied by the deep branch of the ulnar nerve, and adduction of the thumb is unopposed.
49. B   A dermatome is an area of skin supplied by a single segment of the spinal cord.
50. C   The radial nerve was damaged as it lay in the spiral groove on the posterior surface of the shaft of the humerus.
51. B   The supraspinatus muscle is mainly responsible for about the first 15 degrees of abduction of the shoulder joint. From about 15 to 90 degrees, the deltoid muscle is largely responsible. Remember that for a person to raise the arm above 90 degrees the scapula has to rotate, involving the contraction of the trapezius and serratus anterior muscles.
52. A   The rotator cuff is formed by the tendons of the short muscles that cover the anterior, superior, and posterior surfaces of the shoulder joint, namely, subscapularis, supraspinatus, infraspinatus, and teres minor. The tendons are fused to the joint capsule.
53. D
54. C
55. D   The skin on the lateral side of the upper arm covering the upper half of the deltoid muscle is innervated by the supraclavicular nerves (C3 and 4), and the skin covering the lower half of the deltoid muscle is innervated by the upper lateral cutaneous nerve of the arm (C5 and 6) from the axillary nerve.
56. C   The carpal tunnel syndrome is caused by compression of the median nerve as it passes beneath the flexor retinaculum. The median nerve innervates the opponens pollicis, and it gives off digital nerves to the lateral 3½ fingers on the palmar aspect. The muscles of the hypothenar eminence are supplied by the deep branch of the ulnar nerve. The skin of the medial part of the palm is supplied by the palmar cutaneous branch of the ulnar nerve.
57. D   Flexion of the metacarpophalangeal joints is produced by the flexor digitorum superficialis, the flexor digitorum profundus, and the lumbricals and interossei muscles, all of which receive nerve fibers from the medial cord via the median and ulnar nerves. The interphalangeal joints are extended by the lumbricals and interossei muscles, assisted by the extensor digitorum. The lumbricals and the interossei are innervated by the median and ulnar nerves, which have many nerve fibers that originate in the medial cord of the brachial plexus.
58. E   The sweat glands of the upper limb are innervated by sympathetic postganglionic nerve fibers.
59. B   The deep branch of the radial nerve lies within the supinator muscle and is closely related to the neck of the radius. The extensor pollicis longus and the extensor digitorum muscles are innervated by the deep branch of the radial nerve.
60. A   The sweat glands and the blood vessels of the skin over the hypothenar eminence and the palmar surface of the medial 1½ fingers are innervated by sympathetic postganglionic nerve fibers. These fibers travel in the eighth cervical and first thoracic spinal nerves, the medial cord of the brachial plexus, and the ulnar nerve and its palmar cutaneous and digital branches.
61. E
62. A   This patient has a right-sided Erb-Duchenne palsy, that is, a lesion of the fifth and the sixth roots of the brachial plexus. The suprascapular nerve, the nerve to subclavius, and the musculocutaneous and axillary nerves all possess nerve fibers derived from C5 and C6 roots and will therefore be functionless. The following muscles will consequently be paralyzed: (1) supraspinatus (abductor of shoulder) and infraspinatus (lateral rotator of shoulder), (2) the sub-

clavius, (3) the biceps brachii (flexor of elbow and strong supinator of forearm) and the greater part of the brachialis and coracobrachialis, and (4) the deltoid (abductor of shoulder) and the teres minor (lateral rotator of shoulder). Thus, the right upper limb will hang limply by the side (adducted), medially rotated by the unopposed sternocostal part of the pectoralis major; the forearm will be pronated due to loss of the action of the biceps. The position of the upper limb in this condition has been likened to that of a waiter hinting for a tip. In addition, there will be loss of sensation down the lateral side of the right arm.

63. D   This patient has supraspinatus tendinitis. During the middle range of abduction, the tendon of the supraspinatus impinges against the outer border of the acromion. Normally, the large subacromial bursa intervenes and ensures that the movement is relatively free of friction and is painless. In this condition, the bursa has degenerated and the supraspinatus tendon exhibits a localized area of degeneration.

Lower Limb

The primary function of the lower limb is to support the weight of the body and to provide a stable foundation in standing, walking, and running. Each lower limb may be divided into the gluteal region, the thigh, the knee, the leg, the ankle, and the foot.

It is suggested that the lower limb be reviewed in the following order:

1. A brief overview of the bones and major joints, preferably with the help of an articulated skeleton.
2. A consideration of the more important muscles, concentrating on their actions and nerve supply.
3. A brief review of the blood supply and lymphatic drainage.
4. A detailed overview of the nerves and their distribution.

To assist students in the review process, tables have been used extensively.

Bones

Bones of the Pelvic Girdle
The pelvic girdle is composed of four bones: the two hip bones, the sacrum, and the coccyx (see Fig. 4-1). The pelvic girdle provides a strong and stable connection between the trunk and the lower limbs.

Hip Bone
Each hip bone in the child consists of the superior ilium, the posterior and inferior ischium, and the anterior and inferior pubis (Fig. 6-1). At puberty these three bones fuse together to form one large irregular bone. The **acetabulum** is a cup-shaped depression on the outer surface of the hip bone. It is formed by the ilium, the ischium, and the pubis. The acetabulum articulates with the hemispherical head of the femur. The articular surface of the acetabulum is limited to a horseshoe-shaped area and is covered with hyaline cartilage. The **acetabular fossa** is the floor of the acetabulum, which is nonarticular. The **acetabular notch** is situated on the inferior margin of the acetabulum.

The **iliac crest** runs between the **anterior and posterior superior iliac spines.** Below these spines are the corresponding **inferior iliac spines.**

The **ischium** possesses an **ischial spine** and an **ischial tuberosity** (see Fig. 6-1).

The **pubis** is the anterior part of the hip bone; it has a **body** and **superior** and **inferior pubic rami.** The body of the pubis has the **pubic crest** and the **pubic tubercle** and articulates with the pubic bone of the opposite side at the **symphysis pubis.**

In the lower part of the hip bone is a large opening, the **obturator foramen,** which is bounded by the parts of the ischium and pubis (see Fig. 6-1).

Bones of the Thigh
The bones of the thigh consist of the femur and the patella (Fig. 6-2).

Femur
The **head** of the femur forms about two-thirds of a sphere and fits into the acetabulum of the hip bone to form the hip joint. The **fovea capitis** is a small depression in the center of the head for the attachment of the **ligament of the head.** Part of the blood supply to the head of the femur from the obturator artery is conveyed along this ligament and enters the bone at the fovea.

The **neck** connects the head to the shaft and passes downward, backward, and laterally (see Fig. 6-2). The **greater** and **lesser trochanters** are large eminences situated at the junction of the neck and the shaft. Connecting the two trochanters are the **intertrochanteric line** anteriorly, where the iliofemoral ligament is attached, and a prominent **intertrochanteric crest** posteriorly, on which is the **quadrate tubercle.**

The **shaft** is smooth on its anterior surface but has a ridge posteriorly, the **linea aspera,** to which are attached muscles and intermuscular septa. The medial margin of the linea aspera continues below as the **medial supracondylar ridge** to the **adductor tubercle** (see Fig. 6-2) on the medial condyle. The lateral margin becomes continuous below with the **lateral supracondylar ridge.** On the

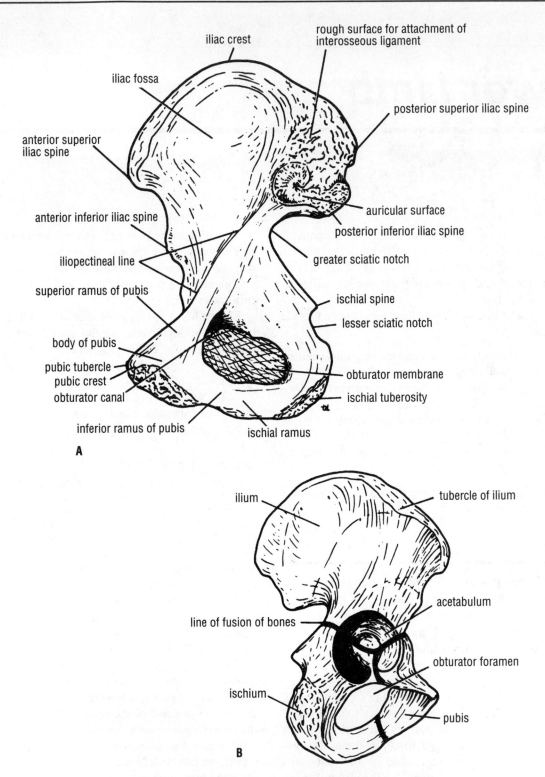

Fig. 6-1.   Medial surface **(A)** and lateral surface **(B)** of right hip bone. Note the lines of fusion between the three bones: the ilium, the ischium, and the pubis.

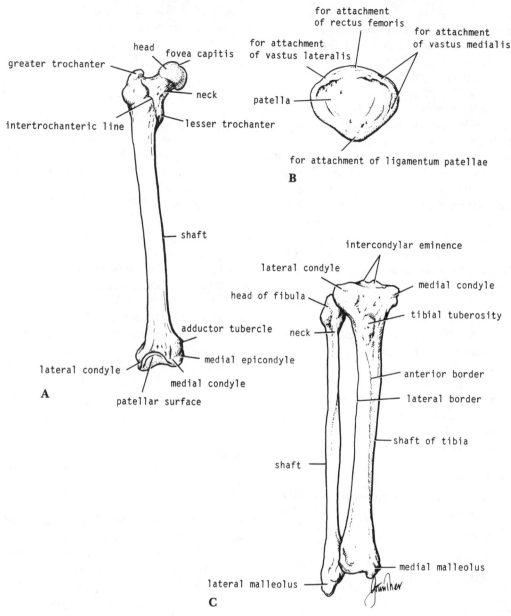

Fig. 6-2. **A.** Anterior surface of right femur. **B.** Anterior surface of right patella. **C.** Anterior surface of right tibia and fibula.

posterior surface of the shaft below the greater trochanter is the **gluteal tuberosity** for the insertion of the gluteus maximus muscle. A flat triangular area on the posterior surface of the lower end of the shaft is called the **popliteal surface.**

The lower end of the femur has **lateral** and **medial condyles,** separated posteriorly by the **intercondylar notch.** The anterior surfaces of the condyles are joined by an articular surface for the patella. The two condyles take part in the formation of the knee joint. Above the condyles are the **medial and lateral epicondyles.** The adductor tubercle is continuous with the medial epicondyle.

Patella

The patella is the largest sesamoid bone (i.e., a bone that develops within a tendon); it lies within the tendon of the quadriceps femoris muscle in front of the knee joint. It is triangular in shape and its apex lies inferiorly; the apex is connected to the tuberosity of the tibia by the ligamentum patellae. The posterior surface articulates with the condyles of the femur.

Bones of the Leg

The bones of the leg are the tibia and the fibula (see Fig. 6-2).

Tibia

The tibia is the large weight-bearing medial bone of the leg. At the upper end are the **lateral and medial condyles,** which articulate with the lateral and medial condyles of the femur, with the **lateral and medial menisci (semilunar cartilages)** intervening. Separating the upper articular surfaces of the tibial condyles is the **intercondylar eminence.** The lateral condyle possesses an oval **articular facet for the head of the fibula** on its lateral aspect.

At the upper end of the anterior border of the shaft of the tibia is the **tuberosity** (see Fig. 6-2), which receives the attachment of the ligamentum patellae. The anterior border is prolonged downward and medially to form the **medial malleolus** below. The lateral border of the tibia gives attachment to the interosseous membrane, which binds the tibia and the fibula together. The lower end of the tibia shows a wide, rough depression on its lateral surface for articulation with the fibula.

Fibula

The fibula provides attachment for muscles. It takes **no part** in the articulation at the knee joint, but below forms part of the ankle joint.

The **head** forms its upper end (see Fig. 6-2); it has a **styloid process** and possesses an **articular surface** for articulation with the lateral condyle of the tibia. The **shaft** is attached to the tibia by the interosseous membrane. The lower end of the fibula forms the **lateral malleolus.**

Bones of the Foot

The bones of the foot are the tarsal bones, the metatarsal bones, and the phalanges (Fig. 6-3).

Tarsal Bones

The tarsal bones are the calcaneum, the talus, the navicular, the cuboid, and the three cuneiform bones.

Calcaneum

The calcaneum is the largest bone of the foot. It articulates above with the talus and in front with the cuboid bone. The posterior surface forms the prominence of the heel. The medial surface possesses a large shelflike ridge, the **sustentaculum tali,** which assists in the support of the talus.

Talus

The talus articulates above at the ankle joint with the tibia and the fibula, below with the calcaneum, and in front with the navicular bone (see Fig. 6-3). It possesses a **head,** a **neck,** and a **body.** Numerous important ligaments are attached to the talus but no muscles are attached to this bone.

Navicular

The navicular lies between the head of the talus and the three cuneiform bones (see Fig. 6-3). The **tuberosity** lies in front of and below the medial malleolus and attaches to the main part of the tibialis posterior tendon.

Cuboid

The cuboid articulates with the anterior end of the calcaneum (see Fig. 6-3). There is a deep groove on its inferior aspect for the tendon of the peroneus longus muscle.

Cuneiform Bones

The three cuneiform bones are small, wedge-shaped bones that articulate proximally with the navicular bone and distally with the first three metatarsal bones. Their wedge shape contributes greatly to the formation and maintenance of the transverse arch of the foot.

Metatarsal Bones and Phalanges

The metatarsal bones and the phalanges resemble the metacarpals and the phalanges of the hand; each possesses a distal **head,** a **shaft,** and a proximal **base** (see Fig. 6-3). There are five metatarsals, and they are numbered from the medial to the lateral side. The **fifth metatarsal** has a prominent **tubercle** on its base, which can be easily palpated along the lateral border of the foot. The tubercle gives attachment to the peroneus brevis tendon.

Each toe has three phalanges except the big toe, which possess only two.

Joints

Hip Joint
Articulation

Between the hemispherical head of the femur and the cup-shaped acetabulum of the hip bone (Fig. 6-4). The articular surface of the acetabulum is horseshoe-shaped and is deficient inferiorly at the **acetabular notch.** The cavity of the acetabulum is deepened by the fibrocartilaginous rim, the **acetabular labrum.** The labrum bridges across the acetabular notch and is called the **transverse acetabular ligament.**

Type

Synovial ball-and-socket joint.

Capsule

Encloses the joint and is attached medially to the acetabular labrum (see Fig. 6-4). Laterally, it is attached to the intertrochanteric line of the femur in front and halfway along the posterior aspect of the neck of the bone behind. It is reinforced by the iliofemoral, pubofemoral, and ischiofemoral ligaments.

Ligaments
Iliofemoral Ligament

This is the strongest and most important ligament of the hip joint (Fig. 6-5). It is shaped like an inverted Y. Its base

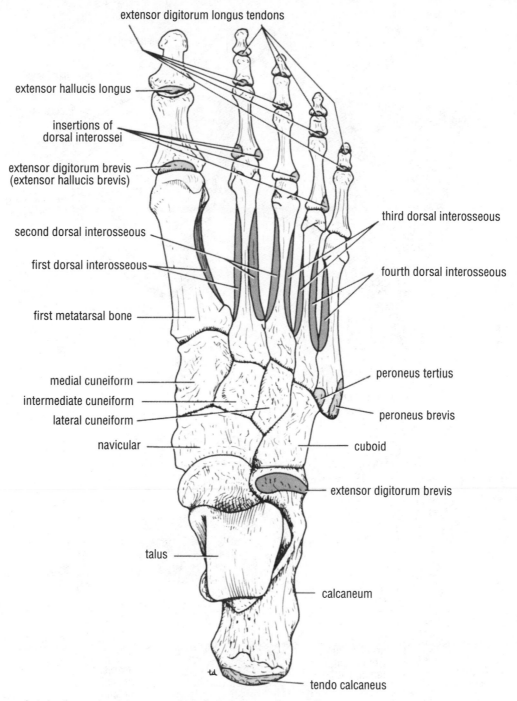

extensor digitorum longus tendons

extensor hallucis longus

insertions of
dorsal interossei

extensor digitorum brevis
(extensor hallucis brevis)

second dorsal interosseous

first dorsal interosseous

first metatarsal bone

medial cuneiform

intermediate cuneiform

lateral cuneiform

navicular

talus

third dorsal interosseous

fourth dorsal interosseous

peroneus tertius

peroneus brevis

cuboid

extensor digitorum brevis

calcaneum

tendo calcaneus

Fig. 6-3.   Bones of right foot, dorsal aspect, showing the muscle attachments.

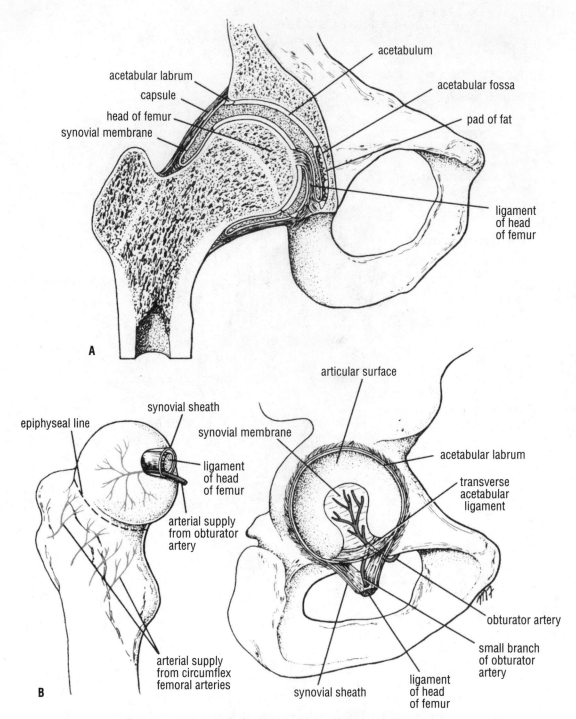

Fig. 6-4. Coronal section of the right hip joint **(A)** and articular surfaces of the right hip joint and arterial supply of head of femur **(B).**

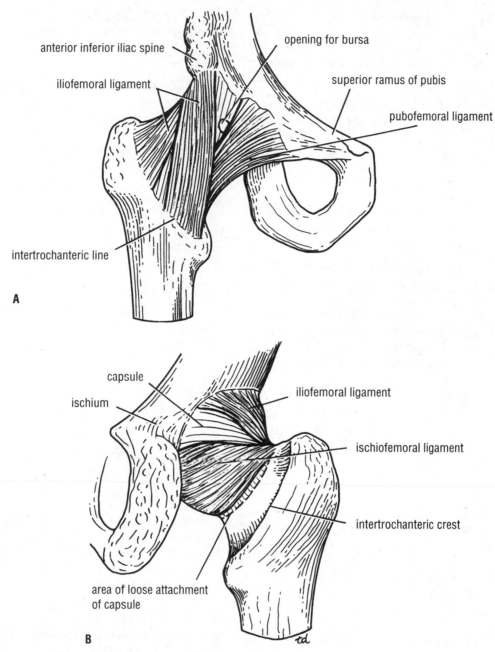

Fig. 6-5.   Anterior aspect **(A)** and posterior aspect **(B)** of right hip joint.

is attached to the anterior inferior iliac spine above, and the two limbs of the Y are attached below to the upper and lower parts of the intertrochanteric line of the femur. It resists hyperextension and lateral rotation of the hip joint.

Pubofemoral Ligament

This is triangular in shape (see Fig. 6-5). The base is attached above to the superior ramus of the pubis, and the apex is attached below to the lower end of the intertrochanteric line. It limits abduction and lateral rotation of the hip joint.

Ischiofemoral Ligament

This is spiral in shape and is attached to the body of the ischium and laterally to the greater trochanter of the femur (see Fig. 6-5). It limits medial rotation of the hip joint.

Ligament of the Head of the Femur

This is flat and triangular in shape (see Fig. 6-4). It is attached by its apex to the fovea capitis of the femur and by its base to the transverse acetabular ligament and the margins of the acetabular notch. The ligament lies within the joint and is ensheathed by synovial membrane. It has a slight limiting action on adduction of the hip joint.

Synovial Membrane

Lines the capsule (see Fig. 6-4). It covers the portion of the femoral neck that lies within the joint capsule. It ensheathes the ligament of the head of the femur and covers the floor of the acetabular fossa. It frequently communicates with the psoas bursa.

Nerve Supply

Femoral, obturator, and sciatic nerves and the nerve to the quadratus femoris.

Movements and the Muscles that Produce Movement

The hip joint has a wide range of movement.

Flexion: Iliopsoas, rectus femoris, sartorius, adductor muscles

Extension (posterior movement of the flexed thigh): Gluteus maximus, hamstring muscles

Abduction: Gluteus medius and minimus, sartorius, tensor fasciae latae, piriformis

Adduction: Adductor longus and brevis, adductor fibers of adductor magnus, pectineus, gracilis

Lateral rotation: Piriformis, obturator internus and externus, superior and inferior gamelli, quadratus femoris, gluteus maximus

Medial rotation: Anterior fibers of gluteus medius and minimus, tensor fasciae latae

Circumduction: A combination of all these movements

Important Relations

Anteriorly: Femoral vessels and nerve

Posteriorly: Sciatic nerve

Knee Joint
Articulation

Above are the condyles of the femur; below are the condyles of the tibia and their menisci (semilunar cartilages) (Fig. 6-6); in front is the articulation between the lower end of the femur and the patella.

Type

Between the femur and the tibia is a synovial joint of the hinge variety; between the patella and the femur is a synovial gliding joint.

Capsule

Encloses the joint except anteriorly where the capsule is deficient. Here, the synovial membrane pouches upward beneath the quadriceps tendon, forming the suprapatellar bursa.

Ligaments
Extracapsular

Ligamentum Patellae. This is a continuation of the tendon of the quadriceps femoris muscle. It is attached above to the lower border of the patella and below to the tubercle of the tibia.

Lateral Collateral Ligament. This is cordlike and attached above to the lateral condyle of the femur and below to the head of the fibula (see Fig. 6-6). It is **separated** from the lateral meniscus by the tendon of the popliteus muscle.

Medial Collateral Ligament. This is a flat band that is attached above to the medial condyle of the femur and below to the medial surface of the shaft of the tibia (see Fig. 6-6). It is **strongly attached to the medial meniscus.**

Oblique Popliteal Ligament. This is a tendinous expansion of the semimembranosus muscle. It strengthens the back of the capsule.

Intracapsular

Cruciate Ligaments. Two very strong ligaments that cross each other within the joint (see Fig. 6-6). They are named anterior and posterior, according to their tibial attachments.

The **anterior cruciate ligament** is attached below to the anterior intercondylar area of the tibia (Fig. 6-7) and passes **upward, backward, and laterally** to be attached to the lateral femoral condyle.

The **posterior cruciate ligament** is attached below to the posterior intercondylar area of the tibia (see Fig. 6-7) and passes **upward, forward, and medially** to be attached to the medial femoral condyle.

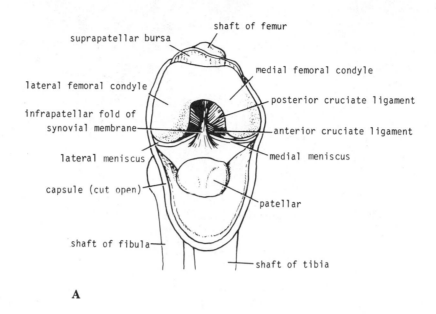

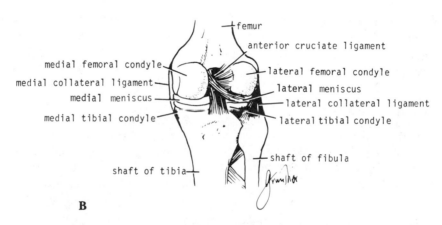

Fig. 6-6.   A. Right knee joint, anterior view, internal aspect: The capsule has been cut and the patella has been turned downward. **B.** Right knee joint, posterior view, internal aspect: The capsule and the synovial membrane have been removed.

Menisci (Semilunar Cartilages)

These menisci are C-shaped sheets of fibrocartilage (see Fig. 6-7). The peripheral convex border of each is thick and attached to the capsule, and the inner concave border is thin and forms a free edge. The upper surfaces are in contact with the femoral condyles, and the lower surfaces are in contact with the tibial condyles. Each meniscus is attached to the upper surface of the tibia by anterior and posterior **horns.** Because the medial meniscus is also attached to the medial collateral ligament, it is relatively immobile and very susceptible to injury. The function of the menisci is to deepen the articular surfaces of the tibial condyles to receive the convex femoral condyles.

Synovial Membrane

Lines the capsule. Anteriorly, it forms a pouch that extends up beneath the quadriceps femoris muscle to form the **suprapatellar bursa.** Posteriorly, the synovial membrane is prolonged downward on the tendon of the popliteus muscle forming the **popliteal bursa.** The synovial membrane is also reflected forward around the front of the cruciate ligaments. As a result, the cruciate ligaments lie **behind** the synovial cavity.

In the anterior part of the lower region of the joint, the synovial membrane is reflected backward from the ligamentum patellae to form the **infrapatellar fold;** the edges of the fold are called the **alar folds.**

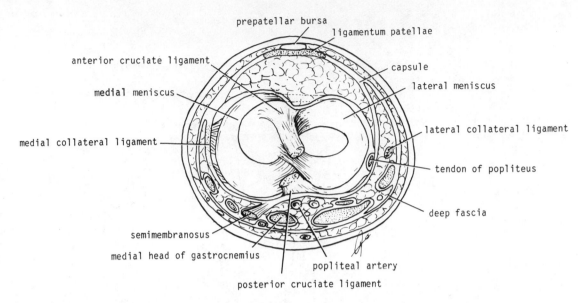

Fig. 6-7. Cross section of right knee joint as seen from above, shows position of ligaments and menisci.

Bursae Related to the Knee Joint
Suprapatellar Bursa
This lies beneath the quadriceps muscle. It is the largest bursa and **always communicates** with the knee joint.

Prepatellar Bursa
This lies between the patella and the skin.

Infrapatellar Bursae
Superficial infrapatellar bursa lies between the ligamentum patellae and the skin. **Deep infrapatellar bursa** lies between the ligamentum patellae and the tibia.

Popliteal Bursa
Surrounds the tendon of the popliteus; it **always communicates** with the joint cavity.

Semimembranosus Bursa
Lies between the tendon of this muscle and the medial condyle of the tibia. It **may communicate** with the joint cavity.

Nerve Supply
Femoral, obturator, common peroneal, and tibial nerves.

Movements and the Muscles that Produce Movement
Flexion: Biceps femoris, semitendinosus, semimembranosus
Extension: Quadriceps femoris
Medial rotation: Sartorius, gracilis, semitendinosus
Lateral rotation: Biceps femoris

The knee joint is most stable when it is in full extension. As the knee joint assumes the position of full extension, me-dial rotation of the femur results in a twisting and tightening of all the major ligaments of the joint. During the movement of flexion, the ligaments are untwisted by the contraction of the **popliteus muscle,** which laterally rotates the femur on the tibia.

Ankle Joint
Articulation
Between the lower end of the tibia, the malleoli above, and the body of the talus below (Fig. 6-8). The **inferior transverse tibiofibular ligament** deepens the socket into which the body of the talus fits snugly.

Type
Synovial hinge joint.

Capsule
Encloses the joint.

Ligaments
Medial (Deltoid) Ligament
Very strong. Attached by its apex to the tip of the medial malleolus (see Fig. 6-8). Below, the deep fibers are attached to the medial surface of the body of the talus; the superficial fibers are attached to the medial side of the talus, the sustentaculum tali, the plantar calcaneonavicular ligament, and the tuberosity of the navicular bone.

Lateral Ligament
Weaker than medial ligament (see Fig. 6-8). There are three bands.
Anterior Talofibular Ligament. Runs from the lateral malleolus to the lateral surface of the talus.

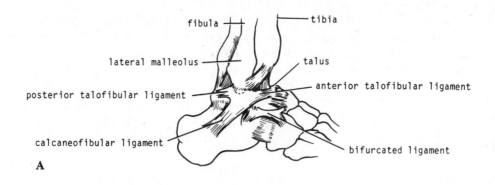

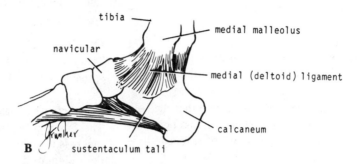

Fig. 6-8.   Right ankle joint. **A.** Lateral aspect. **B.** Medial aspect.

Calcaneofibular Ligament.   Runs from the lateral malleolus to the lateral surface of the calcaneum.
Posterior Talofibular Ligament.   Runs from the lateral malleolus to the posterior tubercle of the talus.

Synovial Membrane
Lines the capsule.

Nerve Supply
Deep peroneal and tibial nerves.

Movements and the Muscles that Produce Movement
Dorsiflexion (toes pointing upward): Tibialis anterior, extensor hallucis longus, extensor digitorum longus, peroneus tertius

Plantar flexion (toes pointing downward): Gastrocnemius, soleus, plantaris, peroneus longus, peroneus brevis, tibialis posterior, flexor digitorum longus, flexor hallucis longus

Important Relations
Anteriorly: Anterior tibial vessels and deep peroneal nerve (see Fig. 6-13)

Posteriorly: Tendo calcaneus (see Fig. 6-15)

Behind the lateral malleolus: Tendons of peroneus longus and brevis (see Fig. 6-15)

Behind the medial malleolus: Posterior tibial vessels, tibial nerve, and the long flexor tendons of the foot (see Fig. 6-15)

Intertarsal Joints
Subtalar Joint
Articulation
Concave inferior surface of the body of the talus and the convex facet on the upper surface of the calcaneum.

Type
Synovial gliding joint.

Talocalcaneonavicular Joint
Articulation
Rounded head of the talus, the upper surface of the sustentaculum tali of the calcaneum, and the posterior concave surface of the navicular bone.

Type
Synovial joint.

Ligaments
Plantar Calcaneonavicular (Spring) Ligament.   Runs from the anterior border of the sustentaculum tali to the inferior surface and the tuberosity of the navicular bone. It supports the head of the talus.

Calcaneocuboid Joint
Articulation
Anterior end of the calcaneum and the posterior surface of the cuboid.

Type
Synovial gliding joint.

Ligaments

Long Plantar Ligament. This strong ligament connects the undersurface of the calcaneus to the cuboid and the bases of the third, fourth, and fifth metatarsal bones.

Short Plantar Ligament. This ligament is wide and strong and connects the undersurface of the calcaneus to the adjoining part of the cuboid.

Movements and the Muscles that Produce Movement

The movements of the subtalar, talocalcaneonavicular, and calcaneocuboid joints are inversion and eversion. Inversion is more extensive than eversion.

Inversion (movement of the foot so that the sole faces medially): Tibialis anterior, extensor hallucis longus, medial tendons of extensor digitorum longus, tibialis posterior

Eversion (opposite movement of the foot so that the sole faces in a lateral direction): Peroneus longus, peroneus brevis, peroneus tertius, lateral tendons of extensor digitorum longus

Cuneonavicular Joint
Articulation
Between the three cuneiform bones and the navicular bone.

Type
Synovial gliding joint.

Cuboideonavicular Joint
The cuboideonavicular joint is a fibrous joint with the bones connected by dorsal, plantar, and interosseous ligaments. A small amount of movement is possible.

Intercuneiform and Cuneocuboid Joints
Intercuneiform and cuneocuboid joints are synovial gliding joints. The bones are connected by dorsal, plantar, and interosseous ligaments.

Tarsometatarsal and Intermetatarsal Joints
Tarsometatarsal and intermetatarsal joints are synovial gliding joints. The bones are connected by dorsal, plantar, and interosseous ligaments.

Metatarsophalangeal and Interphalangeal Joints
Metatarsophalangeal and interphalangeal joints are similar to those of the hand (see p. 132). The movements of abduction and adduction of the toes, performed by the interossei muscles, are small in amount and take place from the **midline of the second digit** (not the third, as in the hand).

Muscles of the Lower Limb

Gluteal Region
The gluteal region is bounded superiorly by the iliac crest and inferiorly by the fold of the buttock (Fig. 6-9). The region is largely made up of the gluteal muscles and a thick layer of superficial fascia.

The muscles of the gluteal region are shown in Table 6-1.

Fascia
Superficial Fascia
This is thick, especially in women, and is impregnated with large quantities of fat.

Deep Fascia
This is continuous below with the fascia lata of the thigh. It splits to enclose the gluteus maximus muscle.

Important Ligaments
The function of the sacrotuberous and sacrospinous ligaments is to stabilize the sacrum and prevent its rotation by the weight of the vertebral column.

Sacrotuberous Ligament
Connects the posterior inferior iliac spine, the lateral part of the sacrum, and the coccyx to the ischial tuberosity (see Fig. 6-9).

Sacrospinous Ligament
Connects the lateral part of the sacrum and coccyx to the spine of the ischium (see Fig. 6-9).

Important Foramina
Greater Sciatic Foramen
This is formed by the conversion of the greater sciatic notch of the hip bone into a foramen by the presence of the sacrotuberous and sacrospinous ligaments.

The following structures pass through the foramen:

Piriformis muscle

Sciatic nerve

Posterior cutaneous nerve of the thigh

Superior and inferior gluteal nerves

Nerves to obturator internus and quadratus femoris

Pudendal nerve

Superior and inferior gluteal arteries and veins

Internal pudendal artery and vein

Lesser Sciatic Foramen
This is formed by the conversion of the lesser sciatic notch of the hip bone into a foramen by the presence of the sacrotuberous and sacrospinous ligaments.

The following structures pass through the foramen:

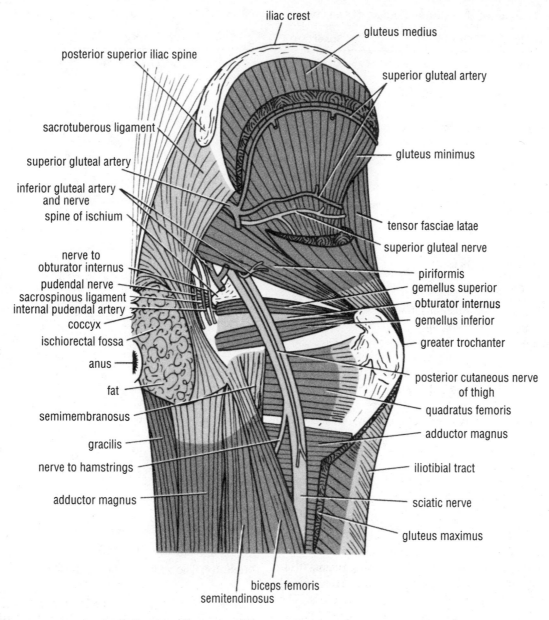

Fig. 6-9. Structures present in right gluteal region; greater part of gluteus maximus and part of gluteus medius have been removed.

Tendon of obturator internus
Nerve to obturator internus
Pudendal nerve
Internal pudendal artery and vein

Thigh

The muscles of the anterior fascial compartment (Fig. 6-10) are shown in Table 6-2; the muscles of the medial fascial compartment are shown in Table 6-3; and the muscles of the posterior fascial compartment (Fig. 6-11) are shown in Table 6-4.

Deep Fascia of the Thigh (Fascia Lata)

The deep fascia encloses the thigh like a trouser leg. Its upper end is attached to the pelvis and its associated ligaments.

Iliotibial Tract

The iliotibial tract is a thickening of the fascia lata on its lateral side. It is attached above to the iliac tubercle and below to the lateral condyle of the tibia. It receives the insertion of the greater part of the gluteus maximus muscle and the tensor fasciae latae muscle.

Table 6-1 Muscles of the Gluteal Region of the Lower Limb

Name of muscle	Origin	Insertion	Nerve supply	Action
Gluteus maximus	Outer surface of ilium, sacrum, coccyx, and sacrotuberous ligament	Iliotibial tract and gluteal tuberosity of femur	Inferior gluteal nerve	Extends and laterally rotates thigh at hip joint; through iliotibial tract, it extends knee joint
Gluteus medius	Outer surface of ilium	Greater trochanter of femur	Superior gluteal nerve	Abducts thigh at hip joint; tilts pelvis when walking
Gluteus minimus	Outer surface of ilium	Greater trochanter of femur	Superior gluteal nerve	Abducts thigh at hip joint; tilts pelvis when walking; anterior fibers medially rotate thigh
Tensor fasciae latae	Iliac crest	Iliotibial tract	Superior gluteal nerve	Assists gluteus maximus in extending the knee joint
Piriformis	Anterior surface of sacrum	Greater trochanter of femur	First and second sacral nerves	Lateral rotator of thigh at hip joint
Obturator internus	Inner surface of obturator membrane	Greater trochanter of femur	Sacral plexus	Lateral rotator of thigh at hip joint
Gemellus superior	Spine of ischium	Greater trochanter of femur	Sacral plexus	Lateral rotator of thigh at hip joint
Gemellus inferior	Ischial tuberosity	Greater trochanter of femur	Sacral plexus	Lateral rotator of thigh at hip joint
Quadratus femoris	Ischial tuberosity	Quadrate tubercle on upper end of femur	Sacral plexus	Lateral rotator of thigh at hip joint

Saphenous Opening

The saphenous opening is a gap in the deep fascia in the front of the thigh just below the inguinal ligament. It allows passage of the great saphenous vein, some small branches of the femoral artery, and lymph vessels. The opening is filled with loose connective tissue called the **cribriform fascia.**

Fascial Compartments of the Thigh

Three fascial septa pass from the inner aspect of the deep fascial sheath of the thigh to the linea aspera of the femur. By this means, the thigh is divided up into three compartments, each having muscles, nerves, and arteries. The compartments are anterior with the femoral nerve, medial (adductor) with the obturator nerve, and posterior with the sciatic nerve.

Femoral Triangle

The triangular area is situated in the upper part of the front of the thigh. Its boundaries are

Superiorly: The inguinal ligament
Laterally: The sartorius muscle
Medially: The adductor longus muscle

The femoral triangle contains the terminal part of the femoral nerve and its branches, the femoral sheath, the femoral artery and its branches, the femoral vein and its tributaries, and the inguinal lymph nodes.

Femoral Sheath

The femoral sheath is a downward protrusion from the abdomen into the thigh of the fascia transversalis and fascia iliaca. The sheath surrounds the femoral blood vessels and lymph vessels for about 1 in. (2.5 cm) below the inguinal ligament. The **femoral artery,** as it enters the thigh beneath the inguinal ligament, occupies the **lateral compartment** of the sheath. The **femoral vein** occupies the **intermediate compartment,** and the lymph vessels (and usually one lymph node) occupy the most **medial compartment.**

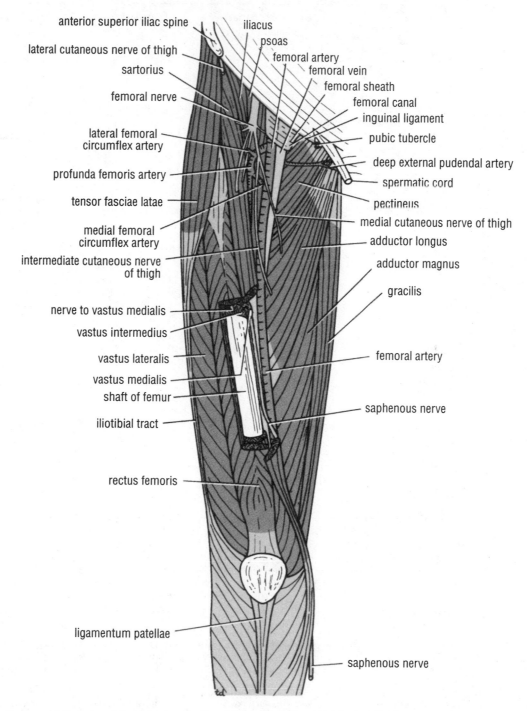

Fig. 6-10. Femoral triangle and adductor (subsartorial) canal in right lower limb.

Table 6-2   Muscles of the Anterior Fascial Compartment of the Thigh

Name of muscle	Origin	Insertion	Nerve supply	Action
Sartorius	Anterior superior iliac spine	Upper medial surface of shaft of tibia	Femoral nerve	Flexes, abducts, laterally rotates thigh at hip joint; flexes and medially rotates leg at knee joint
Iliacus	Iliac fossa of hip bone	With psoas into lesser trochanter of femur	Femoral nerve	Flexes thigh on trunk; if thigh is fixed, it flexes the trunk on the thigh as in sitting up from lying down
Psoas	Twelfth thoracic vertebral body; transverse processes, bodies, and intervertebral discs of the five lumbar vertebrae	With iliacus into lesser trochanter of femur	Lumbar plexus	Flexes thigh on trunk; if thigh is fixed, it flexes the trunk on the thigh as in sitting up from lying down
Pectineus	Superior ramus of pubis	Upper end shaft of femur	Femoral nerve	Flexes and adducts thigh at hip joint
Quadratus femoris				
Rectus femoris	Straight head: anterior inferior iliac spine; reflected head: ilium above acetabulum	Quadriceps tendon into patella	Femoral nerve	Extends leg at knee joint; flexes thigh at hip joint
Vastus lateralis	Upper end and shaft of femur	Quadriceps tendon into patella	Femoral nerve	Extends leg at knee joint
Vastus medialis	Upper end and shaft of femur	Quadriceps tendon into patella	Femoral nerve	Extends leg at knee joint
Vastus intermedius	Shaft of femur	Quadriceps tendon into patella	Femoral nerve	Extends leg at knee joint

Table 6-3   Muscles of the Medial Fascial Compartment of the Thigh

Name of muscle	Origin	Insertion	Nerve supply	Action
Gracilis	Inferior ramus of pubis, ramus of ischium	Upper part of shaft of tibia	Obturator nerve	Adducts thigh at hip joint, flexes leg at knee joint
Adductor longus	Body of pubis	Posterior surface of shaft of femur	Obturator nerve	Adducts thigh at hip joint, assists in lateral rotation
Adductor brevis	Inferior ramus of pubis	Posterior surface of shaft of femur	Obturator nerve	Adducts thigh at hip joint, assists in lateral rotation
Adductor magnus	Inferior ramus of pubis, ramus of ischium, ischial tuberosity	Posterior surface of shaft of femur, adductor tubercle of femur	Obturator nerve: adductor part; sciatic nerve: hamstring part	Adducts thigh at hip joint, assists in lateral rotation, hamstring part extends thigh at hip joint
Obturator externus	Outer surface of obturator membrane	Greater trochanter of femur	Obturator nerve	Lateral rotator of thigh at hip joint

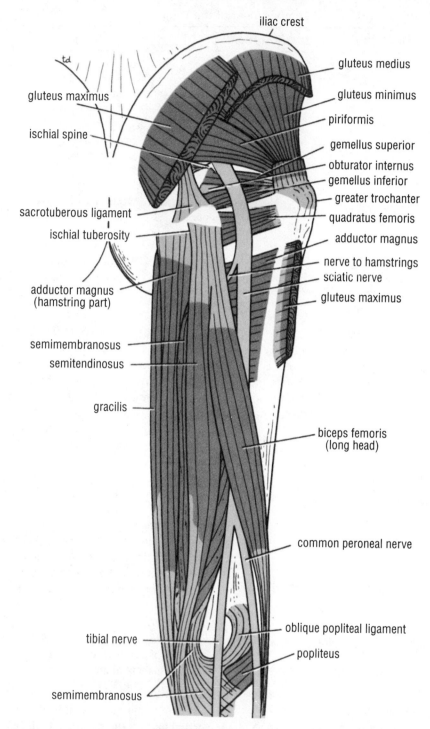

Fig. 6-11. Structures present on posterior aspect of right thigh.

Table 6-4   Muscles of the Posterior Fascial Compartment of the Thigh

Name of muscle	Origin	Insertion	Nerve supply	Action
Biceps femoris	Long head: ischial tuberosity; short head: shaft of femur	Head of fibula	Sciatic nerve (long head: tibial nerve; short head: common peroneal nerve)	Flexes and laterally rotates leg at knee joint, long head also extends thigh at hip joint
Semitendinosus	Ischial tuberosity	Upper part of medial surface of shaft of tibia	Sciatic nerve (tibial portion)	Flexes and medially rotates leg at knee joint, extends thigh at hip joint
Semimembranosus	Ischial tuberosity	Medial condyle of tibia, forms oblique popliteal ligament	Sciatic nerve (tibial portion)	Flexes and medially rotates leg at knee joint, extends thigh at hip joint
Adductor magnus (hamstring portion)	Ischial tuberosity	Adductor tubercle of femur	Sciatic nerve (tibial portion)	Extends thigh at hip joint

Femoral Canal

The femoral canal is the small medial compartment of the femoral sheath for the lymphatics. It is about ½ in. (1.3 cm) long. The canal is a potentially weak area in the wall of the abdomen. A protrusion of peritoneum could be forced down the femoral canal to form a femoral hernia.

Femoral Ring

Femoral ring is the name given to the upper opening of the femoral canal. It is filled by a plug of extra peritoneal fat called the **femoral septum.**

Important Relations

Anteriorly: Inguinal ligament

Posteriorly: Superior ramus of the pubis and the pectineal ligament

Laterally: Femoral vein

Medially: Lacunar ligament (an extension of the inguinal ligament, see p. 52).

Femoral Hernia

1. Protrusion of abdominal parietal peritoneum down through the femoral canal to form the hernial sac.
2. More common in women than in men.
3. The neck of the hernial sac lies below and lateral to the pubic tubercle.
4. The neck of the hernial sac lies at the femoral ring and is related anteriorly to the inguinal ligament, posteriorly to the pectineal ligament, laterally to the femoral vein, and medially to the sharp free edge of the lacunar ligament.

Adductor (Subsartorial) Canal

The adductor canal is an intermuscular cleft situated on the medial aspect of the middle third of the thigh beneath the sartorius muscle. The posterior wall is formed by the adductor magnus muscle, the lateral wall by the vastus medialis, and the anteromedial wall by the sartorius muscle and fascia. It contains the femoral artery and vein, the deep lymph vessels, the saphenous nerve, and the nerve to the vastus medialis.

Knee Region
Popliteal Fossa

The popliteal fossa is a diamond-shaped intermuscular space situated at the back of the knee (Fig. 6-12). It contains the popliteal vessels, the small saphenous vein, the common peroneal and tibial nerves, the posterior cutaneous nerve of the thigh, connective tissue, and lymph nodes.

Boundaries
Laterally

The biceps femoris above and the lateral head of the gastrocnemius and plantaris below.

Medially

The semimembranosus and semitendinosus above and the medial head of the gastrocnemius below.

Leg

The muscles of the anterior fascial compartment (Fig. 6-13) are shown in Table 6-5; the muscles of the lateral fascial compartment (see Fig. 6-13) are shown in Table 6-6; and the muscles of the posterior fascial compartment (Fig. 6-14) are shown in Table 6-7. The muscle on the dorsum of the foot is shown in Table 6-8.

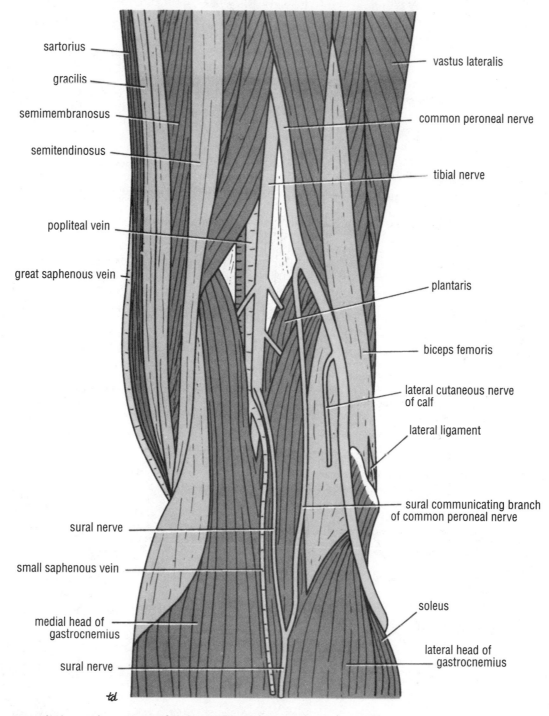

Fig. 6-12.   Boundaries and contents of right popliteal fossa.

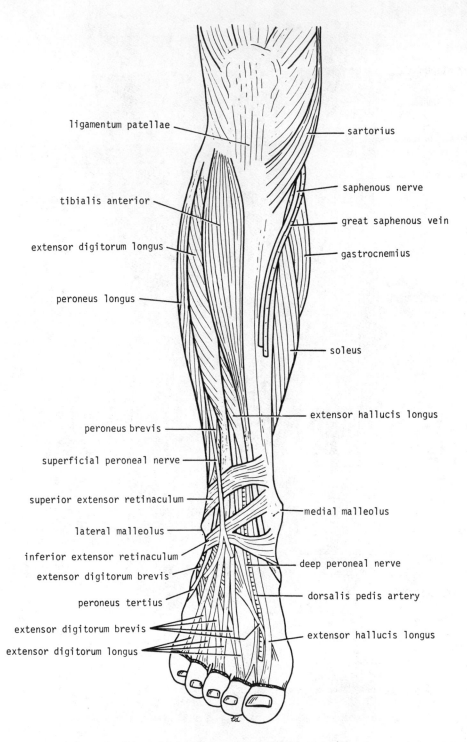

ligamentum patellae

sartorius

tibialis anterior

saphenous nerve

great saphenous vein

extensor digitorum longus

gastrocnemius

peroneus longus

soleus

extensor hallucis longus

peroneus brevis

superficial peroneal nerve

superior extensor retinaculum

medial malleolus

lateral malleolus

inferior extensor retinaculum

deep peroneal nerve

extensor digitorum brevis

peroneus tertius

dorsalis pedis artery

extensor digitorum brevis

extensor digitorum longus

extensor hallucis longus

Fig. 6-13.   Structures present on anterior and lateral aspects of right leg and on dorsum of foot.

Table 6-5 Muscles of the Anterior Fascial Compartment of the Leg

Name of muscle	Origin	Insertion	Nerve supply	Action
Tibialis anterior	Shaft of tibia and interosseous membrane	Medial cuneiform and base of first metatarsal bone	Deep peroneal nerve	Extends* the foot at ankle joint, inverts foot at subtalar and transverse tarsal joints, holds up medial longitudinal arch of foot
Extensor digitorum longus	Shaft of fibula and interosseous membrane	Extensor expansion of lateral four toes	Deep peroneal nerve	Extends toes, dorsiflexes foot at ankle joint
Peroneus tertius	Shaft of fibula and interosseous membrane	Base of fifth metatarsal bone	Deep peroneal nerve	Dorsiflexes foot at ankle joint, everts foot at subtalar and transverse tarsal joints
Extensor hallucis longus	Shaft of fibula and interosseous membrane	Base of distal phalanx of great toe	Deep peroneal nerve	Extends big toe, dorsiflexes foot at ankle joint, inverts foot at subtalar and transverse tarsal joints

*Extension, or dorsiflexion, of the ankle is the movement of the foot away from the ground.

Table 6-6 Muscles of the Lateral Fascial Compartment of the Leg

Name of muscle	Origin	Insertion	Nerve supply	Action
Peroneus longus	Shaft of fibula	Base of first metatarsal and medial cuneiform	Superficial peroneal nerve	Plantar flexes the foot at ankle joint, everts foot at subtalar and transverse tarsal joints, holds up lateral longitudinal arch of foot, supports transverse arch
Peroneus brevis	Shaft of fibula	Base of fifth metatarsal bone	Superficial peroneal nerve	Plantar flexes foot at ankle joint, everts foot at subtalar and transverse tarsal joints, holds up lateral longitudinal arch of foot

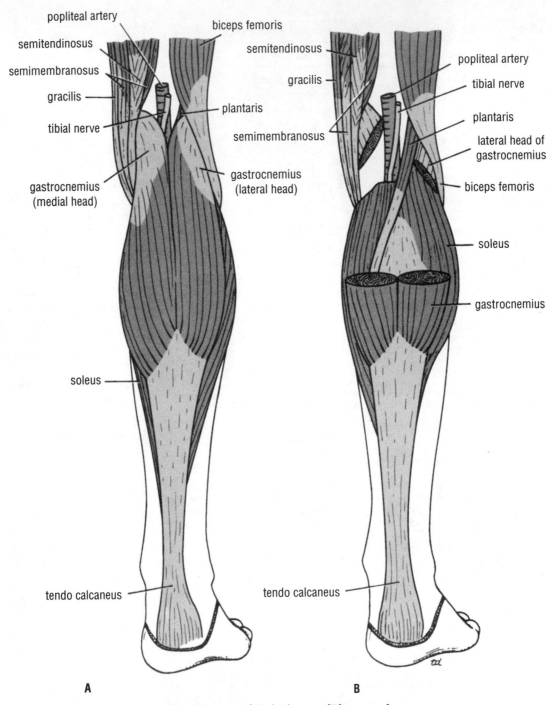

Fig. 6-14. Structures present on posterior aspect of right leg. In **(B)**, part of gastrocnemius has been removed.

Fascial Compartments of the Leg

The deep fascia surrounds the leg and is continuous above with the deep fascia of the thigh. It is attached to the anterior and medial borders of the tibia. Two intermuscular septa pass from its deep aspect to be attached to the fibula. The septa, together with the interosseous membrane, divide the leg into three compartments each having its own muscles, blood supply, and nerve supply. The compartments are anterior with the deep peroneal nerve, lateral (peroneal) with the superficial peroneal nerve, and posterior with the tibial nerve.

Interosseous Membrane

This interosseous membrane binds the tibia and the fibula together and provides attachment for muscles.

Table 6-7 Muscles of the Posterior Fascial Compartment of the Leg

Name of muscle	Origin	Insertion	Nerve supply	Action
Superficial Group				
Gastrocnemius	Medial and lateral condyles of femur	Via tendo calcaneus (Achilles tendon) into calcaneum	Tibial nerve	Plantar flexes foot at ankle joint, flexes knee joint
Plantaris	Lateral supracondylar ridge of femur	Calcaneum	Tibial nerve	Plantar flexes foot at ankle joint, flexes knee joint
Soleus	Shafts of tibia and fibula	Via tendo calcaneus (Achilles tendon) into calcaneum	Tibial nerve	Together with gastrocnemius and plantaris is powerful plantar flexor of ankle joint, provides main propulsive force in walking and running
Deep Group				
Popliteus	Lateral condyle of femur	Shaft of tibia	Tibial nerve	Flexes leg at knee joint, unlocks knee joint by lateral rotation of femur on tibia and thus slackens ligaments of joint
Flexor digitorum longus	Shaft of tibia	Bases of distal phalanges of lateral four toes	Tibial nerve	Flexes distal phalanges of lateral four toes, plantar flexes foot, supports medial and lateral longitudinal arches of foot
Flexor hallucis longus	Shaft of fibula	Base of distal phalanx of big toe	Tibial nerve	Flexes distal phalanx of big toe, plantar flexes foot at ankle joint, supports medial longitudinal arch of foot
Tibialis posterior	Shafts of tibia and fibula and interosseous membrane	Tuberosity of navicular bone and other neighboring bones	Tibial nerve	Plantar flexes foot at ankle joint, inverts foot at subtalar and transverse tarsal joints, supports medial longitudinal arch of foot

Table 6-8   Muscle on the Dorsum of the Foot

Name of muscle	Origin	Insertion	Nerve supply	Action
Extensor digitorum brevis	Calcaneum	By four tendons into the proximal phalanx of big toe (sometimes called extensor hallucis brevis) and long extensor tendons to second, third, and fourth toes	Deep peroneal nerve	Extends first, second, third, and fourth toes

Ankle
Retinacula
The retinacula are thickenings of the deep fascia to keep the long tendons around the ankle joint in position and to act as pulleys (Fig. 6-15).

Superior Extensor Retinaculum
This is attached to the distal ends of the anterior borders of the fibula and the tibia (see Fig. 6-13).

Inferior Extensor Retinaculum
This is a Y-shaped band located in front of the ankle joint (see Fig. 6-13).

Flexor Retinaculum
This extends from the medial malleolus to the medial surface of the calcaneum (see Fig. 6-15). It binds the deep muscles of the back of the leg to the back of the medial malleolus as they pass forward to enter the sole.

Superior Peroneal Retinaculum
This connects the lateral malleolus to the lateral surface of the calcaneum (see Fig. 6-15). It binds the tendons of the peroneus longus and brevis to the back of the lateral malleolus.

Inferior Peroneal Retinaculum
Binds the tendons of the peroneus longus and brevis to the lateral side of the calcaneum (see Fig. 6-15).

Sole of Foot
The muscles of the sole (Figs. 6-16 and 6-17) are usually described in four layers (from inferior to superior). The muscles are shown in Table 6-9.

Deep Fascia
Plantar Aponeurosis
This is a triangular thickening of deep fascia that protects the underlying nerves, blood vessels, and muscles. Its apex is attached to the medial and lateral tubercles of the calcaneum. The base of the aponeurosis divides into five slips that pass into the toes.

Arches of the Foot
There are three bony arches in the sole.

Medial Longitudinal Arch
This is formed by the calcaneum, talus, navicular bone, three cuneiform bones, and the first (medial) three metatarsal bones.
Muscular Support.   Medial part of flexor digitorum brevis, abductor hallucis, flexor hallucis longus, medial part of flexor digitorum longus, flexor hallucis brevis, tibialis anterior, and tendinous extensions of the insertion of the tibialis posterior.
Ligamentous Support.   Plantar and dorsal ligaments, including the important calcaneonavicular (spring) ligament, the medial ligament of the ankle joint, and the plantar aponeurosis.

Lateral Longitudinal Arch
This is formed by the calcaneum, cuboid, and the fourth and fifth metatarsal bones.
Muscular Support.   Abductor digiti minimi, lateral part of the flexor digitorum longus and brevis, and peroneus longus and brevis.
Ligamentous Support.   Long and short plantar ligaments and the plantar aponeurosis.

Transverse Arch
This is formed by the bases of the metatarsal bones, cuboid, and the three cuneiform bones. The wedge shape of the cuneiform bones and the bases of the metatarsal bones play a large part in the support of the transverse arch.
Muscular Support.   Dorsal interossei, transverse head of adductor hallucis, peroneus longus, and peroneus brevis.
Ligamentous Support.   Deep transverse ligaments and very strong plantar ligaments.

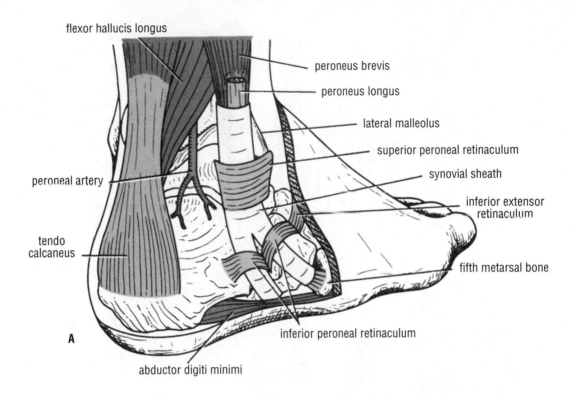

flexor hallucis longus

peroneus brevis

peroneus longus

lateral malleolus

superior peroneal retinaculum

synovial sheath

peroneal artery

inferior extensor retinaculum

tendo calcaneus

fifth metarsal bone

inferior peroneal retinaculum

A

abductor digiti minimi

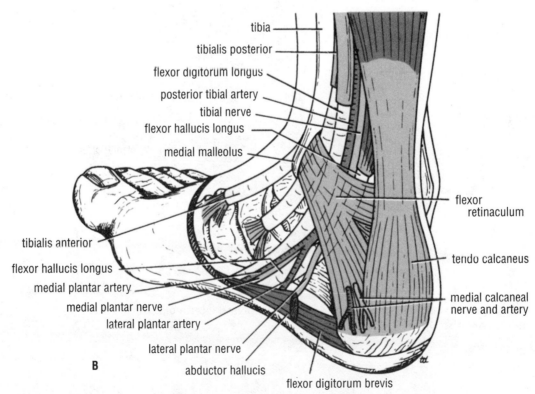

tibia

tibialis posterior

flexor digitorum longus

posterior tibial artery

tibial nerve

flexor hallucis longus

medial malleolus

flexor retinaculum

tibialis anterior

flexor hallucis longus

medial plantar artery

medial plantar nerve

lateral plantar artery

tendo calcaneus

medial calcaneal nerve and artery

lateral plantar nerve

B

abductor hallucis

flexor digitorum brevis

Fig. 6-15. Structures passing behind lateral malleolus **(A)** and medial malleolus **(B)**. Note position of retinacula.

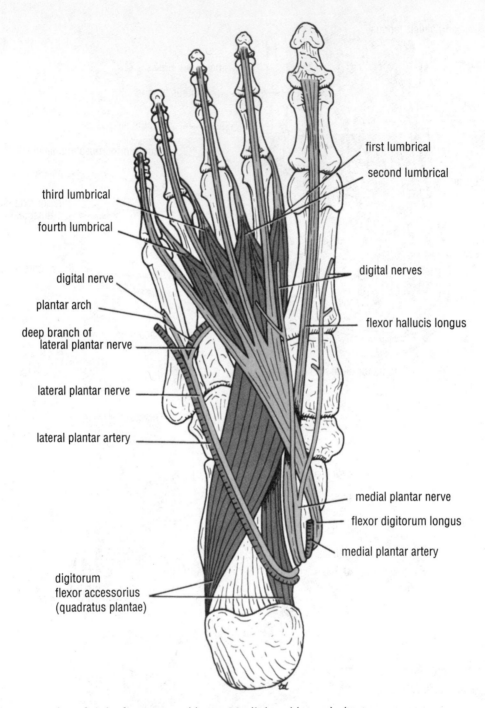

first lumbrical

second lumbrical

third lumbrical

fourth lumbrical

digital nerve

plantar arch

deep branch of
lateral plantar nerve

lateral plantar nerve

lateral plantar artery

digital nerves

flexor hallucis longus

medial plantar nerve

flexor digitorum longus

medial plantar artery

digitorum
flexor accessorius
(quadratus plantae)

Fig. 6-16. Plantar muscles of right foot, second layer. Medial and lateral plantar arteries and nerves are also shown.

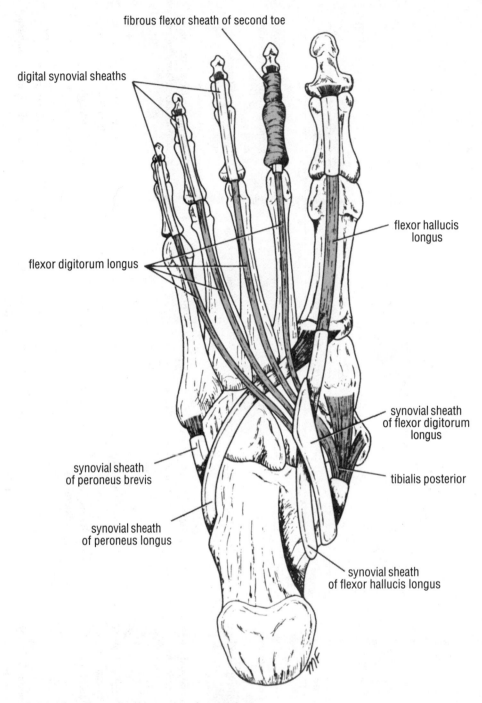

Fig. 6-17. Synovial sheaths of tendons on sole of right foot.

Table 6-9   Muscles of the Sole

Name of muscle	Origin	Insertion	Nerve supply	Action
First layer				
Abductor hallucis	Medial tubercle of calcaneum, flexor retinaculum	Medial side of base of proximal phalanx of big toe	Medial plantar nerve	Flexes and abducts big toe, supports medial longitudinal arch
Flexor digitorum brevis	Medial tubercle of calcaneum	Middle phalanx of four lateral toes	Medial plantar nerve	Flexes lateral four toes, supports medial and lateral longitudinal arches
Abductor digiti minimi	Medial and lateral tubercles of calcaneum	Lateral side of base of proximal phalanx of fifth toe	Lateral plantar nerve	Flexes and abducts fifth toe, supports lateral longitudinal arch
Second layer				
Flexor digitorum accessorius	Medial and lateral sides of calcaneum	Tendon of flexor digitorum longus	Lateral plantar nerve	Assists long flexor tendons to flex lateral four toes
Flexor digitorum longus	Shaft of tibia	Base of distal phalanx of lateral four toes	Tibial nerve	Flexes distal phalanges of lateral four toes, plantar flexes foot, supports longitudinal arches
Lumbricals (4)	Tendons of flexor digitorum longus	Dorsal extensor expansion of lateral four toes	First lumbrical medial plantar, remainder deep branch lateral plantar nerve	Extends toes at interphalangeal joints
Flexor hallucis longus	Shaft of fibula	Base of distal phalanx of big toe	Tibial nerve	Flexes distal phalanx of big toe, plantar flexes foot, supports medial longitudinal arch

Muscle	Origin	Insertion	Nerve	Action
Third layer				
Flexor hallucis brevis	Cuboid, lateral cuneiform bones; tibialis posterior insertion	Medial and lateral sides of base of proximal phalanx of big toe	Medial plantar nerve	Flexes metatarsophalangeal joint of big toe, supports medial longitudinal arch
Adductor hallucis				
Oblique head	Bases of second, third, and fourth metatarsal bones	Lateral side of base of proximal phalanx of big toe	Deep branch lateral plantar nerve	Flexes big toe, supports transverse arch
Transverse head	Plantar ligaments	Lateral side of base of proximal phalanx of big toe	Deep branch lateral plantar nerve	Flexes big toe, supports transverse arch
Flexor digiti minimi brevis	Base of fifth metatarsal bone	Lateral side of base of proximal phalanx of little toe	Lateral plantar nerve	Flexes little toe
Fourth layer				
Interossei				
Dorsal (4)	Adjacent sides of metatarsal bones	Bases of phalanges and dorsal expansion of corresponding toes	Lateral plantar nerve	Abduct toes from second toe, flex metatarsophalangeal joints, extend interphalangeal joints
Plantar (3)	Inferior surfaces of third, fourth, and fifth metatarsal bones	Bases of phalanges and dorsal expansion of corresponding toes	Lateral plantar nerve	Adduct toes to second toe, flex metatarsophalangeal joints, extend interphalangeal joints
Peroneus longus	Shaft of fibula	Base of first metatarsal and medial cuneiform	Superficial peroneal nerve	Plantar flexes the foot at ankle joint, everts foot at subtalar and transverse tarsal joints, holds up lateral longitudinal and transverse arches of foot
Tibialis posterior	Shafts of tibia and fibula and interossecus membrane	Tuberosity of navicular bone and other neighboring bones	Tibial nerve	Plantar flexes foot at ankle joint, inverts foot at subtalar and transverse tarsal joints, supports medial longitudinal arch of foot

Lower Limb

Arteries of the Lower Limb

Femoral Artery

The femoral artery is a continuation of the external iliac artery (Fig. 6-18). It begins behind the inguinal ligament where it lies midway between the anterior superior iliac spine and the symphysis pubis (site for taking femoral pulse). The artery descends through the femoral triangle (see Fig. 6-10) and the adductor canal. It leaves the front of the thigh by passing through the opening in the adductor magnus and enters the popliteal space as the popliteal artery (see Fig. 6-12).

In the femoral triangle, the artery is related laterally to the femoral nerve and medially in the upper part of its course to the femoral vein and the femoral canal.

Branches

1. **Superficial circumflex iliac artery** arises just below the inguinal ligament and runs laterally toward the anterior superior iliac spine.
2. **Superficial epigastric artery** arises just below the inguinal ligament and runs upward to the abdominal wall as far as the umbilicus.
3. **Superficial external pudendal artery.**
4. **Deep external pudendal artery.** This artery and the preceding artery arise just below the inguinal ligament and run medially to supply the skin of the scrotum (or labium majus).
5. **Profunda femoris artery** is a large branch that arises from the femoral artery about 1½ in. (4 cm) below the inguinal ligament (see Fig. 6-10). It supplies structures in the anterior, medial, and posterior fascial compartments of the thigh by means of the following branches:
   a. **Medial and lateral femoral circumflex arteries.**
   b. **Three perforating arteries.**
   c. **Fourth perforating artery,** which is the terminal portion of the profunda artery.
6. **Descending genicular artery.** This small branch arises in the adductor canal.

Trochanteric Anastomosis

The trochanteric anastomosis provides the main blood supply to the head of the femur (in the adult) via the following arteries: (1) superior gluteal artery, (2) inferior gluteal artery, (3) medial femoral circumflex artery, and (4) lateral femoral circumflex artery.

Cruciate Anastomosis

The cruciate anastomosis, together with the trochanteric anastomosis, provides the important connection between the internal iliac and femoral arteries. The following arteries are involved: (1) inferior gluteal artery, (2) medial femoral circumflex artery, (3) lateral femoral circumflex artery, and (4) first perforating artery (a branch of the profunda artery).

Popliteal Artery

The popliteal artery (see Fig. 6-18) is a continuation of the femoral artery and extends from the opening in the adductor magnus to the lower border of the popliteus muscle, where it divides into anterior and posterior tibial arteries. It is deeply placed in the popliteal fossa and lies close to the posterior surface of the femur and the knee joint.

Branches

1. **Muscular branches**
2. **Articular branches** to the knee joint
3. **Terminal branches:** Anterior and posterior tibial arteries

Anastomosis Around the Knee Joint

The arteries involved are (1) the descending genicular artery from the femoral artery, (2) the lateral femoral circumflex artery from the profunda femoris, (3) the articular branches from the popliteal artery, and (4) the branches from the anterior and posterior tibial arteries.

Anterior Tibial Artery

The anterior tibial artery arises at the bifurcation of the popliteal artery in the popliteal fossa (see Fig. 6-18). It passes forward between the tibia and the fibula through the upper part of the interosseus membrane to enter the anterior compartment of the leg. It descends with the deep peroneal nerve to the front of the ankle joint where it becomes the dorsalis pedis artery (see Fig. 6-13).

At the ankle, the anterior tibial artery lies midway between the malleoli and has the tendon of the extensor hallucis longus on its medial side and the tendons of extensor digitorum longus on its lateral side (site for taking the anterior tibial pulse).

Branches

1. **Muscular branches**
2. **Anastomotic branches,** which anastomose with branches of other arteries around the knee and ankle joints

Dorsalis Pedis Artery

The dorsalis pedis artery begins in front of the ankle joint midway between the malleoli; it is a continuation of the anterior tibial artery (see Figs. 6-13 and 6-18). The dorsalis pedis artery ends by entering the sole through the proximal part of the space between the first and second metatarsal bones. Having passed between the two heads of the first dorsal interosseous muscle, it joins the lateral plantar artery and completes the plantar arch (see Fig. 6-16).

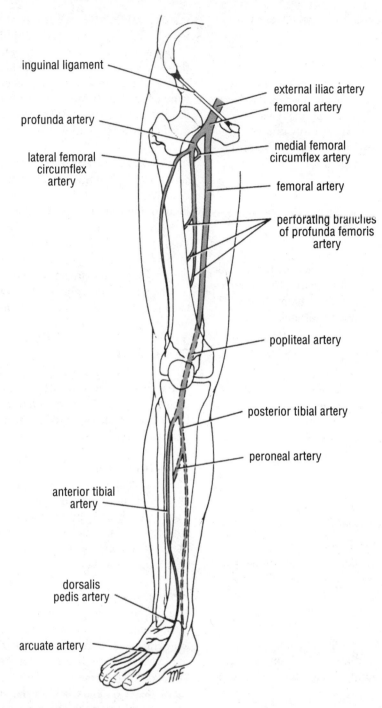

Fig. 6-18.   Major arteries of the lower limb.

At first the artery is superficial, having the tendons of the extensor digitorum longus on its lateral side and the tendon of the extensor hallucis longus on its medial side (site for taking the dorsalis pedis pulse).

Branches
1. **Lateral tarsal artery.** Supplies the dorsum of the foot
2. **Arcuate artery.** Runs laterally across the bases of the metatarsal bones and gives off branches to the toes

3. **First dorsal metatarsal artery.** Supplies both sides of the big toe

Posterior Tibial Artery
The posterior tibial artery arises at the bifurcation of the popliteal artery in the popliteal fossa (see Fig. 6-18). It descends in the posterior compartment of the leg, accompanied by the tibial nerve. The artery terminates behind the medial malleolus by dividing into medial and lateral plan-

tar arteries. The pulse may be felt midway between the medial malleolus and the heel.

Branches

1. **Peroneal artery.** This large artery arises close to the origin of the posterior tibial artery. It descends in close association with the flexor hallucis longus muscle to the region of the ankle. It gives off **muscular branches,** a **nutrient artery** to the fibula, and **anastomotic branches** around the ankle joint.
2. **Muscular branches.**
3. **Nutrient artery** to the tibia.
4. **Anastomotic branches** around the ankle joint.
5. **Medial and lateral plantar arteries.**

Medial Plantar Artery

This is the smaller of the terminal branches of the posterior tibial artery (see Fig. 6-16). It runs forward along the medial border of the foot with the medial plantar nerve. It gives off many muscular and cutaneous branches.

Lateral Plantar Artery

This is the larger of the terminal branches of the posterior tibial artery (see Fig. 6-16). It runs forward deep to the abductor hallucis and the flexor digitorum brevis with the lateral plantar nerve. It ends by curving medially to form the **plantar arch** by anastomosing with the dorsalis pedis artery. The plantar arch gives off perforating and metatarsal arteries; the metatarsal arteries give rise to digital arteries.

Veins of the Lower Limb

The superficial veins lie in the superficial fascia and are of **great clinical importance.** The deep veins accompany the main arteries.

Superficial Veins
Dorsal Venous Network

The dorsal venous network lies on the dorsum of the foot (Fig. 6-19). It is drained on the medial side by the great saphenous vein and on the lateral side by the small saphenous vein.

Great Saphenous Vein

The great saphenous vein arises from the medial side of the dorsal venous network of the foot (see Fig. 6-19) and ascends directly **in front of** the medial malleolus. Accompanied by the saphenous nerve, it ascends the leg in the superficial fascia. It passes behind the knee and curves forward around the medial side of the thigh. It then passes through the saphenous opening in the deep fascia and joins the femoral vein about 1½ in. (4 cm) below and lateral to the pubic tubercle. The great saphenous vein possess numerous valves and is connected to the small saphenous vein by branches that pass behind the knee. A number of **perforating veins** connect the great saphenous vein with the deep veins along the medial side of the calf.

The great saphenous vein receives the following small tributaries near its termination: the **superficial circumflex iliac vein,** the **superficial epigastric vein,** and the **superficial external pudendal vein.**

Small Saphenous Vein

The small saphenous vein arises from the lateral side of the dorsal venous network of the foot (see Fig. 6-19). It ascends **behind** the lateral malleolus in company with the sural nerve. It passes up the back of the leg and pierces the deep fascia to enter the popliteal fossa. It drains into the popliteal vein. The small saphenous vein communicates with the deep veins and the great saphenous vein.

The superficial veins of the lower limbs are commonly used for blood transfusion; they are common sites for varicosities.

Deep Veins
Venae Comitantes

The deep veins accompany the respective arteries as venae comitantes. The venae comitantes of the anterior and posterior tibial arteries unite in the popliteal fossa to form the popliteal vein.

Popliteal Vein

The popliteal vein is formed by the union of the venae comitantes of the anterior and posterior tibial arteries (see Fig. 6-12). It ends by passing through the opening in the adductor magnus to become the femoral vein. The popliteal vein receives numerous tributaries, including the small saphenous vein.

Femoral Vein

The femoral vein is a continuation of the popliteal vein at the opening in the adductor magnus. It ascends through the adductor canal and the femoral triangle, accompanied by the femoral artery. In the femoral sheath, it lies on the medial side of the femoral artery and lateral to the femoral canal. As it ascends behind the inguinal ligament, it becomes continuous with the external iliac vein. The femoral vein receives the great saphenous vein and veins that correspond to branches of the femoral artery.

Lymphatic Drainage of the Lower Limb

The **superficial lymph vessels** ascend the limb in the superficial fascia with the superficial veins. The **deep lymph vessels** lie deep to the deep fascia and follow the deep ar-

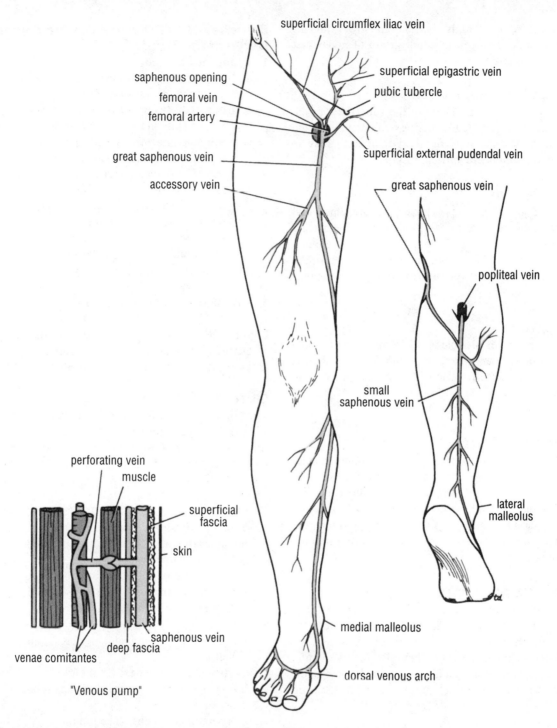

Fig. 6-19. Superficial veins of right lower limb. Note importance of valved perforating veins in the "venous pump."

teries and veins. All the lymph vessels of the lower limb ultimately drain into the deep inguinal group of nodes that are situated in the groin.

Superficial Inguinal Nodes

The superficial inguinal nodes lie in the superficial fascia just below the inguinal ligament; they all drain into the deep inguinal nodes. The superficial inguinal nodes may be divided into horizontal and vertical groups.

Horizontal Group

The horizontal group receives lymph from the superficial lymph vessels of the anterior abdominal wall below the level of the umbilicus, the perineum, the external genitalia of both sexes (but not the testes), and the lower half of the anal canal. It also receives lymph from the skin of the buttocks.

Vertical Group

The vertical group lies alongside the terminal part of the great saphenous vein and receives the majority of the superficial lymph vessels of the lower limb, except from the back and lateral side of the calf and lateral side of the foot (which drain into popliteal nodes).

Deep Inguinal Nodes

The deep inguinal nodes are usually three in number and lie along the medial side of the femoral vein and in the femoral canal. They receive all the lymph from the superficial inguinal nodes and from all the deep structures of the lower limb. The efferent lymph vessels pass up through the femoral canal into the abdominal cavity and drain into the external iliac nodes.

Popliteal Lymph Nodes

Situated in the popliteal fossa, the popliteal lymph nodes receive superficial lymph vessels that accompany the small saphenous vein from the lateral side of the foot, and the back and lateral side of the calf. They also receive lymph from the deep structures of the leg below the knee. The efferent vessels from these nodes drain upward to the deep inguinal nodes.

Nerves of the Lower Limb

Femoral Nerve

The femoral nerve arises from the lumbar plexus (L2, 3, and 4). It enters the thigh behind the inguinal ligament and lies lateral to the femoral vessels and the femoral sheath in the femoral triangle (see Fig. 6-10). It quickly terminates by dividing into anterior and posterior divisions.

Branches of the Divisions of the Femoral Nerve in the Thigh

1. **Cutaneous branches.**
   a. **Medial cutaneous nerve of the thigh** supplies the skin on the medial side of the thigh.
   b. **Intermediate cutaneous nerve of the thigh** supplies the skin on the anterior surface of the thigh.
   c. **Saphenous nerve** descends through the femoral triangle and the adductor canal, crossing the femoral artery. The nerve emerges on the medial side of the knee joint between the tendons of the sartorius and gracilis. It accompanies the great saphenous vein down the medial side of the leg and in front of the medial malleolus. It passes along the medial border of the foot as far as the ball of the big toe.
2. **Muscular branches** to the sartorius, pectineus, quadriceps femoris.
3. **Articular branches** to the hip and knee joints.

Figure 6-20 summarizes the branches of the femoral nerve.

Dermatomal charts for the anterior and posterior surfaces of the body are shown in Figures 1-7 and 1-8.

Obturator Nerve

The obturator nerve arises from the lumbar plexus (L2, 3, and 4). Having run forward on the lateral wall of the pelvis, it reaches the obturator canal (upper part of the obturator foramen). The obturator nerve divides into anterior and posterior divisions.

Branches of the Obturator Nerve in the Thigh

1. **Anterior division** descends into the thigh anterior to the obturator externus and adductor brevis.
   a. **Muscular branches:** Gracilis, adductor brevis, adductor longus, and sometimes pectineus.
   b. **Cutaneous branch:** Skin on medial side of thigh.
   c. **Articular branch:** Hip joint.
2. **Posterior division** descends through obturator externus and passes behind adductor brevis and in front of adductor magnus.
   a. **Muscular branches:** Obturator externus, adductor magnus (adductor part), and sometimes adductor brevis.
   b. **Articular branch:** Knee joint.

Figure 6-21 summarizes the branches of the obturator nerve.

Dermatomal charts for the anterior and posterior surfaces of the body are shown in Figures 1-7 and 1-8.

Sciatic Nerve

The sciatic nerve arises from the sacral plexus (L4, 5; S1, 2, and 3). It passes out of the pelvis into the gluteal region through the greater sciatic foramen (see Fig. 6-9). The nerve appears below the piriformis and is covered by the

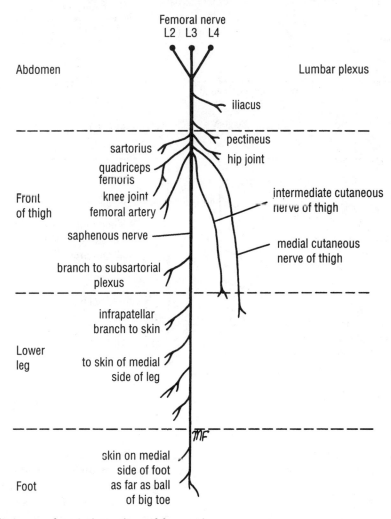

Fig. 6-20. Summary diagram of main branches of femoral nerve.

gluteus maximus muscle. It descends through the gluteal region and enters the posterior compartment of the thigh. In the lower third of the thigh (occasionally at a higher level), it ends by dividing into the tibial and common peroneal nerves (see Fig. 6-11).

Branches of the Sciatic Nerve

1. **Muscular branches:** Biceps femoris (long head), semitendinosus, semimembranosus, hamstring part of adductor magnus
2. **Articular branches:** Hip joint
3. **Terminal branches:** Tibial nerve, common peroneal nerve

Tibial Nerve

The tibial nerve descends through the popliteal fossa and the posterior compartment of the leg. It lies deep to the gastrocnemius and soleus muscles and reaches the interval between the medial malleolus and the heel. It is covered by the flexor retinaculum and divides into medial and lateral plantar nerves.

Branches of the Tibial Nerve

1. **Cutaneous branches**
   a. **Sural nerve** (joined by communicating branch of common peroneal nerve) supplies the skin of the calf, the back of the leg, the lateral border of the foot, and the lateral side of the little toe (see Fig. 6-12).
   b. **Medial calcaneal nerve** supplies the skin over the medial surface of the heel.
2. **Muscular branches:** Gastrocnemius, plantaris, soleus, popliteus, flexor digitorum longus, flexor hallucis longus, tibialis posterior.
3. **Articular branches:** Knee joint, ankle joint.
4. **Medial plantar nerve** runs forward deep to the abductor hallucis with the medial plantar artery (see Fig. 6-16).
   a. **Cutaneous branch:** Medial part of sole and medial 3½ toes and nail beds.

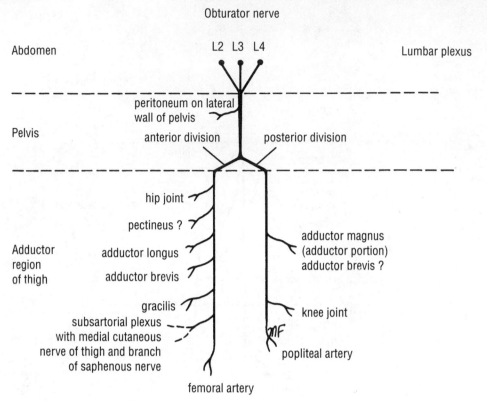

Fig. 6-21.   Summary diagram of main branches of obturator nerve.

b. **Muscular branch:** Abductor hallucis, flexor digitorum brevis, flexor hallucis brevis, first lumbrical.
5. **Lateral plantar nerve** runs forward deep to the abductor hallucis and flexor digitorum brevis in company with the lateral plantar artery (see Fig. 6-16).
   a. **Cutaneous branch:** Lateral part of sole and lateral 1½ toes and nail beds.
   b. **Muscular branch:** Flexor digitorum accessorius, abductor digiti minimi, flexor digiti minimi brevis, adductor hallucis, interosseous muscles, second lumbrical, third lumbrical, fourth lumbrical.

Common Peroneal Nerve

The common peroneal nerve descends through the popliteal fossa (see Fig. 6-12). It then passes laterally around the neck of the fibula, pierces the peroneus longus muscle, and divides into the superficial peroneal nerve and the deep peroneal nerve.

Branches of the Common Peroneal Nerve

1. **Cutaneous branches**
   a. **Sural communicating branch** (see Fig. 6-12) joins the sural nerve (see Branches of the Tibial Nerve).
   b. **Lateral cutaneous nerve of the calf** supplies the skin on the lateral side of the back of the leg.
2. **Muscular branch:** Short head of biceps femoris.
3. **Articular branch:** Knee joint.
4. **Superficial peroneal nerve** descends between the peroneus longus and brevis in the lateral fascial compartment and becomes subcutaneous (see Fig. 6-13).
   a. **Cutaneous branch:** To skin on front of lower leg and dorsum of foot, except cleft between big and second toes (innervated by deep peroneal nerve).
   b. **Muscular branch:** Peroneus longus and brevis.
5. **Deep peroneal nerve** descends in the anterior fascial compartment deep to the extensor digitorum longus muscle and on the interosseous membrane (see Fig. 6-13). It is accompanied by the anterior tibial vessels. On the dorsum of the foot it divides into medial and lateral terminal branches.
   a. **Cutaneous branch** supplies adjacent sides of the big and second toes.
   b. **Muscular branch:** Tibialis anterior, extensor digitorum longus, peroneus tertius, extensor hallucis longus, extensor digitorum brevis.
   c. **Articular branch:** Ankle and tarsal joints.

Figures 6-22 and 6-23 summarize the branches of the sciatic nerve.

Dermatomal charts for the anterior and posterior surfaces of the body are shown in Figures 1-7 and 1-8.

Injuries to the Nerves of the Lower Limb
Femoral Nerve Lesions

Stab or gunshot wounds in the thigh are common causes for femoral nerve lesions. It is rare for the nerve to be completely divided. The quadriceps femoris muscle is paralyzed and the knee cannot be extended. There is sensory loss over the medial side of the lower part of the leg and along the medial border of the foot as far as the ball of the big toe.

Sciatic Nerve Lesions

Badly placed intramuscular injections in the gluteal region, fracture dislocations of the hip joint, and penetrating wounds may damage the sciatic nerve. The hamstring muscles are paralyzed so that flexion of the knee is greatly weakened. All the muscles below the knee are paralyzed, and the weight of the foot causes it to assume the plantar-flexed position, or **foot drop.** There is loss of skin sensation below the knee, except for a narrow part of the leg and the medial border of the foot as far as the ball of the big toe.

Fig. 6-22. Summary diagram, showing origin of sciatic nerve and main branches of common peroneal nerve.

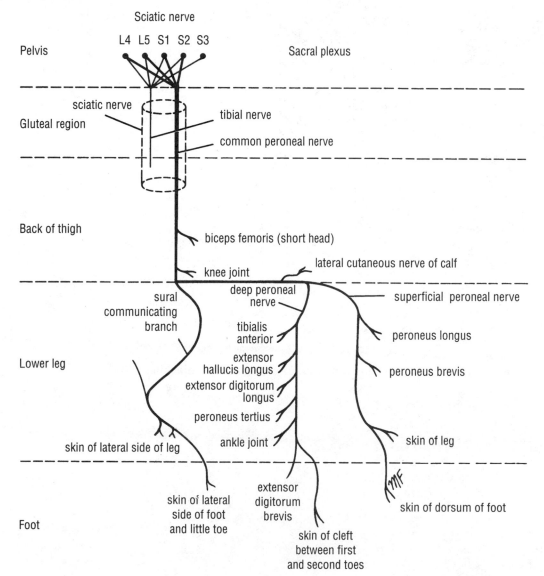

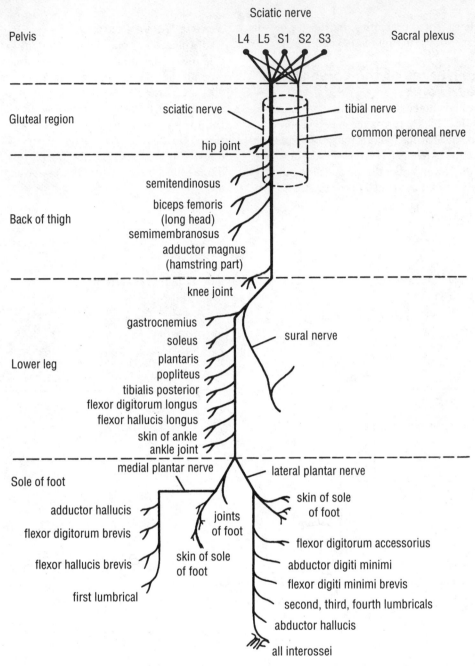

Fig. 6-23.   Summary diagram, showing origin of sciatic nerve and main branches of tibial nerve.

National Board Type Questions

Match the numbered structures shown in the anteroposterior radiograph of the hip region with the appropriate lettered structures (A–E) listed below.

Match the numbered structures shown in the anteroposterior radiograph of the knee joint with the appropriate lettered structures (A–E).

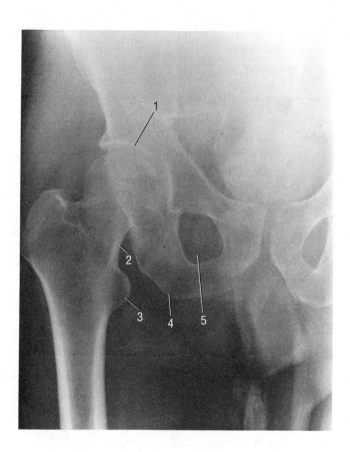

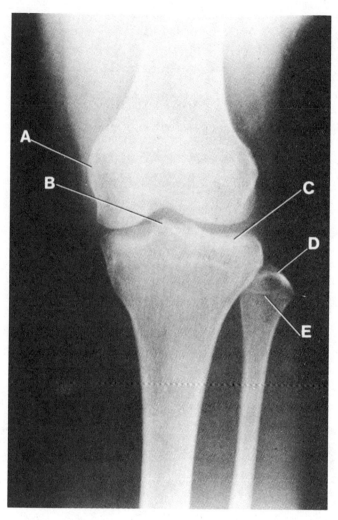

1. Structure 1.
2. Structure 2.
3. Structure 3.
4. Structure 4.
5. Structure 5.

A. Neck of femur. 2
B. Ischial tuberosity. 4
C. Acetabulum. 1
D. Obturator foramen. 5
E. Lesser trochanter. 3

6. Head of fibula. = C
7. Medial condyle of femur. A
8. Site of epiphyseal line. E
9. Lateral condyle of tibia. C
10. Intercondylar eminence. B

Match the numbered structures shown in the lateral radiograph of the ankle region with the appropriate lettered structures (A–F) listed below.

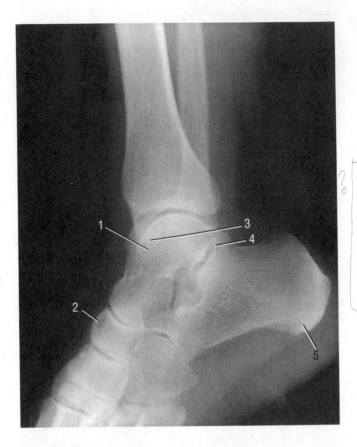

11. Structure 1.   A. Calcaneum. 5
12. Structure 2.   B. Lateral malleolus.
13. Structure 3.   C. Body of talus.
14. Structure 4.   D. Navicular.
15. Structure 5.   E. Medial malleolus.
                          F. None of the above.

Match each statement on the left with the correct answer on the right. Answers may be used more than once.

16. The ____E____ limits flexion   A. Iliofemoral ligament.
    of the thigh at the hip joint   B. Anterior abdominal
    (with knee extended).   wall.
17. The ____D____ limits   C. Ischiofemoral
    abduction of the thigh at the   ligament.
    hip joint.   D. Pubofemoral
18. The ____A____ limits exten-   ligament.
    sion of the thigh at the hip   E. None of the above.
    joint.
19. The ____B____ limits flexion
    of the thigh at the hip joint
    (with knee flexed).

Match the muscles on the left with their appropriate nerve supply on the right.

20. Long head of the biceps   A. Obturator nerve.
    femoris. C   B. Femoral nerve.
21. Gracilis. A   C. Tibial portion of sciatic
22. Gluteus maximus. E   nerve.
23. Sartorius. B   D. Common peroneal
24. Hamstring portion of   portion of sciatic nerve.
    adductor magnus. C   E. Inferior gluteal nerve.
25. Adductor longus. A

Match the skin areas listed on the left with their appropriate lymphatic drainage on the right.

26. A malignant melanoma   A. Horizontal group of
    (skin cancer) in the   superficial inguinal
    center of the buttock.   nodes.
27. Nail bed of the big toe.   B. Vertical group of super-
28. Skin of the lateral side   ficial inguinal nodes.
    of the foot.   C. Popliteal nodes.
29. Skin of the medial side   D. Internal iliac nodes.
    of the knee.   E. None of the above.
30. Skin around the anus.

In each of the following questions, answer:

    A. If only (1) is correct
    B. If only (2) is correct
    C. If both (1) and (2) are correct
    D. If neither (1) nor (2) is correct

31. (1) The sciatic nerve enters the gluteal region through
    the lesser sciatic foramen.
    (2) The common peroneal nerve can be palpated as it
    winds around the lateral side of the neck of the
    fibula.
32. (1) The pulsations of the dorsalis pedis artery can be
    felt on the dorsum of the foot between the tendons of
    the extensor digitorum longus and the extensor hallu-
    cis longus.
    (2) The pulsations of the femoral artery may be felt at
    the midpoint of a line joining the pubic symphysis to
    the anterior superior iliac spine.
33. After ligation of the femoral artery above the origin of
    the profunda femoris artery, blood would still reach
    the leg via the:
    (1) Inferior gluteal artery.
    (2) Perforating arteries.
34. The calcaneum of the foot articulates with:
    (1) The medial cuneiform.
    (2) The fibula.
35. The iliotibial tract receives the tendinous insertions of
    the:
    (1) Pectineus muscle.
    (2) Upper two-thirds of gluteus maximus muscle.

36. Muscle(s) of the thigh that cross(es) two joints:
    (1) Rectus femoris. T
    (2) Long head of biceps femoris. T
37. The popliteal artery:
    (1) Lies superficial to the popliteal vein. F
    (2) Terminates by dividing into the peroneal and tibial arteries. T
38. Tendon(s) of the flexor digitorum longus muscle:
    (1) Receive the insertion of quadratus plantae (flexor digitorum accessorius) muscle.
    (2) Pierce the tendons of flexor digitorum brevis muscle.
39. The gluteus maximus muscle:
    (1) Inserts into the lesser trochanter of the femur. F
    (2) Covers the posterior surface of the sciatic nerve. T
40. The adductor canal:
    (1) Contains the nerve to vastus medialis and the saphenous nerve. T
    (2) Begins at the apex of the femoral triangle. T

Select the **best** response.

41. If the dorsalis pedis artery is severed just proximal to its medial and lateral tarsal branches, blood may reach the dorsum of the foot through which of the following vessels?
    A. Peroneal artery.
    B. Posterior tibial artery.
    C. Medial plantar artery.
    D. Lateral plantar artery.
    E. All of the above.
42. In the child, the chief arterial supply to the head of the femur is derived from the:
    A. Obturator artery
    B. Internal pudendal artery.
    C. Branches from the medial and lateral circumflex femoral arteries.
    D. Deep circumflex iliac artery.
    E. Superficial circumflex iliac artery.
43. The femoral ring is the:
    A. Opening in the deep fascia of the thigh for the great saphenous vein.
    B. Opening in the adductor magnus for the femoral artery.
    C. Proximal opening in the femoral canal.
    D. Compartment in the femoral sheath for the femoral artery.
    E. Compartment in the femoral sheath for the femoral nerve.
44. Following a football injury, an orthopedic surgeon noted that the right tibia could be moved anteriorly with excessive freedom when the knee was flexed. Which ligament is most likely to be torn?
    A. Lateral collateral ligament.
    B. Posterior cruciate ligament.
    C. Anterior cruciate ligament.
    D. Medial collateral ligament.
    E. Patellar ligament.
45. A femoral hernia has the following characteristics **except:**
    A. It is more common in women.
    B. The swelling occurs below and lateral to the pubic tubercle.
    C. It descends through the femoral canal.
    D. Its neck is related immediately laterally to the femoral artery.
    E. Its neck is related medially to the sharp edge of the lacunar ligament.
46. Which ligament contributes to the formation of the greater sciatic foramen?
    A. Dorsal sacroiliac.
    B. Interosseous sacroiliac.
    C. Uterosacral.
    D. Sacrospinous. & sacrotuberous
    E. Arcuate.
47. The gastrocnemius and soleus muscles have all the following features in common **except:**
    A. Tibial nerve supply.
    B. Are found in the posterior compartment of the leg.
    C. Arise from the femoral condyles and flex the knee joint.
    D. Insert via the tendo calcaneus.
    E. Plantar flex the ankle joint.
48. All the following statements about the sartorius muscle are correct **except:**
    A. It flexes the leg at the knee joint.
    B. It flexes the thigh at the hip joint.
    C. It laterally rotates the thigh at the hip joint.
    D. It adducts the thigh at the hip joint.
    E. It attaches to the anterior superior iliac spine.
49. In order to lift the left foot off the ground while walking, which of the following muscles plays an important role?
    A. Left gluteus medius.
    B. Left gluteus maximus.
    C. Right adductor longus.
    D. Right gluteus medius.
    E. None of the above.
50. Rupture of the tendo calcaneus results in inability to:
    A. Dorsiflex the foot.
    B. Evert the foot.
    C. Invert the foot.
    D. Plantar flex the foot.
    E. None of the above.
51. A sprained ankle that results from excessive eversion would most likely demonstrate a torn:
    A. Talofibular ligament.
    B. Tendo calcaneus.
    C. Deltoid ligament.

D. Interosseous ligament.

E. Peroneal retinaculum.

52. Which nerve provides the most extensive cutaneous supply to the medial side of the foot?

A. Medial cutaneous nerve of the thigh.

B. Lateral plantar nerve.

C. Saphenous nerve.

D. Sural nerve.

E. Superficial peroneal nerve.

53. If the foot is permanently dorsiflexed and everted, which nerve might be injured?

A. Deep peroneal nerve.

B. Superficial peroneal nerve.

C. Common peroneal nerve.

D. Tibial nerve.

E. Obturator nerve.

54. The femoral nerve arises from the following segments of the spinal cord:

A. L2 and 3.

B. L4, 5; S1, 2, and 3.

C. L2, 3, and 4.

D. L1 and 2.

E. L5; S1, 2, and 3.

55. The dermatome present over the lateral side of the foot is:

A. S5.

B. L3.

C. S1.

D. L4.

E. L5.

56. The following structures are transmitted through the lesser sciatic foramen **except:**

A. The tendon of obturator internus muscle.

B. Internal pudendal vessels.

C. The nerve to obturator internus muscle.

D. Pudendal nerve.

E. Inferior gluteal artery.

57. The femoral sheath is formed by the following layer(s) of fascia:

A. Pectineus fascia.

B. Fascia iliaca and fascia transversalis.

C. Fascia lata and the membranous layer of superficial fascia.

D. Psoas fascia and the fatty layer of superficial fascia.

E. Processus vaginalis.

58. The following facts concerning the great saphenous vein are true **except:**

A. Arises on the dorsum of the foot.

B. Enters the leg by passing anterior to the medial malleolus.

C. Drains into the femoral vein about 1½ in. (3.8 cm) below and lateral to the pubic tubercle.

D. Is accompanied by the saphenous nerve.

E. Has no communication with the deep veins of the leg.

59. Which of the following muscles everts the foot?

A. Tibialis posterior.

B. Flexor hallucis longus.

C. Peroneus longus.

D. Tibialis anterior.

E. Flexor digitorum longus.

60. The lateral meniscus of the knee joint:

A. Has a thick inner border.

B. Is strongly attached around its circumference to the tibia.

C. Is more frequently torn than the medial meniscus.

D. Is strongly attached to the lateral collateral ligament.

E. Its anterior horn is attached to the tibia in front of the intercondylar eminence.

61. The muscles that are attached to the greater trochanter of the femur include the following **except:**

A. Gluteus medius.

B. Piriformis.

C. Gluteus minimus.

D. Iliopsoas.

E. Obturator externus.

62. A patient in the supine position with the hip and knee joints extended is asked to abduct the lower limb against resistance provided by the physician. This exercise tests which of the following muscles?

A. Semitendinosus.

B. Gluteus medius.

C. Pectineus.

D. Gracilis.

E. Semimembranosus.

63. The following facts concerning the obturator nerve are true **except:**

A. Originates from the lumbar plexus.

B. Enters the thigh immediately beneath the inguinal ligament.

C. Innervates the adductor muscles of the thigh.

D. Divides into an anterior and a posterior division.

E. Supplies the skin on the medial side of the thigh.

64. Injury to the common peroneal nerve would result in an:

A. Inability to invert the foot.

B. Inability to plantar flex the ankle.

C. Inability to evert the foot.

D. Inability to plantar flex the big toe.

E. Inability to feel skin sensation on the medial side of the leg.

65. After lesion of the tibial part of the sciatic nerve, some active flexion may still be possible at the knee joint. The muscles responsible for this remaining flexion include the:

A. Short head of the biceps femoris.

B. Gastrocnemius.

C. Plantaris.

D. Popliteus.

E. Long head of biceps femoris.

In the next section, study each case history. Select the **best** answer to each question following it.

A 69-year-old man was walking down a flight of steps when he slipped and fell. On impact his left foot hit the ground and was forcibly inverted and medially rotated. He stated that he could feel something give on the lateral side of the dorsum of the foot. Although shocked by the fall and the resulting foot pain, he was able to walk home. Within 2 hours the left foot and ankle had swollen considerably. The swelling made examination by his physician difficult, but it was clear that movements of the ankle were possible although inversion was extremely painful. On standing the patient could bear his body weight without too much discomfort. The physician ordered an anteroposterior and a lateral radiograph of the ankle region; the results showed nothing abnormal.

66. This patient was diagnosed as having an acute sprain of the lateral ankle. Which ligament is most likely to have been damaged?
    A. The deltoid ligament.
    B. The anterior talofibular ligament.
    C. Plantar calcaneonavicular ligament.
    D. Anterior ligament of inferior tibiofibular joint.
    E. Posterior talofibular ligament.

A 29-year-old woman was involved in an automobile accident. Her car skidded into a tree, and she was thrown forward, striking her right knee on the dashboard. On examination in the emergency department, she was found to have a posterior fracture dislocation of her right hip joint.

67. What bone or bones are likely to have been fractured?
    A. Greater tuberosity of the femur.
    B. Floor of the acetabulum.
    C. Posterior rim of the acetabulum and fracture of the femoral head.
    D. Lesser trochanter of the femur.
    E. Anterior inferior iliac spine.
68. What anatomical structure is likely to have been compromised in a hip dislocation of this type?
    A. The tendon of obturator internus.
    B. Sciatic nerve.
    C. The quadratus femoris muscle.
    D. The gluteus maximus muscle.
    E. Superior gluteal nerve.

Answers and Explanations

1. C
2. A
3. E
4. B
5. D
6. D
7. A
8. E
9. C
10. B
11. C
12. D

13. E
14. B
15. A
16. E
17. D
18. A
19. B
20. C
21. A
22. E
23. B
24. C
25. A
26. A   Remember that the skin of the back below the level of the iliac crests drains into the horizontal group of superficial inguinal nodes.
27. B
28. C
29. D
30. A   Remember that the mucous membrane of the lower half of the anal canal has the same lymphatic drainage.
31. B   The sciatic nerve enters the gluteal region through the greater sciatic foramen.
32. C   These are important answers.
33. C   The inferior gluteal artery is a branch of the internal iliac artery, and the perforating arteries are branches of the profunda femoris. The latter arteries also anastomose with arteries around the knee joint.
34. D   The calcaneum articulates with the talus and the cuboid.
35. B   The pectineus muscle is inserted into the upper part of the shaft of the femur.
36. C
37. D   The popliteal artery lies deep to the popliteal vein against the femur. The popliteal artery terminates by dividing into the anterior and posterior tibial arteries.
38. C
39. B   The gluteus maximus muscle is mainly inserted into the iliotibial tract, and only a small part is inserted into the back of the upper part of the shaft of the femur.
40. C
41. E
42. A   The nutrient artery, a branch of the obturator artery, reaches the femoral head in a child along the ligament of the head and enters the bone at the fovea capitis. The femoral head is separated from the arteries supplying the neck of the femur by the epiphyseal cartilage.
43. C
44. C   The anterior cruciate ligament is attached to the tibia in the anterior intercondylar area and passes upward, backward, and laterally to be attached to the lateral femoral condyle.
45. D   The neck of a femoral hernia is related immediately laterally to the femoral vein.
46. D   Both the sacrospinous and the sacrotuberous ligaments contribute to the formation of the greater sciatic foramen.
47. C   The soleus does not arise from the femoral condyles.

Lower Limb

48. D   The sartorius muscle is an abductor of the hip joint.
49. D   The right gluteus medius, together with the right gluteus minimus muscles, tilt the pelvis so that the left lower limb is raised, permitting the left foot to be advanced forward clear of the ground.
50. D   The soleus and gastrocnemius muscles are attached to the calcaneum via the tendo calcaneus. These muscles plantar flex the ankle joint. The plantaris is only a weak plantar flexor muscle.
51. C
52. C   The saphenous nerve supplies the skin of the medial border of the foot as far as the ball of the big toe. The medial side of the big toe is supplied by the superficial peroneal nerve.
53. D
54. C
55. C
56. E   The inferior gluteal artery emerges from the greater sciatic foramen.
57. B
58. E   The great saphenous vein has numerous communications with the deep veins of the leg through the valved perforating veins.
59. C

60. E
61. D   The iliopsoas muscle is inserted into the lesser trochanter of the femur.
62. B   The gluteus medius muscle is a strong abductor of the hip joint.
63. B   The obturator nerve enters the thigh through the obturator canal.
64. C   The plantar flexors of the ankle joint and the invertors of the foot are mainly supplied by the tibial nerve. The skin on the medial side of the leg receives its nerve supply from the saphenous nerve, a branch of the femoral nerve.
65. A   The short head of the biceps femoris is supplied by the common peroneal nerve.
66. B   The anterior talofibular ligament is very commonly damaged, more so than those ligaments on the medial side of the joint. Excessive inversion of the foot with plantar flexion of the ankle or attempted medial rotation of the ankle is the common cause.
67. C   The head of the femur is driven backward by the force of the accident through the posterior rim of the acetabulum that is fractured; the head of the femur may also be fractured.
68. B   The sciatic nerve is commonly damaged in this type of fracture dislocation.

Head and Neck

The head and neck is the most complicated area of gross anatomy to understand and learn. However, from the practical point of view and for examination purposes, only certain areas have to be committed to memory. For example, the distribution of the cranial nerves is most important, whereas the precise and detailed relationships of different structures in the neck can be briefly reviewed.

It is suggested that the head and neck be reviewed in the following order:

1. A brief overview of the skull, including the mandible and the temporomandibular joint. (The availability of a dry skull would be of great help.)
2. The important muscles of the neck should be reviewed; know the boundaries of the triangles and their associated fasciae.
3. A brief overview of the muscles of the scalp and face should be undertaken; they are all supplied by the seventh cranial nerve.
4. A brief review of the blood vessels and lymphatic drainage.
5. A detailed overview of the cranial nerves and their distribution.
6. Review the important branches of the cervical plexus of nerves.
7. Understand the distribution of the autonomic nervous system.
8. Review the parts of the digestive and respiratory systems in the head and neck.
9. Review the pituitary, thyroid, and parathyroid glands.
10. Review the gross parts of the eye and the ear.

To assist students in the review process, tables have been used extensively.

Skull

Newborn Skull
Fontanelles
The **anterior fontanelle** is diamond-shaped and lies between the two halves of the frontal bone and the two parietal bones. The fibrous membrane forming the floor of the anterior fontanelle is replaced by bone and is closed by 18 months of age. The heart rate, the intracranial pressure, and the degree of hydration can be examined by palpating the anterior fontanelle. The **posterior fontanelle** is triangular in shape and lies between the two parietal bones and the occipital bone. The posterior fontanelle is usually closed by the end of the first year.

Tympanic Part of the Temporal Bone
The tympanic part is merely a C-shaped ring at birth, compared with a C-shaped curved plate in the adult. This means that the tympanic membrane is much nearer the surface in the newborn.

Mastoid Process
Not present at birth, the mastoid process develops during the first 2 years.

Adult Skull
Anterior View
The frontal bone, or forehead bone, curves downward to make the upper margin of the orbits (Fig. 7-1). The **superciliary arches** can be seen on either side.

The **orbital margins** are bounded by the frontal bone superiorly, the zygomatic bone laterally, the maxilla inferiorly, and the processes of the maxilla and the frontal bone medially.

Within the **frontal bone,** just above the orbital margins, are two hollow spaces lined with mucous membrane called the **frontal air sinuses.** These communicate with the nasal cavity and serve as voice resonators.

The two **nasal bones** form the bridge of the nose. Their lower borders, with the maxillae, make the **anterior nasal aperture.** The nasal cavity is divided in two by the bony nasal septum, which is largely formed by the **vomer.** The **superior and middle conchae** are shelves of bone that project into the nasal cavity from the **ethmoid** on each side; the **inferior conchae** are separate bones.

The two **maxillae** form the upper jaw (see Fig. 7-1), the anterior part of the hard palate, part of the lateral walls of

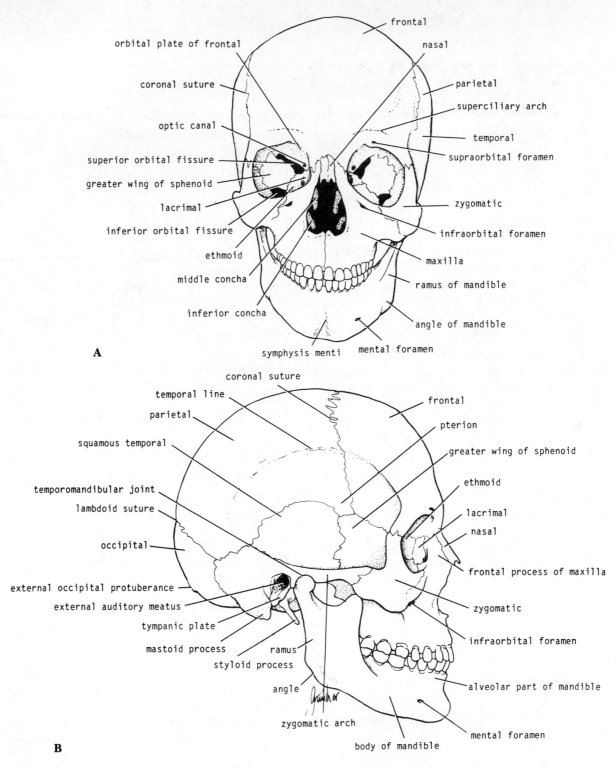

Fig. 7-1.   A. Anterior aspect of skull. **B.** Lateral aspect of skull.

the nasal cavities, and part of the floors of the orbital cavities. The two bones meet in the midline at the **intermaxillary suture** and form the lower margin of the nasal aperture. Below the orbit the maxilla is perforated by the **infraorbital foramen.** The alveolar process projects downward, and together with the opposite side forms the **alveolar arch,** which carries the upper teeth. Within each maxilla is a large pyramid-shaped cavity lined with mucous membrane called the **maxillary sinus.** This communicates with the nasal cavity and serves as a voice resonator.

The **zygomatic bone** forms the prominence of the cheek and part of the lateral wall and floor of the orbital cavity. Medially, it articulates with the maxilla, and laterally, it articulates with the zygomatic process of the temporal bone to form the **zygomatic arch.**

Lateral View

The **frontal bone** forms the anterior part of the side of the skull and articulates with the parietal bone at the **coronal suture** (see Fig. 7-1).

The **parietal bones** form the sides and roof of the cranium and articulate with each other in the midline at the **sagittal suture.** Behind, they articulate with the occipital bone at the **lambdoid suture.**

The skull is completed at the side by the squamous part of the **occipital bone;** parts of the **temporal bone,** namely, the **squamous, tympanic, mastoid process, styloid process,** and **zygomatic process;** and the **greater wing of the sphenoid.** Note the position of the **external auditory meatus.** The ramus and body of the mandible lie inferiorly.

Note that the thinnest part of the lateral wall of the skull is where the anteroinferior corner of the parietal bone articulates with the greater wing of the sphenoid; this point is referred to as the **pterion.** Clinically, the pterion is a very important area because it overlies the anterior division of the **middle meningeal artery and vein.**

Inferior View

If the mandible is discarded, the anterior part of the inferior aspect of the skull is seen to be formed by the **hard palate.** The **palatal processes of the maxillae** and the **horizontal plates of the palatine bones** can be identified.

Above the posterior edge of the hard palate are the **choanae** (posterior nasal apertures). These are separated from each other by the posterior margin of the **vomer** and are bounded laterally by the **medial pterygoid plates** of the sphenoid bone. The inferior end of the medial pterygoid plate is prolonged as a curved spike of bone, the **pterygoid hamulus.**

The following bony structures should be identified: the **lateral pterygoid plate,** the **infratemporal fossa,** the **tuberosity of the maxilla,** the **petrous part of the temporal bone,** the **styloid process,** the **mastoid process,** the **occipital condyles,** the **external occipital protuberance,**

and the **superior nuchal line.** Identify also the **mandibular fossa of the temporal bone** and the **articular tubercle,** which form the upper articular surfaces for the temporomandibular joint.

The following fissures or foramina should also be located because they allow passage of the cranial nerves and other important structures from the skull: the **foramen ovale,** the **foramen spinosum,** the **carotid canal,** the **jugular foramen,** the **foramen lacerum,** the **stylomastoid foramen,** and the **hypoglossal canal.** In addition, note the important **foramen magnum** and the opening of the bony part of the **auditory tube.**

Superior View of the Base of the Skull

The base of the skull is divided up into three cranial fossae: anterior, middle, and posterior (Fig. 7-2). The anterior cranial fossa is separated from the middle cranial fossa by the lesser wing of the sphenoid, and the middle cranial fossa is separated from the posterior cranial fossa by the petrous part of the temporal bone.

Anterior Cranial Fossa

The anterior cranial fossa lodges the frontal lobes of the cerebral hemispheres. Note the perforations of the **cribriform plate of the ethmoid** that transmit the olfactory nerves.

Middle Cranial Fossa

The lateral parts of the middle cranial fossa lodge the temporal lobes of the cerebral hemispheres; the middle part is raised and formed by the **body of the sphenoid bone** (see Fig. 7-2). In front is the **sulcus chiasmatis,** which is related to the optic chiasma and leads laterally to the **optic canal** on each side. The optic canal transmits the optic nerve and the ophthalmic artery. Posterior to the sulcus is an elevation, the **tuberculum sellae.** Behind the elevation is a deep depression, the **sella turcica,** which lodges the **hypophysis cerebri.** The sella turcica is bounded posteriorly by a square plate of bone called the **dorsum sellae.** The superior angles of the dorsum sellae have two tubercles, the **posterior clinoid processes,** which give attachment to the tentorium cerebelli. The cavernous sinus is directly related to the side of the body of the sphenoid.

Note that the sphenoid bone resembles a bat having a centrally placed **body** with **greater** and **lesser wings** that are outstretched on each side. The body of the sphenoid contains the **sphenoid air sinuses,** which are lined with mucous membrane and communicate with the nasal cavity; they serve as voice resonators.

The **superior orbital fissure** is a slitlike opening between the lesser and greater wings of the sphenoid. It transmits the lacrimal, the frontal, the trochlear, the oculomotor, the nasociliary, and the abducent nerves, together with the superior ophthalmic vein.

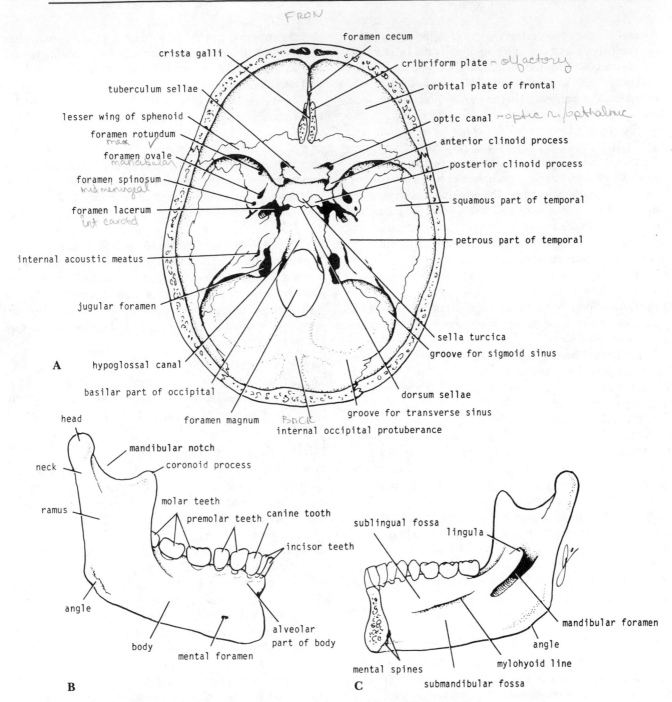

Fig. 7-2. **A.** Internal surface of base of skull. **B.** Lateral aspect of mandible. **C.** Medial aspect of mandible.

The **foramen rotundum,** which is situated behind the medial end of the superior orbital fissure, perforates the greater wing of the sphenoid (see Fig. 7-2) and transmits the maxillary division of the trigeminal nerve.

The **foramen ovale** lies posterolateral to the foramen rotundum (see Fig. 7-2). It perforates the greater wing of the sphenoid and transmits the sensory and motor roots of the mandibular division of the trigeminal nerve. It also transmits the lesser petrosal nerve (see p. 242).

The **foramen spinosum** is small and lies posterolateral to the foramen ovale (see Fig. 7-2). It perforates the greater wing of the sphenoid and transmits the middle meningeal artery.

The **foramen lacerum** is large and irregular in shape and lies between the apex of the petrous part of the temporal bone and the sphenoid bone (see Fig. 7-2). It is largely filled by cartilage and fibrous tissue. It allows the passage of the internal carotid artery from the carotid canal into the cranial cavity.

Posterior Cranial Fossa

The posterior cranial fossa is very deep and lodges parts of the hindbrain, namely, the cerebellum, the pons, and the medulla oblongata. The **foramen magnum** (see Fig. 7-2) occupies the central area of the floor and transmits the medulla oblongata, its meninges, the ascending spinal parts of the accessory nerves, and the two vertebral arteries.

The **hypoglossal canal** is situated close to the anterolateral boundary of the foramen magnum; it transmits the hypoglossal nerve.

The **jugular foramen** lies between the petrous part of the temporal bone and the condylar part of the occipital bone. It transmits the glossopharyngeal, the vagus, and the accessory nerves. It is here that the sigmoid sinus leaves the skull to become the internal jugular vein.

The **internal acoustic meatus** pierces the posterior surface of the petrous part of the temporal bone. It transmits the vestibulocochlear nerve and the facial nerve (see Fig. 7-2).

Table 7-1 provides a summary of the more important openings in the base of the skull and the structures that pass through them.

Mandible

The mandible, or lower jaw, is the largest and strongest bone of the face (see Fig. 7-2). It articulates with the skull at the **temporomandibular joint.** The mandible consists of a horseshoe-shaped **body** and a pair of **rami.** The body of the mandible meets the ramus on each side at the **angle.**

The **body of the mandible,** on its external surface in the midline, has a faint ridge, the **symphysis menti.** The **mental foramen** can be seen below the second premolar tooth (see Fig. 7-2); it transmits one of the terminal branches of the inferior alveolar nerve.

On the medial surface of the body of the mandible is the **submandibular fossa** for the submandibular salivary gland. In front of this fossa is the **sublingual fossa** for the sublingual salivary gland. Between the two fossae runs an oblique ridge called the **mylohyoid line** (see Fig. 7-2) for the attachment of the mylohyoid muscle.

The upper part of the body is called the **alveolar part;** in the adult it contains 16 sockets for the roots of the lower teeth.

The **ramus of the mandible** is vertical and has an anterior **coronoid process** and a posterior **condyloid**

Table 7-1 Summary of the More Important Openings in the Base of the Skull and the Structures that Pass Through Them

Opening in skull	Bone of skull	Structures transmitted
Anterior Cranial Fossa		
Perforations in cribriform plate	Ethmoid	Olfactory nerves
Middle Cranial Fossa		
Optic canal	Lesser wing of sphenoid	Optic nerve, ophthalmic artery
Superior orbital fissure	Between lesser and greater wings of sphenoid	Lacrimal, frontal, trochlear, oculomotor, nasociliary, and abducent nerves; superior ophthalmic vein
Foramen rotundum	Greater wing of sphenoid	Maxillary division of the trigeminal nerve
Foramen ovale	Greater wing of sphenoid	Mandibular division of the trigeminal nerve, lesser petrosal nerve
Foramen spinosum	Greater wing of sphenoid	Middle meningeal artery
Foramen lacerum	Between petrous part of temporal and sphenoid	Internal carotid artery
Posterior Cranial Fossa		
Foramen magnum	Occipital	Medulla oblongata, spinal part of accessory nerve, and right and left vertebral arteries
Hypoglossal canal	Occipital	Hypoglossal nerve
Jugular foramen	Between petrous part of temporal and condylar part of occipital	Glossopharyngeal, vagus, and accessory nerves; sigmoid sinus becomes internal jugular vein
Internal acoustic meatus	Petrous part of temporal	Vestibulocochlear and facial nerves

process (see Fig. 7-2) or **head;** the two processes are separated by the **mandibular notch.** Below the condyloid process is a short **neck.** The **mandibular foramen** lies on the medial surface of the ramus; it transmits the inferior alveolar nerve. The foramen leads into the **mandibular canal** and opens onto the lateral surface at the **mental foramen.**

The condyloid process (or head) of the mandible articulates with the temporal bone at the temporomandibular joint.

Temporomandibular Joint
Articulation
Above, the mandibular fossa and the articular tubercle of the temporal bone. Below, the head of the mandible (Fig. 7-3).

Type
Synovial joint. The fibrocartilaginous disc divides the joint into upper and lower cavities.

Capsule
Encloses the joint.

Ligaments
Lateral Temporomandibular Ligament
This is attached above to the articular tubercle at the root of the zygomatic arch and below to the neck of the mandible. The fibers extend downward and backward. The ligament limits the posterior movement of the mandible.

Sphenomandibular Ligament
This is attached above to the spine of the sphenoid and below to the bony projection (lingula) of the mandibular foramen. Its function is unknown.

Stylomandibular Ligament
This is attached to the styloid process above and to the angle of the mandible below. Its function is unknown.

Articular Disc
This is an oval disc of fibrocartilage (see Fig. 7-3). It is attached in front to the tendon of the lateral pterygoid muscle and posteriorly by fibrous tissue to the head of the mandible. The circumference of the disc is attached to the capsule. The disc permits gliding movement in the upper part of the joint and hinge movement in the lower part of the joint.

Synovial Membrane
Lines the capsule in the upper and lower cavities.

Nerve Supply
Auriculotemporal and masseteric nerves, branches of the mandibular division of the trigeminal nerve.

Movements and the Muscles that Produce Movement (see Fig. 7-3)
Protrusion: The head of the mandible and the articular disc move forward in the upper part of the joint. Lateral pterygoid muscle.

Retraction: The head of the mandible and the articular disc move backward in the upper part of the joint. Posterior fibers of the temporalis muscle.

Depression of mandible (mouth is opened): The head of the mandible rotates on the undersurface of the articular disc around a horizontal axis. Digastrics, geniohyoids, and mylohyoid muscles.

Elevation of mandible (mouth is closed): The head of the mandible rotates on the undersurface of the articular disc. At the same time, the posterior fibers of the temporalis muscle pull back the head of the mandible, and the articular disc is pulled backward by fibroelastic tissue, which connects the disc to the temporal bone posteriorly. Temporalis, masseter, and medial pterygoid muscles.

Lateral chewing movements: Alternate protruding and retracting the mandible on each side.

The muscles of mastication are shown in Table 7-2. All the muscles are developed from the first pharyngeal arch and are therefore innervated by the mandibular division of the trigeminal nerve (fifth cranial nerve).

Muscles of the Neck

The superficial muscles of the side of the neck (Fig. 7-4) are shown in Table 7-3; the suprahyoid and infrahyoid muscles are shown in Table 7-4; the anterior and lateral vertebral muscles are shown in Table 7-5; and the suboccipital muscles are shown in Table 7-6.

Key Neck Muscles
Sternocleidomastoid Muscle
When the sternocleidomastoid muscle (see Fig. 7-4) contracts, it is easily seen as an oblique band crossing the side of the neck from the sternoclavicular joint to the mastoid process of the skull. It divides the neck into anterior and posterior triangles. The anterior border covers the carotid arteries, the internal jugular vein, the vagus nerve, and the deep cervical lymph nodes; it also overlaps the thyroid gland. The muscle is covered superficially by skin, fascia, the platysma muscle, and the external jugular vein. The great auricular nerve and the transverse cutaneous nerve (branches of the cervical plexus) also cross the outer surface of the muscle. The deep surface of the posterior border is related to the cervical plexus, the phrenic nerve, and the upper part of the brachial plexus. The origin, insertion, nerve supply, and action of the sternocleidomastoid muscle are shown in Table 7-3.

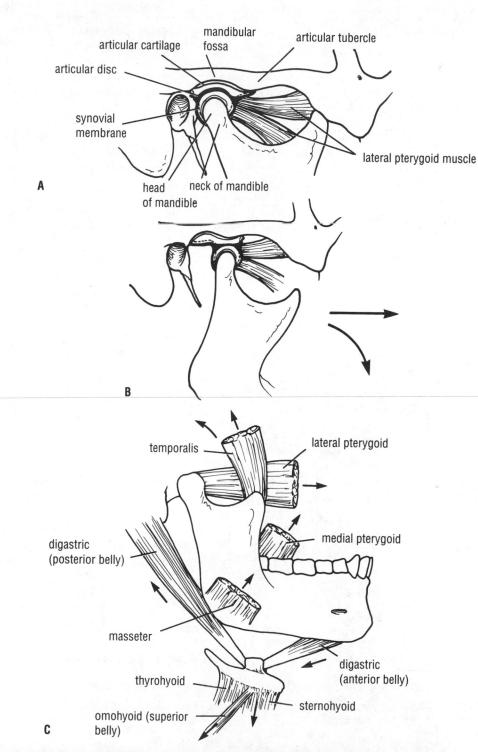

Fig. 7-3. Temporomandibular joint (**A**) with mouth closed and (**B**) with mouth open. Note the position of the head of the mandible and articular disc in relation to the articular tubercle in each case. **C.** The attachment of the muscles of mastication to the mandible; the arrows indicate the direction of their actions.

Table 7-2 Muscles of Mastication

Name of muscle	Origin	Insertion	Nerve supply	Action
Masseter	Zygomatic arch	Lateral surface of ramus of mandible	Mandibular division of trigeminal nerve	Raises mandible to occlude teeth in mastication
Temporalis	Floor of temporal fossa and covering fascia	Coronoid process of mandible	Mandibular division of trigeminal nerve	Anterior and superior fibers elevate the mandible, posterior fibers retract the mandible
Lateral pterygoid	Greater wing of sphenoid and lateral pterygoid plate	Neck of mandible and articular disc of temporomandibular joint	Mandibular division of trigeminal nerve	Pulls neck of mandible and disc forward
Medial pterygoid	Tuberosity of maxilla and lateral pterygoid plate	Medial surface of angle of mandible	Mandibular division of trigeminal nerve	Raises mandible

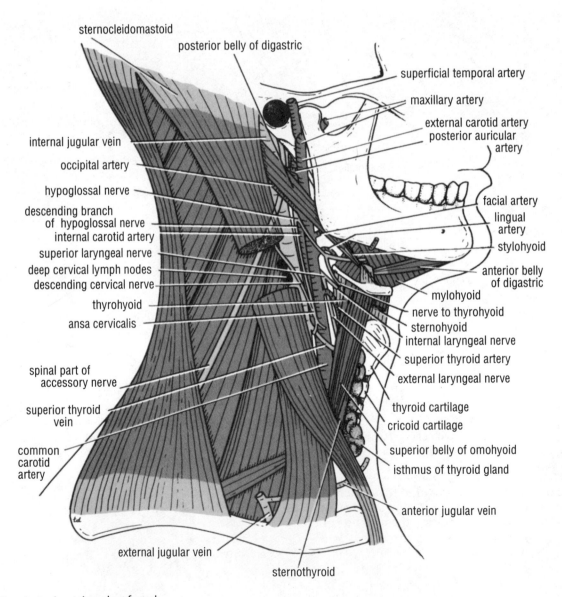

Fig. 7-4. Anterior triangle of neck.

Table 7-3 Superficial Muscles of the Side of the Neck

Name of muscle	Origin	Insertion	Nerve supply	Action
Platysma	Deep fascia of upper part of chest	Lower margin of body of mandible and angle of mouth	Facial nerve	Depresses mandible and angle of mouth
Trapezius	Occipital bone, ligamentum nuchae, spines of all thoracic vertebrae	Upper fibers into lateral third of clavicle, middle fibers and lower fibers into spine of scapula	Spinal part of accessory nerve and C3 and 4 spinal nerves	Upper fibers elevate the scapula, middle fibers pull scapula medially, lower fibers pull medial border of scapula downward
Sternocleidomastoid	Manubrium sterni and medial third of clavicle	Mastoid process of temporal bone and occipital bone	Spinal part of accessory nerve and C2 and 3 spinal nerves	Muscles of the two sides acting together extend the head and flex the neck, one muscle alone rotates head to opposite side

Scalenus Anterior Muscle

The scalenus anterior muscle is a key muscle to the understanding of the root of the neck (Fig. 7-5). It is deeply placed and descends almost vertically from the vertebral column to the first rib.

Important Relations

Anteriorly: Related to the carotid arteries, the vagus nerve, the internal jugular vein, and the deep cervical lymph nodes. The transverse cervical and suprascapular arteries and the prevertebral layer of deep cervical fascia bind the phrenic nerve to the muscle.

Posteriorly: Related to the pleura, the origin of the brachial plexus, and the second part of the subclavian artery. The scalenus medius muscle lies behind the muscle.

Medially: Related to the vertebral artery and vein and the sympathetic trunk. On the left side, the medial border is related to the thoracic duct.

Laterally: Related to the emerging branches of the cervical plexus, the roots of the brachial plexus, and the third part of the subclavian artery.

The origin, insertion, nerve supply, and action of the scalenus anterior muscle are shown in Table 7-5.

Cervical Fascia
Superficial Cervical Fascia

Superficial cervical fascia is a thin layer of connective tissue that unites the dermis of the skin to the deep fascia. It contains adipose tissue and encloses the platysma muscle. Also embedded in it are cutaneous nerves, superficial veins, and the superficial lymph nodes.

Deep Cervical Fascia

Deep cervical fascia supports the muscles, vessels, and viscera of the neck (Fig. 7-6). In certain areas it is condensed to form well-defined fibrous sheets called the **investing layer,** the **pretracheal layer,** and the **prevertebral layer.** It is also condensed to form the **carotid sheath** (see Fig. 7-6).

Investing Layer

The investing layer is a thick layer that encircles the neck, splitting to enclose the trapezius and sternocleidomastoid muscles (see Fig. 7-6). It is attached posteriorly to the ligamentum nuchae and above to the mandible, the zygomatic arch, and the superior nuchal line of the occipital bone. Below, it splits into two layers that are attached to the anterior and posterior borders of the manubrium and the clavicle. It also splits to enclose the parotid and the submandibular salivary glands.

Pretracheal Layer

The pretracheal layer is a thin layer that is attached above to the laryngeal cartilages (see Fig. 7-6). Below, it extends into the thorax to blend with the fibrous pericardium. Laterally, it is attached to the investing layer of deep cervical fascia beneath the sternocleidomastoid muscle. It surrounds the thyroid and parathyroid glands, forming a sheath for them, and encloses the infrahyoid muscles.

Prevertebral Layer

The prevertebral layer is a thick layer that passes like a septum across the neck behind the pharynx and the esophagus and in front of the prevertebral muscles and the vertebral column (see Fig. 7-6). It continues posteriorly to be attached to the ligamentum nuchae. It forms the fascial floor

Table 7-4   Suprahyoid Muscles and Infrahyoid Muscles

Name of muscle	Origin	Insertion	Nerve supply	Action
Suprahyoid muscles				
Digastric				
Posterior belly	Mastoid process of temporal bone	Intermediate tendon that is bound to hyoid bone	Facial nerve	Depresses mandible or elevates hyoid bone
Anterior belly	Lower border of mandible near midline	Intermediate tendon as above	Mylohyoid nerve, mandibular division of trigeminal nerve	Depresses mandible or elevates hyoid bone
Stylohyoid	Styloid process of temporal bone	Body of hyoid bone	Facial nerve	Elevates hyoid bone
Mylohyoid	Mylohyoid line on inner surface of body of mandible	Body of hyoid bone and raphe that extends from mandible to hyoid bone	Mandibular division of trigeminal nerve (inferior alveolar nerve)	Elevates floor of mouth and hyoid bone or depresses mandible
Geniohyoid	Inferior mental spine on back of symphysis menti of mandible	Body of hyoid bone	C1 through hypoglossal nerve	Elevates hyoid bone or depresses mandible
Infrahyoid muscles				
Sternohyoid	Manubrium sterni and medial end of clavicle	Body of hyoid bone	Ansa cervicalis (C1, 2, 3)	Depresses hyoid bone
Sternothyroid	Manubrium sterni	Oblique line on lamina of thyroid cartilage	Ansa cervicalis (C1, 2, 3)	Depresses larynx
Thyrohyoid	Oblique line on lamina of thyroid cartilage	Body of hyoid bone	C1 through hypoglossal nerve	Depresses hyoid bone or elevates larynx
Omohyoid				
Inferior belly	Upper margin of scapula and suprascapular ligament	Intermediate tendon bound to clavicle and first rib	Ansa cervicalis (C1, 2, 3)	Depresses hyoid bone
Superior belly	Body of hyoid bone	Intermediate tendon bound to clavicle and first rib	Ansa cervicalis (C1, 2, 3)	Depresses hyoid bone

of the posterior triangle and extends laterally over the first rib into the axilla to form the important **axillary sheath** (see p. 134).

Carotid Sheath

The carotid sheath is a local condensation of the prevertebral, the pretracheal, and the investing layers of deep fascia that surround the **common** and **internal carotid arteries**, the **internal jugular vein**, the **vagus nerve**, and the **deep cervical lymph nodes** (see Fig. 7-6).

Cervical Ligaments

The muscles of the head and neck are associated with important ligaments.

Stylohyoid Ligament

The stylohyoid ligament connects the styloid process to the lesser cornu of the hyoid bone (the ligament is the remains of the second pharyngeal arch).

Stylomandibular Ligament

The stylomandibular ligament connects the styloid process to the angle of the mandible (the ligament is a thickening of the investing layer of deep fascia).

Sphenomandibular Ligament

The sphenomandibular ligament connects the spine of the sphenoid bone to the lingula of the mandible (the ligament is the remains of the first pharyngeal arch).

Table 7-5 Anterior and Lateral Vertebral Muscles

Name of muscle	Origin	Insertion	Nerve supply	Action
Longus colli	Anterior surface of vertebrae between atlas and third thoracic vertebra	Same as origin	Anterior rami of cervical nerves	Flexes cervical part of vertebral column
Longus capitis	Transverse processes of lower cervical vertebrae	Occipital bone	Anterior rami of cervical nerves	Flexes head
Rectus capitis anterior	Front of lateral mass of atlas	Occipital bone	Cervical plexus	Flexes head
Rectus capitis lateralis	Transverse process of atlas	Occipital bone	Cervical plexus	Lateral flexion of head
Scalenus anterior	Transverse processes of third, fourth, fifth, and sixth cervical vertebrae	First rib	Cervical spinal nerves	Elevates first rib, laterally flexes and rotates cervical part of vertebral column
Scalenus medius	Transverse processes of upper six cervical vertebrae	First rib	Cervical spinal nerves	Elevates first rib, laterally flexes and rotates cervical part of vertebral column

Pterygomandibular Ligament

The pterygomandibular ligament connects the hamular process of the medial pterygoid plate to the posterior end of the mylohyoid line of the mandible. It gives attachment to the superior constrictor and the buccinator muscles.

Triangles of the Neck

The sternocleidomastoid muscle divides the neck into anterior and posterior triangles (see Fig. 7-6).

Anterior Triangle

The anterior triangle is bounded above by the body of the mandible, the sternocleidomastoid muscle posteriorly, and the midline anteriorly (see Fig. 7-6). It is further subdivided into the **carotid triangle,** the **digastric triangle,** the **submental triangle,** and the **muscular triangle.** The position and boundaries of these triangles are shown in Figure 7-6.

Posterior Triangle

The posterior triangle is bounded posteriorly by the trapezius muscle, anteriorly by the sternocleidomastoid muscle, and inferiorly by the clavicle (see Fig. 7-6). It is subdivided by the inferior belly of the omohyoid muscle into the large **occipital triangle** above and a small **supraclavicular triangle** below (see Fig. 7-6).

Suboccipital Triangle

The suboccipital triangle is deeply placed in the upper part of the neck just below the skull. It is covered by the semispinalis capitis muscle. The three sides of the triangle are formed by the rectus capitis posterior major and the superior and inferior oblique muscles. In the floor of the triangle are the posterior arch of the atlas, the vertebral artery, and the posterior ramus of the first cervical nerve.

Table 7-6 Suboccipital Muscles

Name of muscle	Origin	Insertion	Nerve supply	Action
Rectus capitis posterior major	Spine of axis	Occipital bone	Posterior ramus of first cervical nerve	Extension of head
Rectus capitis posterior minor	Posterior arch of atlas	Occipital bone	Posterior ramus of first cervical nerve	Extension of head
Obliquus capitis inferior	Spine of axis	Transverse process of atlas	Posterior ramus of first cervical nerve	Rotates face to same side
Obliquus capitis superior	Transverse process of atlas	Occipital bone	Posterior ramus of first cervical nerve	Extends head to same side

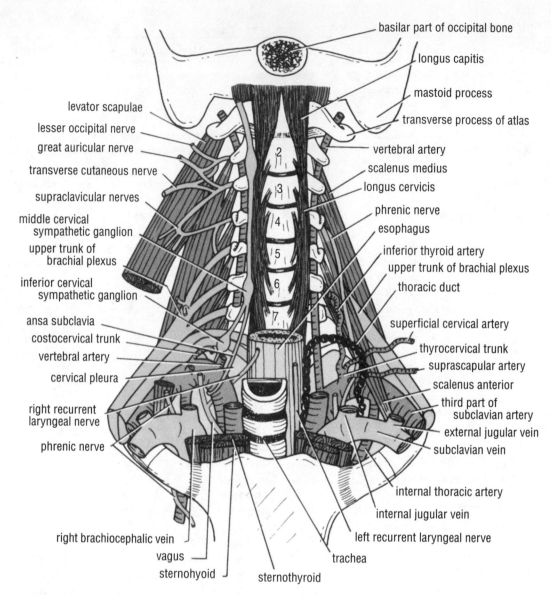

Fig. 7-5.   Prevertebral region and root of neck.

Muscles of the Scalp and Face

The muscles of the scalp and face are all derived from the second pharyngeal arch and are therefore supplied by the facial nerve (seventh cranial nerve).

Muscles of the Scalp

The scalp consists of five layers, the first three of which are bound together and move as a whole on the skull.

1. Skin
2. Connective tissue of superficial fascia
3. Aponeurosis of occipitofrontalis muscle
4. Loose connective tissue (permits layers 1–3 to move on layer 5)
5. Periosteum of skull bones

Note that the first letter of each layer combines to spell the acronym **SCALP.**

The origin, insertion, nerve supply, and action of the scalp muscles are shown in Table 7-7.

Muscles of Facial Expression

Situated in the superficial fascia, the muscles of facial expression arise from the skull and are inserted into the skin (Fig. 7-7). The muscles serve as sphincters and dilators to the orbit, nose, and mouth. They also modify the expression of the face.

The origin, insertion, nerve supply, and action of the muscles of facial expression are shown in Table 7-8. Try using the various muscles yourself while looking in the mirror. **It is unnecessary, for examination purposes, to know the precise attachments of these muscles.**

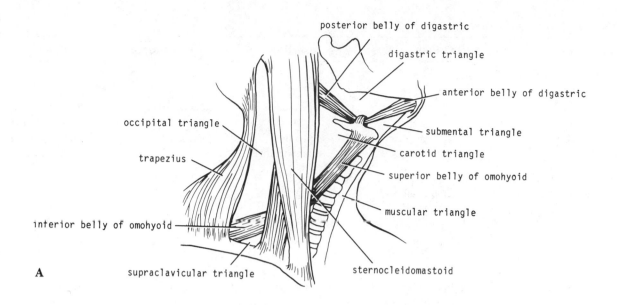

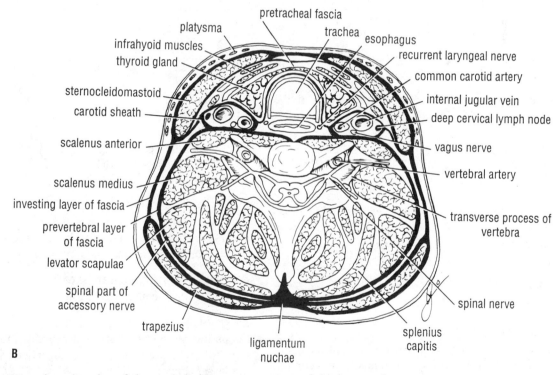

Fig. 7-6. **A.** Muscular triangles of the neck. **B.** The arrangement of the layers of deep fascia in the neck.

Arteries of the Head and Neck

Common Carotid Artery

The right common carotid artery arises from the brachiocephalic artery behind the right sternoclavicular joint (Fig. 7-8). The left artery arises from the arch of the aorta. The common carotid artery runs upward in the carotid sheath through the neck under cover of the anterior border of the sternocleidomastoid muscle. At the upper border of the thyroid cartilage it divides into the external and internal carotid arteries. In the carotid sheath, the artery is related laterally to the internal jugular vein; the vagus nerve lies between these two structures.

Carotid Sinus

The carotid sinus is a localized dilatation of the common carotid artery at its point of division (see Fig. 7-8). It is supplied by the glossopharyngeal nerve. The carotid sinus

Table 7-7   Muscles of the Scalp and the External Ear

Name of muscle	Origin	Insertion	Nerve supply	Action
Muscle of Scalp				
Occipitofrontalis				
Occipital bellies	Occipital bone	Epicranial aponeurosis	Facial nerve	Moves scalp on skull and raises eyebrows
Frontal bellies	Skin and fascia of eyebrow	Epicranial aponeurosis	Facial nerve	Moves scalp on skull and raises eyebrows
Muscles of External Ear				
Auricularis anterior	Epicranial aponeurosis	Auricle	Facial nerve	Small amount of auricular movement in some individuals
Auricularis superior	Epicranial aponeurosis	Auricle	Facial nerve	Small amount of auricular movement in some individuals
Auricularis posterior	Epicranial aponeurosis	Auricle	Facial nerve	Small amount of auricular movement in some individuals

is a pressoreceptor that assists in the regulation of the blood pressure.

Carotid Body

The carotid body is a small structure that lies posterior to the point of division of the common carotid artery. It is supplied by the glossopharyngeal nerve. The carotid body is a chemoreceptor, being sensitive to excess carbon dioxide and reduced oxygen tensions in the blood. It assists in regulating the heart and respiratory rates.

Branches

1. External carotid artery
2. Internal carotid artery

External Carotid Artery

The external carotid artery begins at the level of the upper border of the thyroid cartilage as one of the terminal branches of the common carotid artery (see Fig. 7-8). It ascends to terminate in the parotid salivary gland behind the neck of the mandible. Here, it divides into the superficial temporal and maxillary arteries.

As it emerges from under cover of the anterior border of the sternocleidomastoid muscle, the external carotid artery lies within the carotid triangle, where its pulsations can easily be felt.

Branches

1. Superior thyroid artery
2. Ascending pharyngeal artery
3. Lingual artery
4. Facial artery
5. Occipital artery
6. Posterior auricular artery
7. Superficial temporal artery
8. Maxillary artery

Superior Thyroid Artery

Curves downward to the upper pole of the thyroid gland (see Fig. 7-8). It is accompanied by the **external laryngeal nerve,** which supplies the cricothyroid muscle.

Ascending Pharyngeal Artery

Ascends along the pharyngeal wall and supplies it.

Lingual Artery

Loops upward and forward and supplies the tongue (see Fig. 7-8). The artery is crossed superficially by the **hypoglossal nerve.**

Facial Artery

Loops upward on the lateral surface of the pharynx close to the tonsil (see Fig. 7-8). It grooves the submandibular salivary gland, bends around the lower border of the mandible, and ascends over the face close to the anterior border of the masseter muscle. Its pulsations can be felt against the mandible. The artery ascends lateral to the mouth and terminates at the medial angle of the orbit.

Branches of the facial artery supply the tonsil, the submandibular salivary gland, and the muscles and skin of the face.

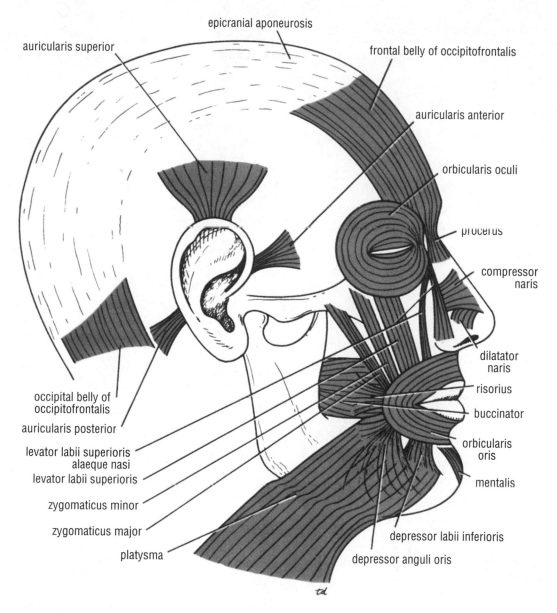

Fig. 7-7. Muscles of facial expression.

Occipital Artery
Supplies the back of the scalp (see Fig. 7-8).

Posterior Auricular Artery
Supplies the auricle and the scalp.

Superficial Temporal Artery
Ascends over the zygomatic arch, where it may be palpated just in front of the auricle (see Fig. 7-8). It is accompanied by the auriculotemporal nerve. The artery supplies the scalp.

Maxillary Artery
Runs forward medial to the neck of the mandible (see Fig. 7-8); it leaves the infratemporal fossa through the pterygo-

maxillary fissure to enter the pterygopalatine fossa. Here it splits up into branches that follow the branches of the maxillary division of the trigeminal nerve.

Branches
1. Branches supply the upper and lower jaws, the muscles of mastication, the nose, the palate, and the meninges.
2. The **middle meningeal artery** ascends between the roots of the auriculotemporal nerve to enter the skull through the foramen spinosum. The artery runs laterally within the skull and divides into anterior and posterior branches. The **anterior branch** is particularly important since it lies close to the motor area of the cerebral cortex. Accompanied by its vein, it grooves or tunnels through the upper part of the greater wing of

Table 7-8   Muscles of Facial Expression

Name of muscle	Origin	Insertion	Nerve supply	Action
Muscles of Eyelids				
Orbicularis oculi				
Orbital part	Frontal bone, maxillary bone, medial palpebral ligament	No interruption, forms concentric loops	Facial nerve	Pulls skin of forehead, temple, and cheek like a purse string: "screws up the eye"
Palpebral part	Medial palpebral ligament	Lateral palpebral ligament	Facial nerve	Closes eyelids
Lacrimal part	Lacrimal bone	Both eyelids	Facial nerve	Empties tears from lacrimal sac
Corrugator supercilii	Medial part of superciliary arch of frontal bone	Skin of eyebrow	Facial nerve	Pulls eyebrows medially
Muscles of Nostrils				
Compressor naris	Frontal process of maxilla	Via aponeurosis into muscle of opposite side	Facial nerve	Compresses nasal aperture
Dilator naris	Maxilla	Ala of nose	Facial nerve	Widens nasal aperture
Procerus	Nasal bone and lateral nasal cartilage	Skin between eyebrows	Facial nerve	Wrinkles skin at root of nose
Muscles of Lips and Cheeks				
Sphincter muscle of lips				
Orbicularis oris	Maxilla and mandible, skin of lips, some fibers from buccinator muscle	Surrounds orifice of mouth	Facial nerve	Compresses lips together
Dilator muscles of lips (levator labii superioris alaeque nasi, levator labii superioris, zygomaticus minor, zygomaticus major, levator anguli oris, risorius, depressor anguli oris, depressor labii inferioris, mentalis)	Bones and fascia around oral aperture	Substance of lips	Facial nerve	Separates lips
Buccinator	Alveolar margins of maxilla and mandible; pterygomandibular ligament	Fibers decussate and enter upper and lower lips	Facial nerve	Compresses cheeks and lips against teeth
Platysma	Deep fascia of upper part of chest	Lower margin of body of mandible and angle of mouth	Facial nerve	Depresses mandible and draws down angle of mouth

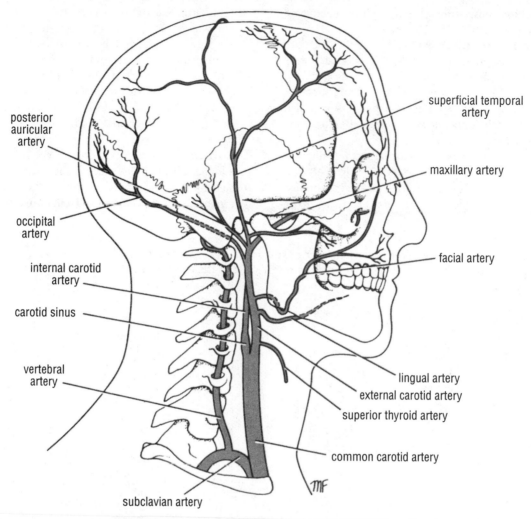

posterior
auricular
artery

superficial temporal
artery

maxillary artery

occipital
artery

facial artery

internal carotid
artery

carotid sinus

vertebral
artery

lingual artery

external carotid artery

superior thyroid artery

common carotid artery

subclavian artery

Fig. 7-8. Main arteries of the head and neck. Note that for clarity, the thyrocervical trunk, the costocervical trunk, and the internal thoracic artery, branches of the subclavian artery, have not been shown.

the sphenoid bone and the thin anterior inferior angle of the parietal bone. Here it is prone to damage following a blow to the lateral side of the head. Since the blood vessels lie between the meningeal layer of dura (dura proper) and the periosteal layer of dura (periosteum of skull), the resulting hemorrhage would be extradural.

Internal Carotid Artery

The internal carotid artery begins at the level of the upper border of the thyroid cartilage (see Fig. 7-8). It ascends the neck in the carotid sheath with the internal jugular vein and the vagus nerve and passes deep to the parotid salivary gland.

The internal carotid artery enters the cranial cavity by passing forward through the carotid canal in the petrous part of the temporal bone. It passes upward and then runs forward in the cavernous sinus. At the anterior end of the sinus, the artery bends upward through the roof and medial to the anterior clinoid process. The artery then inclines backward, lateral to the optic chiasma, to terminate by dividing into the anterior and middle cerebral arteries.

Branches

There are **no** branches in the neck. However, many important branches are given off in the skull.

1. The **ophthalmic artery** arises from the internal carotid artery as it emerges from the cavernous sinus. The ophthalmic artery passes forward into the orbital cavity through the optic canal, below and lateral to the optic nerve. It gives off the **central artery of the retina,** which enters the optic nerve and runs forward to enter the eyeball. The central artery is an end artery and is the only blood supply to the retina.

2. The **posterior communicating artery** runs backward to join the posterior cerebral artery.

3. The **anterior cerebral artery** is a terminal branch of the internal carotid artery (Fig. 7-9). It passes forward between the cerebral hemispheres and winds around the corpus callosum to supply the medial and superolateral surfaces of the cerebral hemisphere. It is joined to the artery of the opposite side by the **anterior communicating artery.**

4. The **middle cerebral artery** is the largest terminal branch of the internal carotid artery (see Fig. 7-9). It runs laterally in the lateral cerebral sulcus. It supplies the entire lateral surface of the cerebral hemisphere ex- cept for the narrow strip along the superolateral margin, which is supplied by the anterior cerebral artery, the occipital pole, and the inferolateral surface of the hemisphere (which are supplied by the posterior cerebral artery). This artery thus supplies all the motor area of the cerebral cortex except the leg area. It also gives off central branches that supply central masses of gray matter and the internal capsule.

Circulus Arteriosus (Circle of Willis)

Lies in the subarachnoid space at the base of the brain. It is formed by the anastomosis between the branches of the two internal carotid arteries and the two vertebral arteries

Fig. 7-9. Arteries and cranial nerves seen on inferior surface of brain. In order to show the course of the middle cerebral artery, the anterior pole of the left temporal lobe has been removed.

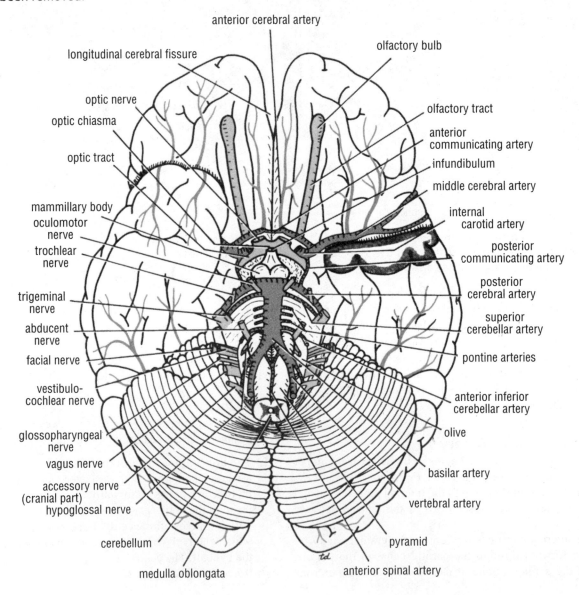

(see Fig. 7-9). The anterior communicating, the anterior cerebral, the internal carotid, the posterior communicating, the posterior cerebral, and the basilar (formed by the junction of the two vertebral arteries) are all arteries that contribute to the circle. Cortical and central branches arise from the circle and supply the brain.

Subclavian Arteries

Right Subclavian Artery

The right subclavian artery arises from the brachiocephalic artery, behind the right sternoclavicular joint (see Fig. 7-8). It arches upward and laterally over the pleura and between the scalenus anterior and medius muscles. At the outer border of the first rib it becomes the axillary artery.

Left Subclavian Artery

The left subclavian artery arises from the arch of the aorta in the thorax. It ascends to the root of the neck and then arches laterally in a manner similar to the right subclavian artery.

The scalenus anterior muscle is used to divide the subclavian artery into three parts.

First Part of the Subclavian Artery

The first part extends from the origin of the subclavian artery to the medial border of the scalenus anterior. This part of the artery gives off the vertebral artery, the thyrocervical trunk, and the internal thoracic artery.

Branches

1. The **vertebral artery** ascends the neck through the foramina in the transverse processes of the **upper** six cervical vertebrae (see Fig. 7-8). It passes medially above the posterior arch of the atlas and then ascends through the foramen magnum into the skull. On reaching the anterior surface of the medulla oblongata at the level of the lower border of the pons, it joins the vessel of the opposite side to form the basilar artery.

    The **basilar artery** (see Fig. 7-9) ascends in a groove on the anterior surface of the pons, giving off branches to the pons, cerebellum, and internal ear, and finally divides into the two posterior cerebral arteries.

    On each side, the **posterior cerebral artery** (see Fig. 7-9) curves laterally and backward around the midbrain. Cortical branches supply the inferolateral surfaces of the temporal lobe and the visual cortex on the lateral and medial surfaces of the occipital lobe.

    a. **Branches of the vertebral artery in the neck:**
       Spinal and muscular arteries
    b. **Branches of the vertebral artery in the skull**
       i. Meningeal
       ii. Anterior and posterior spinal
       iii. Posterior inferior cerebellar
       iv. Medullary arteries

2. **Thyrocervical trunk.** This short trunk gives off three terminal branches (see Fig. 7-5).
    a. The **inferior thyroid artery** ascends to the level of the cricoid cartilage and curves medially and downward behind the carotid sheath to reach the posterior surface of the thyroid gland. Here it is closely related to the **recurrent laryngeal nerve.** The inferior thyroid artery supplies the thyroid gland, the inferior parathyroid gland, and neighboring structures.
    b. The **superficial cervical artery** runs laterally over the phrenic nerve and crosses the brachial plexus.
    c. The **suprascapular artery** runs laterally over the phrenic nerve and crosses the brachial plexus. It follows the suprascapular nerve into the supraspinous fossa of the scapula and takes part in the anastomosis around the scapula.

3. The **internal thoracic artery** enters the thorax behind the first costal cartilage and in front of the pleura (see Fig. 7-5). It is crossed by the phrenic nerve and then descends vertically, a fingerbreadth lateral to the sternum. In the sixth intercostal space, it divides into the superior epigastric and musculophrenic arteries (see p. 24).

Second Part of the Subclavian Artery

The second part lies behind the scalenus anterior muscle.

Branch

The **costocervical trunk** runs backward over the dome of the pleura and divides into the **superior intercostal artery,** which supplies the first and second intercostal spaces, and the **deep cervical artery,** which supplies the deep muscles of the neck.

Third Part of the Subclavian Artery

The third part extends from the lateral border of the scalenus anterior (see Fig. 7-5) across the posterior triangle of the neck to the lateral border of the first rib. With the nerves of the brachial plexus, it is surrounded by the **axillary sheath** of fascia.

Branches

The third part of the subclavian artery usually has no branches. Occasionally, the superficial cervical arteries, the suprascapular arteries, or both arise from this part of the subclavian artery.

Veins of the Head and Neck

The veins of the head and neck may be divided into (1) the veins of the brain, the venous sinuses, the diploic veins, and the emissary veins; and (2) the veins of the scalp, the face, and the neck.

Veins of the Brain

The veins of the brain are thin walled and have no valves. They are made up of the **cerebral veins,** the **cerebellar veins,** and the **veins of the brain stem.** All these veins drain into neighboring venous sinuses.

Venous Sinuses

The venous sinuses are placed between the periosteal and meningeal layers of the dura mater (Fig. 7-10). They have thick fibrous walls but possess no valves. They receive tributaries from the brain, the skull bones, the orbit, and the internal ear.

Superior Sagittal Sinus

The superior sagittal sinus lies in the upper fixed border of the falx cerebri (see Fig. 7-10). It runs backward and usually becomes continuous with the right transverse sinus. The sinus communicates on each side with the **venous lacunae.** Numerous arachnoid villi and granulations project into the lacunae.

Inferior Sagittal Sinus

The inferior sagittal sinus lies in the lower free margin of the falx cerebri (see Fig. 7-10). It runs backward and joins the **great cerebral vein** to form the straight sinus.

Straight Sinus

The straight sinus lies at the junction of the falx cerebri with the tentorium cerebelli. Formed by the union of the inferior sagittal sinus with the great cerebral vein, it usually drains into the left transverse sinus.

Transverse Sinus

The right transverse sinus begins as a continuation of the superior sagittal sinus; the left transverse sinus is usually a continuation of the straight sinus (see Fig. 7-10). Each sinus lies in the lateral attached margin of the tentorium cerebelli. They end on each side by becoming the sigmoid sinus.

Sigmoid Sinuses

The sigmoid sinuses are a direct continuation of the transverse sinuses. Each sinus curves downward behind the

Fig. 7-10.   Interior of skull, showing dura mater and its contained venous sinuses. Note connections of veins of scalp and veins of face with venous sinuses.

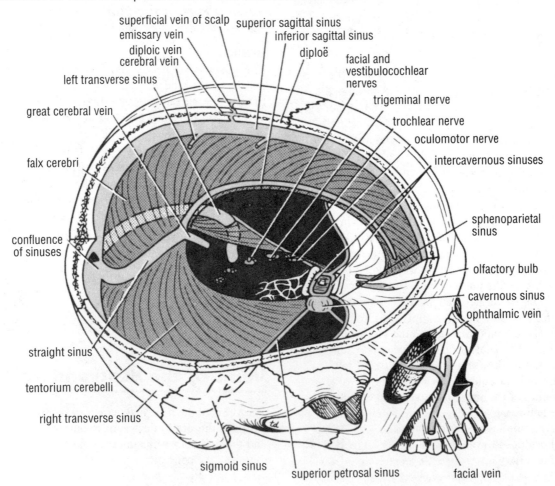

mastoid antrum and leaves the skull through the jugular foramen to become the internal jugular vein.

Occipital Sinus

The occipital sinus lies in the attached margin of the falx cerebelli. It communicates with the vertebral veins through the foramen magnum and with the transverse sinuses.

Cavernous Sinuses

Each cavernous sinus lies on the lateral side of the body of the sphenoid bone (see Fig. 7-28). Anteriorly, the sinus receives the inferior ophthalmic vein and the central vein of the retina. The sinus drains posteriorly into the transverse sinus through the superior petrosal sinus. **Intercavernous sinuses** connect the two cavernous sinuses through the sella turcica.

Important Structures Associated with the Cavernous Sinuses

1. The internal carotid artery and the sixth cranial nerve travel through it (see Fig. 7-28).
2. In the lateral wall, the third and fourth cranial nerves and the ophthalmic and maxillary divisions of the fifth cranial nerve (see Fig. 7-28).
3. The hypophysis cerebri lies medially in the sella turcica (see Fig. 7-28).
4. The veins of the face are connected with the cavernous sinus via the facial vein and inferior ophthalmic vein (important route for spread of infection from the face).

Superior and Inferior Petrosal Sinus

The petrosal sinuses run along the upper and lower borders of the petrous part of the temporal bone.

Diploic Veins

The diploic veins occupy channels within the bones of the vault of the skull.

Emissary Veins

The emissary veins are valveless veins that pass through the skull bones and connect the veins of the scalp to the venous sinuses (important route for spread of infection).

Veins of the Face and the Neck
Facial Vein

The facial vein is formed at the medial angle of the eye by the union of the **supraorbital** and **supratrochlear veins** (Fig. 7-11). It is connected through the ophthalmic veins with the cavernous sinus. The facial vein descends down the face with the facial artery and passes lateral to the mouth. It crosses the lower margin of the mandible and is joined by the anterior division of the retromandibular vein and drains into the internal jugular vein.

Superficial Temporal Vein

The superficial temporal vein is formed on the side of the scalp (see Fig. 7-11). It follows the superficial temporal

artery and the auriculotemporal nerve and enters the parotid salivary gland. Here it joins the maxillary vein to form the retromandibular vein.

Maxillary Vein

The maxillary vein is formed in the infratemporal fossa from the pterygoid venous plexus (see Fig. 7-11). The maxillary vein joins the superficial temporal vein to form the retromandibular vein.

Retromandibular Vein

The retromandibular vein is formed by the union of the superficial temporal and the maxillary veins (see Fig. 7-11). On leaving the parotid salivary gland, it divides into an anterior branch that joins the facial vein and a posterior branch that joins the **posterior auricular vein** to form the external jugular vein.

External Jugular Vein

The external jugular vein is formed behind the angle of the jaw by the union of the posterior auricular vein with the posterior division of the retromandibular vein (see Fig. 7-11). It descends across the sternocleidomastoid muscle, beneath the platysma muscle, and drains into the subclavian vein behind the middle of the clavicle.

Tributaries

1. **Posterior external jugular vein** from the back of the scalp
2. **Superficial cervical vein** from skin and fascia over the posterior triangle
3. **Suprascapular vein** from the suprascapular fossa
4. **Anterior jugular vein**

Anterior Jugular Vein

The anterior jugular vein starts just below the chin and descends in the neck close to the midline (see Fig. 7-11). Just above the sternum, it is joined to the opposite vein by the **jugular arch.** The anterior jugular vein joins the external jugular vein deep to the sternocleidomastoid muscle.

Internal Jugular Vein

The internal jugular is a large vein that drains blood from the brain, the face, the scalp, and the neck (see Fig. 7-11). It starts above as a continuation of the sigmoid venous sinus. It leaves the skull through the jugular foramen and descends through the neck in the carotid sheath lateral to the vagus nerve and the internal and common carotid arteries. It ends by joining the subclavian vein to form the brachiocephalic vein behind the medial end of the clavicle. It is closely related to the **deep cervical lymph nodes** throughout its course.

Tributaries

1. **Inferior petrosal sinus**
2. **Facial vein**

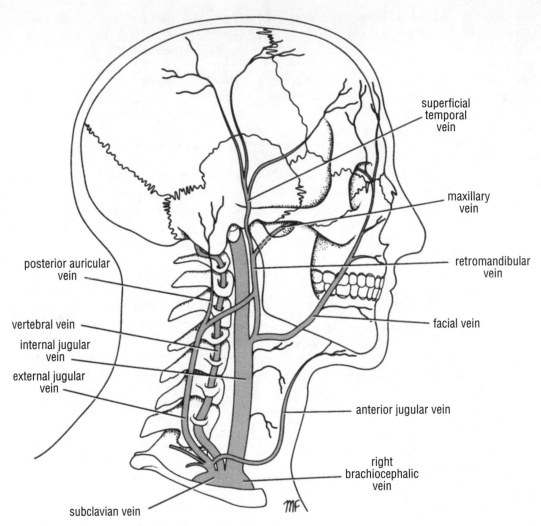

Fig. 7-11.   Main veins of the head and neck.

3. **Pharyngeal veins**
4. **Lingual vein**
5. **Superior thyroid vein**
6. **Middle thyroid vein**

Subclavian Vein

The subclavian vein is a continuation of the axillary vein at the outer border of the first rib (see Fig. 7-11). It joins the internal jugular vein to form the brachiocephalic vein. It receives the external jugular vein and it often receives the **thoracic duct** on the left side and the **right lymphatic duct** on the right side.

Lymphatic Drainage of the Head and Neck

The lymph nodes are arranged in two groups (Fig. 7-12): the **regional group,** which surrounds the neck below the chin like a collar, and the **deep vertical group,** which is embedded in the carotid sheath.

Regional Lymph Nodes
Occipital Nodes
Occipital nodes are situated at the apex of the posterior triangle over the occipital bone. They receive lymph from the back of the scalp.

Mastoid Nodes

Mastoid nodes lie over the mastoid process. They receive lymph from the scalp above the ear, the auricle, and the external auditory meatus.

Parotid Nodes

Parotid nodes are situated on or within the parotid salivary gland. They receive lymph from the scalp above the parotid gland, the eyelids, the auricle, and the external auditory meatus.

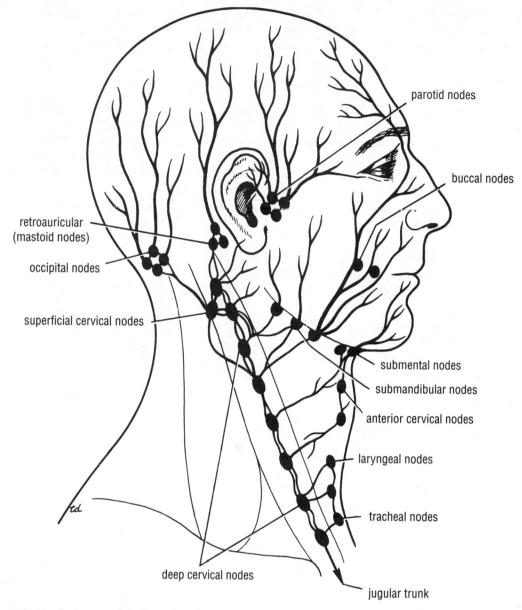

Fig. 7-12. Lymphatic drainage of the head and neck.

Buccal Nodes
Buccal nodes are one or two in number and they lie on the buccinator muscle. They drain lymph from the face and anterior part of the scalp, which passes to the submandibular nodes.

Submandibular Nodes
Submandibular nodes lie superficial to the submandibular salivary gland just below the body of the mandible. They receive lymph from the front of the scalp; the nose; the cheek; the upper and lower lip (except the central part of the lower lip); the frontal, maxillary, and ethmoid sinuses; the upper and lower teeth (except the lower incisors); the anterior two-thirds of the tongue (except the tip); the floor of the mouth; and the vestibule and the gums.

Submental Nodes
Submental nodes lie in the submental triangle just below the chin. They receive lymph from the tip of the tongue, the floor of the anterior part of the mouth, the incisor teeth, the central part of the lower lip, and the skin over the chin.

Anterior Cervical Nodes
Anterior cervical nodes lie along the course of the anterior jugular veins. They receive lymph from the skin of the front of the neck.

Superficial Cervical Nodes

Superficial cervical nodes lie along the external jugular vein. They receive lymph from the skin over the angle of the jaw and the lower part of the parotid gland, as well as the lobe of the ear.

Retropharyngeal Nodes

Retropharyngeal nodes lie between the pharynx and the vertebral column. They receive lymph from the nasopharynx, the auditory tube, and the vertebral column.

Laryngeal Nodes

Laryngeal nodes lie in front of the larynx. They receive lymph from the larynx.

Tracheal (Paratracheal) Nodes

Tracheal nodes lie alongside the trachea. They receive lymph from the trachea and thyroid gland.

Deep Cervical Lymph Nodes

Deep cervical lymph nodes are arranged in a vertical chain along the course of the internal jugular vein within the carotid sheath (see Fig. 7-12). They receive lymph from all the regional nodes previously discussed.

The efferent lymph vessels from all the deep cervical lymph nodes join to form the **jugular trunk,** which drains into the thoracic duct or the right lymphatic duct.

Jugulodigastric Node

The jugulodigastric node is located behind the angle of the jaw and drains the tonsil.

Jugulo-omohyoid Node

The jugulo-omohyoid node is located about halfway down the neck; it is mainly associated with drainage of the tongue.

Cranial Nerves

There are 12 pairs of cranial nerves:

1. Olfactory (sensory)
2. Optic (sensory)
3. Oculomotor (motor)
4. Trochlear (motor)
5. Trigeminal (mixed)
6. Abducent (motor)
7. Facial (mixed)
8. Vestibulocochlear (sensory)
9. Glossopharyngeal (mixed)
10. Vagus (mixed)
11. Accessory (motor)
12. Hypoglossal (motor)

Olfactory Nerves

The olfactory nerves arise from olfactory nerve cells in the olfactory mucous membrane lining the upper part of the nasal mucous membrane above the superior concha (Fig. 7-13). Bundles of the nerves pass through the openings in the cribriform plate of the ethmoid bone to enter the **olfactory bulb** inside the skull. The olfactory bulb is connected to the **olfactory area of the cerebral cortex** by the olfactory tract.

Optic Nerve

The optic nerve axons arise from the cells of the **ganglionic layer of the retina** (see Fig. 7-13). The optic nerve leaves the eyeball in the orbital cavity and passes through the optic canal to enter the middle cranial fossa. It unites with the optic nerve of the opposite side to form the **optic chiasma.**

In the chiasma, the fibers from the medial (nasal) half of each retina cross the midline and enter the **optic tract** of the opposite side, whereas the fibers from the lateral (temporal) half of each retina pass posteriorly in the optic tract of the same side (see Fig. 7-13). Most of the fibers of the optic tract terminate by synapsing with nerve cells in the **lateral geniculate body,** which is a small projection from the posterior part of the thalamus. A few fibers pass to the pretectal nucleus and superior colliculus of the midbrain and are concerned with light reflexes.

The axons of the nerve cells of the lateral geniculate body pass posteriorly as the **optic radiation** and terminate in the **visual cortex** of the cerebral hemisphere.

Oculomotor Nerve

The oculomotor nerve emerges on the anterior surface of the midbrain (Fig. 7-14). It passes forward in the lateral wall of the cavernous sinus. The nerve divides into a superior and an inferior ramus that enter the orbital cavity through the superior orbital fissure.

The oculomotor nerve supplies:

1. The levator palpebrae superioris, the superior rectus, the medial rectus, the inferior rectus, and the inferior oblique muscles (extrinsic muscles of the eye).
2. Parasympathetic nerve fibers to the constrictor pupillae of the iris and the ciliary muscles (intrinsic muscles of the eye). The fibers synapse in the **ciliary ganglion** and reach the eyeball in the **short ciliary nerves.**

The oculomotor nerve is therefore entirely motor and is responsible for lifting the upper eyelid; turning the eye upward, downward, and medially; constricting the pupil; and accommodating the eye.

Trochlear Nerve

The trochlear nerve is the most slender of the cranial nerves. Having decussated (crossed) the nerve of the opposite side, it leaves the posterior surface of the midbrain (see Fig. 7-14). It runs forward in the lateral wall of the cavernous sinus and enters the orbit through the superior orbital fissure.

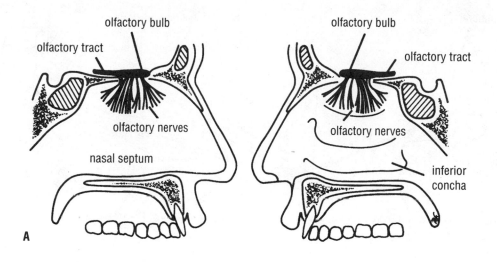

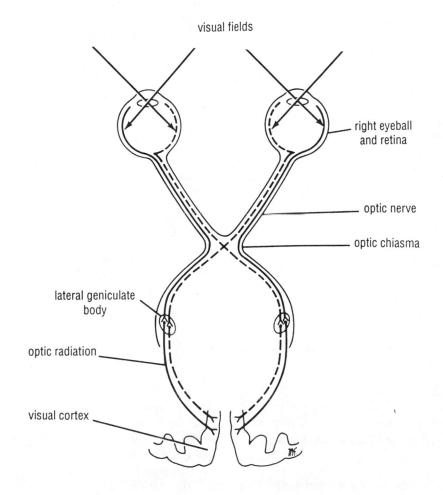

Fig. 7-13. **A.** Distribution of olfactory nerves on nasal septum and lateral wall of nose. **B.** The optic nerve and its connections.

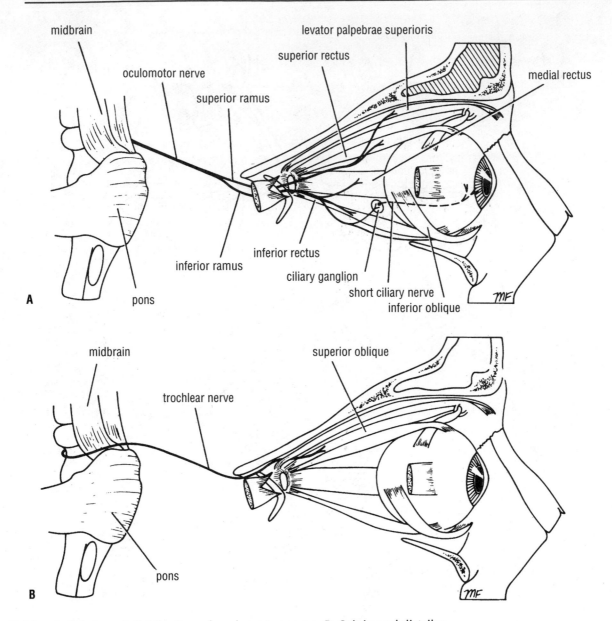

Fig. 7-14.   **A.** Origin and distribution of oculomotor nerve. **B.** Origin and distribution of trochlear nerve.

The trochlear nerve supplies the superior oblique muscle of the eyeball (extrinsic muscle). It is entirely motor and assists in turning the eye downward and laterally.

Trigeminal Nerve

The trigeminal nerve is the largest cranial nerve (Fig. 7-15). It leaves the anterior aspect of the pons as a small motor root and a large sensory root. The nerve passes forward out of the posterior cranial fossa to reach the middle cranial fossa. Here the sensory root expands to form the **trigeminal ganglion.** The trigeminal ganglion lies within a pouch of dura mater called the **trigeminal cave.** The ophthalmic, maxillary, and mandibular nerves arise from the anterior border of the ganglion.

Ophthalmic Nerve

The ophthalmic nerve (see Fig. 7-15) is entirely sensory. It runs forward in the lateral wall of the cavernous sinus and divides into three branches (the lacrimal, frontal, and nasociliary nerves), which enter the orbital cavity through the superior orbital fissure.

Branches

1. The **lacrimal nerve** runs forward on the upper border of the lateral rectus muscle. It is joined by the zygomaticotemporal branch of the maxillary nerve, which contains the parasympathetic secretomotor fibers to the lacrimal gland. The lacrimal nerve enters the lacrimal

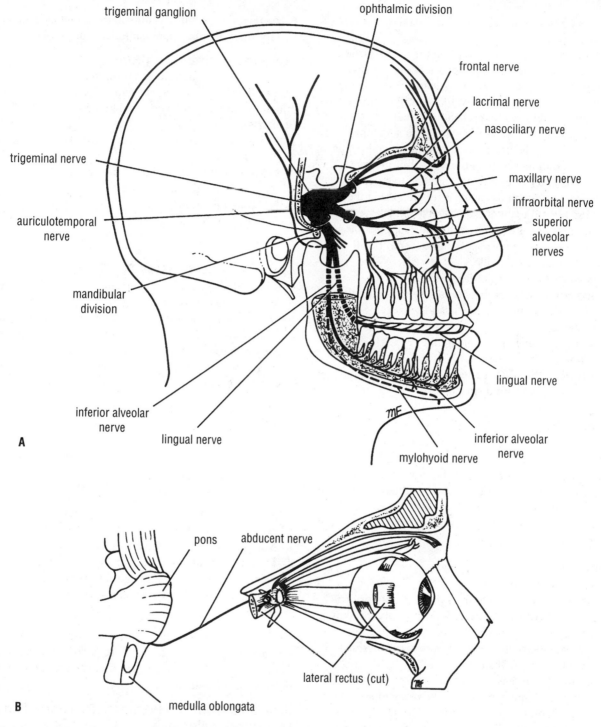

Fig. 7-15. **A.** Distribution of trigeminal nerve. **B.** Origin and distribution of abducent nerve.

gland and gives branches to the conjunctiva and skin of the upper eyelid.

2. The **frontal nerve** runs forward on the upper surface of the levator palpebrae superioris and divides into the **supraorbital** and **supratrochlear nerves.** These nerves leave the orbital cavity and supply the frontal air sinus and the skin of the forehead and scalp.

3. The **nasociliary nerve** crosses the optic nerve and runs forward on the upper border of the medial rectus muscle. It continues as the **anterior ethmoid nerve** through the anterior ethmoidal foramen to enter the cranial cavity. It then descends through a slit at the side of the crista galli to enter the nasal cavity. It gives off two **internal nasal branches** and then supplies the skin of the tip of the nose with the **external nasal nerve.** Its branches include:
   a. **Sensory fibers** to the ciliary ganglion
   b. **Long ciliary nerves** that contain sympathetic fibers to the dilator pupillae muscle and sensory fibers to the cornea
   c. **Infratrochlear nerve** that supplies the skin of the eyelids
   d. **Posterior ethmoidal nerve** that is sensory to the ethmoid and sphenoid sinuses

Maxillary Nerve

The maxillary nerve (see Fig. 7-15) is purely sensory. It leaves the skull through the foramen rotundum and crosses the pterygopalatine fossa to enter the orbit through the inferior orbital fissure. It continues as the **infraorbital nerve** in the infraorbital groove and emerges on the face through the infraorbital foramen. It gives sensory fibers to the skin of the face and side of the nose.

Branches

1. **Meningeal branches.**
2. **Zygomatic.** It divides into the **zygomaticotemporal** and **zygomaticofacial nerves,** which supply the skin of the face. The zygomaticotemporal branch gives parasympathetic secretomotor fibers to the lacrimal gland via the lacrimal nerve.
3. **Ganglionic branches.** Two short nerves that suspend the pterygopalatine ganglion in the pterygopalatine fossa. They contain sensory fibers that have passed through the ganglion from the nose, the palate, and the pharynx. They also contain postganglionic parasympathetic fibers that are going to the lacrimal gland.
4. **Posterior superior alveolar nerve** supplies the maxillary sinus and the upper molar teeth and adjoining parts of the gum and the cheek.
5. **Middle superior alveolar nerve** supplies the maxillary sinus, the upper premolar teeth, the gums, and the cheek.

6. **Anterior superior alveolar nerve** supplies the maxillary sinus and the upper canine and incisor teeth.

Pterygopalatine Ganglion

This is a parasympathetic ganglion that is suspended from the maxillary nerve in the pterygopalatine fossa. It is secretomotor to the lacrimal and nasal glands (see p. 239).
Branches
1. **Orbital branches** enter the orbit through the inferior orbital fissure.
2. **Greater and lesser palatine nerves** supply the palate, the tonsil, and the nasal cavity.
3. **Nasal branches.**
4. **Pharyngeal branch** supplies the roof of the nasopharynx.

Mandibular Nerve (see Fig. 7-15)

The mandibular nerve is motor and sensory. The sensory root leaves the trigeminal ganglion and passes through the foramen ovale to enter the infratemporal fossa. The motor root also passes through the foramen ovale and joins the sensory root to form the trunk of the mandibular nerve. The nerve trunk then divides into a small anterior and a large posterior division.

Branches from the Main Trunk of the Mandibular Nerve

1. **Meningeal branch.**
2. **Nerve to the medial pterygoid muscle** supplies not only the medial pterygoid but the tensor veli palatini muscle.

Branches from the Anterior Division of the Mandibular Nerve

1. **Masseteric nerve** to the masseter muscle
2. **Deep temporal nerves** to the temporalis muscle
3. **Nerve to the lateral pterygoid muscle**
4. **Buccal nerve** to the skin and mucous membrane of the cheek

Note that the buccal nerve *does not* supply the buccinator muscle (facial nerve). Note also that the buccal nerve is the *only sensory* branch of the anterior division of the mandibular nerve.

Branches from the Posterior Division of the Mandibular Nerve

1. **Auriculotemporal nerve** supplies the skin of the auricle, the external auditory meatus, the temporomandibular joint, and the scalp. Also conveys postganglionic parasympathetic secretomotor fibers from the otic ganglion to the parotid salivary gland.
2. **Lingual nerve** descends in front of the inferior alveolar nerve and enters the mouth. It runs forward on the side of the tongue, crossing the submandibular duct. In its course it is joined by the **chordi tympani** nerve. It

supplies the mucous membrane of the anterior two-thirds of the tongue and the floor of the mouth. It gives off **preganglionic parasympathetic secretomotor fibers** to the submandibular ganglion.

3. **Inferior alveolar nerve** enters the mandibular canal to supply the teeth of the lower jaw. It emerges through the mental foramen (mental nerve) to supply the skin of the chin. Before entering the canal it gives off the **mylohyoid nerve** that supplies the mylohyoid muscle and the anterior belly of the digastric muscle.

4. **Communicating branch** frequently runs from the inferior alveolar nerve to the lingual nerve.

Note that the branches of the posterior division of the mandibular nerve are sensory except for the nerve to the mylohyoid muscle.

Otic Ganglion

This parasympathetic ganglion is located medial to the mandibular nerve just below the skull and is adherent to the nerve to the medial pterygoid muscle. The preganglionic fibers originate in the glossopharyngeal nerve and reach the ganglion via the lesser petrosal nerve (see p. 242). The postganglionic secretomotor fibers reach the parotid salivary gland via the auriculotemporal nerve.

Submandibular Ganglion

This parasympathetic ganglion lies deep to the submandibular salivary gland and is attached to the lingual nerve by small nerves. Preganglionic parasympathetic fibers reach the ganglion from the facial nerve via the chorda tympani and the lingual nerves. Postganglionic secretomotor fibers pass to the submandibular and sublingual salivary glands.

Abducent Nerve

The abducent nerve is a small nerve that leaves the anterior surface of the brain between the pons and the medulla oblongata (see Fig. 7-15). It passes forward through the cavernous sinus with the internal carotid artery. The nerve enters the orbit through the superior orbital fissure and supplies the lateral rectus muscle. The abducent nerve is therefore responsible for turning the eye laterally.

Facial Nerve

The facial nerve has a motor root and a sensory root (**nervus intermedius**) (Fig. 7-16). The nerve emerges on the anterior surface of the brain between the pons and the medulla oblongata. It enters the internal acoustic meatus with the vestibulocochlear nerve. At the bottom of the meatus, the facial nerve enters the facial canal and passes laterally through the inner ear. On reaching the medial wall of the tympanic cavity, the nerve swells to form the sensory **geniculate ganglion** (see Fig. 7-16). The nerve now bends sharply backward above the promontory. At the posterior

wall of the tympanic cavity, the facial nerve bends down on the medial side of the aditus of the mastoid antrum. The nerve descends behind the pyramid and emerges from the temporal bone through the stylomastoid foramen.

Important Branches of the Facial Nerve

1. **Greater petrosal nerve** arises from the nerve at the geniculate ganglion (see Fig. 7-16). It contains preganglionic parasympathetic fibers that synapse in the pterygopalatine ganglion. The postganglionic fibers are secretomotor to the lacrimal gland and glands of the nose and palate. The greater petrosal nerve also contains taste fibers from the palate.

2. **Nerve to stapedius** supplies the stapedius muscle within the pyramid (see Fig. 7-16).

3. **Chorda tympani** arises from the facial nerve in the facial canal in the posterior wall of the tympanic cavity (see Fig. 7-16). It runs forward over the medial surface of the upper part of the tympanic membrane and leaves the tympanic cavity through the **petrotympanic fissure.** It is now in the infratemporal fossa and joins the lingual nerve. The nerve contains preganglionic parasympathetic secretomotor fibers to the submandibular and sublingual salivary glands. It also contains taste fibers from the anterior two-thirds of the tongue and the floor of the mouth.

4. **Posterior auricular, posterior belly of digastric, and stylohyoid nerves** (see Fig. 7-16) are muscular branches given off as the facial nerve emerges from the stylomastoid foramen.

5. **Five terminal branches to the muscles of facial expression.** These are the **temporal, zygomatic, buccal, mandibular,** and **cervical branches** (see Fig. 7-16). The facial nerve lies within the parotid salivary gland after it leaves the stylomastoid foramen; it is located between the superficial and deep parts of the gland. Here it gives off the terminal branches that emerge from the anterior border of the gland and pass to the muscles of the face and scalp. **Note that the buccal branch supplies the buccinator muscle.** The cervical branch supplies the platysma and the depressor anguli oris muscles.

Vestibulocochlear Nerve

The vestibulocochlear nerve is a sensory nerve that consists of two sets of fibers, **vestibular** and **cochlear** (Fig. 7-17). The fibers emerge from the anterior surface of the brain between the pons and the medulla oblongata and enter the internal acoustic meatus.

Vestibular Fibers

The vestibular fibers are the central processes of the nerve cells of the **vestibular ganglion** situated in the internal acoustic meatus (see Fig. 7-17). The vestibular fibers origi-

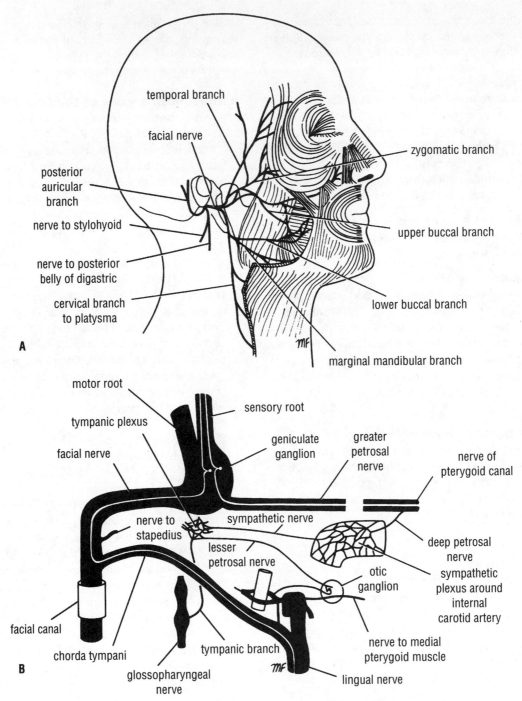

Fig. 7-16. **A.** Distribution of facial nerve. **B.** Branches of facial nerve within petrous part of temporal bone; the taste fibers are shown in white. The glossopharyngeal nerve is also shown.

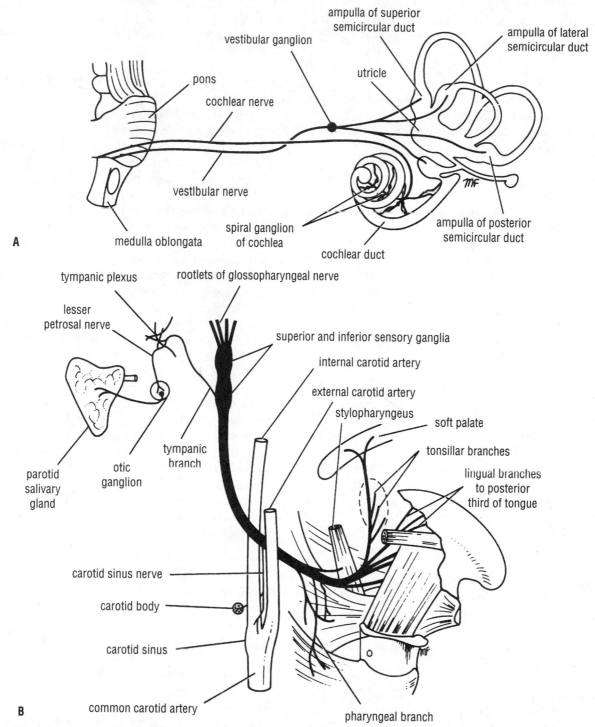

Fig. 7-17.   **A.** Origin and distribution of vestibulocochlear nerve. **B.** Distribution of glossopharyngeal nerve.

nate from the vestibule and the semicircular canals and are therefore concerned with the sense of position and movement of the head.

Cochlear Fibers

The **cochlear fibers** are the central processes of nerve cells of the **spiral ganglion of the cochlea** (see Fig. 7-17). The cochlear fibers originate in the **spiral organ of Corti** and are therefore concerned with the sense of hearing.

Glossopharyngeal Nerve

The glossopharyngeal nerve is a motor and sensory nerve (see Fig. 7-17). It emerges from the anterior surface of the medulla oblongata between the olive and the inferior cerebellar peduncle. It leaves the skull through the jugular foramen; the **superior and inferior sensory ganglia** are located on the nerve as it passes through the foramen. The nerve descends in the neck and winds around the posterior border of the stylopharyngeus muscle to be distributed to the pharynx and the tongue.

Important Branches of the Glossopharyngeal Nerve

1. **Tympanic branch** passes to the **tympanic plexus** in the tympanic cavity. Preganglionic parasympathetic fibers for the parotid salivary gland now leave the plexus as the **lesser petrosal nerve** and they synapse in the otic ganglion (see Fig. 7-17).
2. **Carotid branch** contains sensory fibers from the carotid sinus (pressoreceptor mechanism for the regulation of blood pressure) and the carotid body (chemoreceptor mechanism for the regulation of the heart rate and respiration) (see Fig. 7-17).
3. **Nerve to the stylopharyngeus muscle.**
4. **Pharyngeal branches** (see Fig. 7-17) to the **pharyngeal plexus.** The plexus gives sensory fibers to the pharynx and also receives branches from the vagus nerve and the sympathetic trunk.
5. **Lingual branch** (see Fig. 7-17) to the mucous membrane of the posterior third of the tongue (including the vallate papillae).

Vagus Nerve

The vagus nerve is both a motor and a sensory nerve (Fig. 7-18). It emerges from the anterior surface of the medulla oblongata between the olive and the inferior cerebellar peduncle. The nerve leaves the skull through the jugular foramen and has **superior** and **inferior sensory ganglia.** Below the inferior ganglion, the cranial root of the accessory nerve joins the vagus nerve and is distributed mainly in its pharyngeal and recurrent laryngeal branches.

The vagus nerve descends vertically in the neck within the carotid sheath with the internal jugular vein and the internal and common carotid arteries. It descends through the thorax in the mediastinum, passing behind the root of the lung, and enters the abdomen through the esophageal opening in the diaphragm.

Important Branches of the Vagus Nerve in the Neck

1. **Meningeal and auricular branches.**
2. **Pharyngeal branch** contains nerve fibers from the cranial root of the accessory nerve. It joins the **pharyngeal plexus** and supplies all the muscles of the pharynx except the stylopharyngeus and all the muscles of the soft palate except the tensor veli palatini.
3. **Superior laryngeal nerve** (see Fig. 7-18) divides into internal and external laryngeal nerves. The **internal laryngeal nerve** is sensory to the mucous membrane of the piriform fossa and the larynx down as far as the vocal folds. The **external laryngeal nerve** is motor and located close to the superior thyroid artery; it supplies the cricothyroid muscle.
4. **Recurrent laryngeal nerve** (see Fig. 7-18). On the right side, **it hooks around the first part of the subclavian artery** and ascends in the groove between the trachea and the esophagus. The nerve is closely related to the inferior thyroid artery. It supplies all the muscles of the larynx except the cricothyroid muscle, the mucous membrane of the larynx below the vocal folds, and the mucous membrane of the upper part of the trachea. On the left side, the **nerve hooks around the arch of the aorta** and ascends into the neck between the trachea and the esophagus.
5. **Cardiac branches.** Two or three branches arise in the neck. They descend into the thorax and end in the cardiac plexus (see Fig. 7-18).

The vagus nerve has the most extensive distribution of all the cranial nerves. Its distribution in the thorax is described on page 42. For abdominal distribution, see Figure 7-18.

Accessory Nerve

The accessory nerve is a motor nerve with cranial and spinal roots (Fig. 7-19).

Cranial Root

The cranial root emerges from the anterior surface of the medulla oblongata between the olive and the inferior cerebellar peduncle (see Fig. 7-19).

Spinal Root

The spinal root arises from nerve cells in the anterior gray column of the upper five segments of the cervical part of the spinal cord (see Fig. 7-19). The nerve ascends alongside the spinal cord and enters the skull through the foramen magnum.

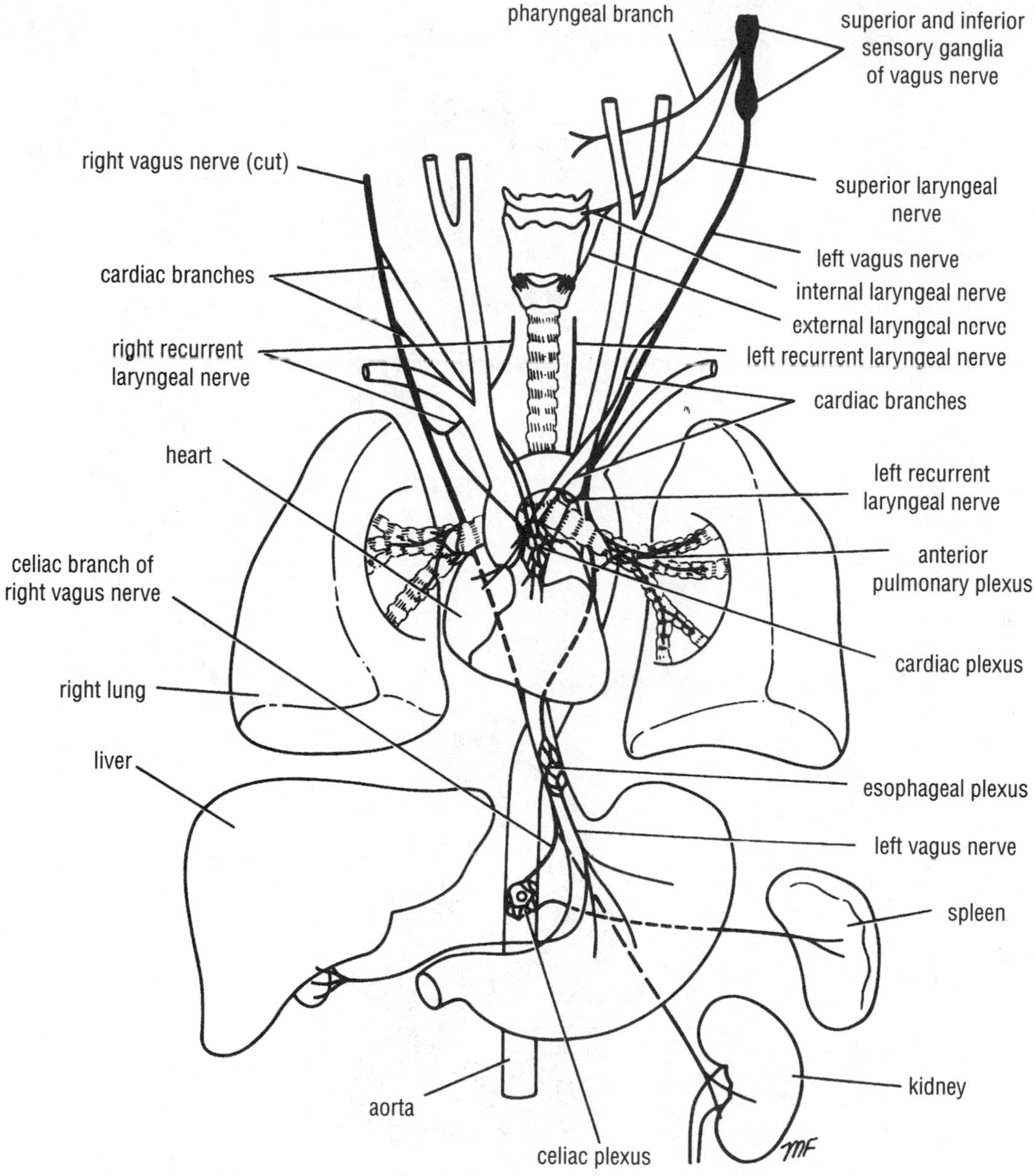

Fig. 7-18. Distribution of vagus nerve.

The two roots unite and leave the skull via the jugular foramen. The roots then separate and the cranial root joins the vagus nerve and is distributed in its pharyngeal and recurrent laryngeal branches to the muscles of the soft palate, pharynx, and larynx. The spinal root runs downward and laterally and enters the deep surface of the sternocleidomastoid muscle, which it supplies. The nerve then crosses the posterior triangle of the neck to supply the trapezius muscle.

The accessory nerve thus brings about movements of the soft palate, the pharynx, and the larynx and controls the movements of two large muscles of the neck.

Hypoglossal Nerve

The hypoglossal nerve is a motor nerve that emerges on the anterior surface of the medulla oblongata between the pyramid and the olive (see Fig. 7-19). It leaves the skull through the hypoglossal canal. The nerve descends in the neck to reach the lower border of the posterior belly of the digastric muscle. Here it turns forward and crosses the internal and external carotid arteries and the loop of the lingual artery. The nerve passes forward deep to the mylohyoid muscle on the side of the tongue. It lies below the submandibular duct and the lingual nerve. In the upper part

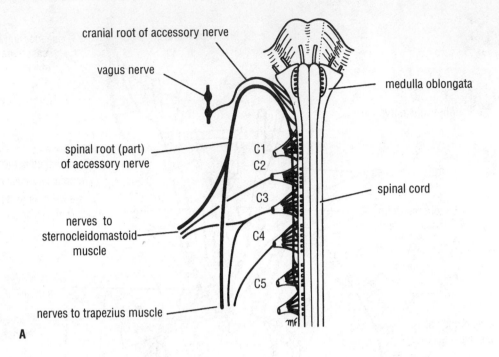

cranial root of accessory nerve

vagus nerve

medulla oblongata

spinal root (part) of accessory nerve

C1
C2

C3

spinal cord

nerves to sternocleidomastoid muscle

C4

C5

nerves to trapezius muscle

A

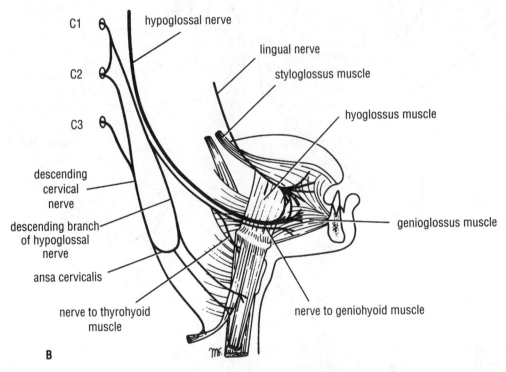

C1

hypoglossal nerve

lingual nerve

C2

styloglossus muscle

C3

hyoglossus muscle

descending cervical nerve

descending branch of hypoglossal nerve

ansa cervicalis

genioglossus muscle

nerve to thyrohyoid muscle

nerve to geniohyoid muscle

B

Fig. 7-19.   **A.** Origin and distribution of accessory nerve. **B.** Distribution of hypoglossal nerve.

of its course, the hypoglossal nerve is joined by C1 fibers from the cervical plexus.

Important Branches of the Hypoglossal Nerve
1. **Meningeal branch.**
2. **Descending branch** (C1 fibers). This passes downward and joins the **descending cervical nerve** (C2 and 3) to form a loop called the **ansa cervicalis.** Branches from the loop supply the omohyoid, the sternohyoid, and the sternothyroid muscles.
3. **Nerve to the thyrohyoid muscle** (C1).
4. **Muscular branches** to all the muscles of the tongue except the palatoglossus (pharyngeal plexus).
5. **Nerve to the geniohyoid** (C1).

The hypoglossal nerve proper thus supplies the majority of the muscles of the tongue and controls the shape and movements of the tongue.

The openings in the skull through which the cranial nerves leave the cranial cavity are shown in Table 7-1.

Cervical Plexus

The cervical plexus is formed by the anterior rami of the first four cervical spinal nerves (see Fig. 7-5). The rami join to form loops that lie in front of the origins of the levator scapulae and scalenus medius muscles. The cervical plexus supplies the skin and muscles of the head, neck, and shoulders.

Branches
1. **Lesser occipital nerve** (C2) supplies the back of the scalp and the auricle.
2. **Greater occipital nerve** (C2–3) ascends over back of scalp and supplies it as far forward as the vertex of the skull.
3. **Transverse cervical nerve** (C2–3) supplies the skin over the front of the neck.
4. **Supraclavicular nerves** (C3–4). The medial, intermediate, and lateral branches supply the skin over the shoulder region. Clinically these nerves are important since pain may be referred along them from the phrenic nerve.

Important Branch
Phrenic Nerve
A large nerve, the phrenic nerve is the **only** motor nerve supply to the diaphragm (see Fig. 7-5). The nerve runs vertically down in the neck on the anterior surface of the scalenus anterior muscle, crossing the muscle from its lateral to its medial border. The nerve enters the thorax by passing anterior to the subclavian artery. The further distribution of the phrenic nerve in the thorax and the abdomen is described on page 42.

Brachial Plexus

The roots and trunks of this important plexus are found in the posterior triangle of the neck. The plexus is described on page 153.

Autonomic Nervous System in the Head and Neck

Sympathetic Part
Sympathetic Trunks
In the neck, the sympathetic trunk on each side extends upward to the skull and downward to the first rib, where it becomes continuous with the thoracic part of the trunk. The trunk lies behind the carotid sheath; it possesses three ganglia: the superior, middle, and inferior cervical (see Fig. 7-5).

Superior Cervical Ganglion
Lies immediately below the skull.
Branches
1. **Internal carotid nerve** accompanies the internal carotid artery and forms the **internal carotid plexus.**
2. **Gray rami communicantes** pass to the upper four cervical spinal nerves.
3. **Arterial branches** to the common and external carotid arteries.
4. **Cranial nerve branches** to the ninth, tenth, and twelfth cranial nerves.
5. **Pharyngeal branches** join pharyngeal branches of the ninth and tenth cranial nerves to form the **pharyngeal plexus.**
6. **Superior cardiac branch** descends into the thorax to join the cardiac plexus.

Middle Cervical Ganglion
Lies at level of cricoid cartilage.
Branches
1. **Gray rami communicantes** to the fifth and sixth cervical spinal nerves.
2. **Thyroid branches.**
3. **Middle cardiac branch** descends into the thorax to join the cardiac plexus.

Inferior Cervical Ganglion
In the majority of people, the inferior cervical ganglion is fused to the first thoracic ganglion to form the **stellate ganglion.** It is located between the transverse process of the seventh cervical vertebra and the neck of the first rib.
Branches
1. **Gray rami communicantes** to the seventh and eighth cervical spinal nerves.
2. **Arterial branches** to the subclavian and vertebral arteries.

Head and Neck

3. **Inferior cardiac branch** descends into the thorax to join the cardiac plexus.

Ansa Subclavia

The portion of the sympathetic trunk that connects the middle to the inferior cervical ganglia is in the form of two or more bundles (see Fig. 7-5). The anterior bundle crosses anterior to the first part of the subclavian artery and is called the **ansa subclavia.**

Parasympathetic Part

The cranial part of the craniosacral outflow of the parasympathetic part of the autonomic nervous system is located in the nuclei of the **oculomotor (third),** the **facial (seventh),** the **glossopharyngeal (ninth),** and the **vagus (tenth) cranial nerves.**

The parasympathetic nucleus of the oculomotor nerve is called the **Edinger-Westphal nucleus,** those of the facial the **lacrimatory** and **superior salivary nuclei,** that of the glossopharyngeal the **inferior salivary nucleus,** and that of the vagus the **dorsal nucleus of the vagus.** The axons of these connector nerve cells are myelinated preganglionic fibers that emerge from the brain within the cranial nerves.

The preganglionic fibers synapse in peripheral ganglia located close to the viscera they innervate. The cranial parasympathetic ganglia are the **ciliary,** the **pterygopalatine,** the **submandibular,** and the **otic.** In certain locations, the ganglion cells are placed in nerve plexuses, such as the **cardiac plexus,** the **pulmonary plexus,** the **myenteric plexus (Auerbach's plexus),** and the **mucosal plexus (Meissner's plexus);** the last two plexuses are found in the gastrointestinal tract. The postganglionic fibers are non-myelinated and short in length.

Review of the Digestive System in the Head and Neck

Mouth

The mouth is divided into the **vestibule,** which lies between the lips and the cheeks externally and the gums and the teeth internally, and the **mouth cavity,** which lies within the gums and the teeth (Figs. 7-20 and 7-21).

The **cheek** is made up of the buccinator muscle, is covered on the outside by skin, and is lined by mucous membrane. Opposite the upper second molar, the **duct of the parotid gland** opens into the vestibule on a small papilla.

Roof of Mouth

The roof of the mouth is formed by the hard palate anteriorly and the soft palate posteriorly.

Floor of Mouth

The floor of the mouth is formed largely by the anterior two-thirds of the tongue and by mucous membrane extending from the tongue to the gums. The **frenulum of the tongue** connects the undersurface of the tongue to the floor of the mouth (see Fig. 7-21). On each side of the frenulum is the **papilla and orifice of the duct of the submandibular salivary gland** (see Fig. 7-21). Lateral to this is the **sublingual fold,** produced by the underlying sublingual salivary gland. The ducts of the gland are 18 to 20 in number and open on the summit of the fold.

Sensory Innervation of the Mouth

Roof: Greater palatine and nasopalatine nerves (maxillary nerve)

Floor: Lingual nerve (common sensation) and chorda tympani (taste)

Cheek: Buccal nerve (mandibular nerve)

Teeth

There are two sets of teeth: deciduous (20) and permanent (32).

Tongue

The tongue is a mass of striated muscle covered with mucous membrane and divided by a **median fibrous septum.**

Muscles

1. **Intrinsic:** Longitudinal, transverse, and vertical
2. **Extrinsic:** Genioglossus, hyoglossus, styloglossus, and palatoglossus

The origin, insertion, nerve supply, and action of the tongue muscles are summarized in Table 7-9.

Mucous Membrane

On the upper surface of the tongue between the anterior two-thirds and the posterior third is a V-shaped groove, the **sulcus terminalis.** Three types of papillae are present on the upper surface of the anterior two-thirds: **filiform, fungiform,** and **vallate.** The vallate papillae are 10 to 12 in number and are situated in a row in front of the sulcus terminalis.

Blood Supply of Tongue

Lingual artery, tonsillar branch of facial artery, and ascending pharyngeal artery.

Lymph Drainage of Tongue

Tip: submental lymph nodes; sides of anterior two-thirds: submandibular lymph nodes; and posterior third: deep cervical lymph nodes.

Sensory Innervation

Anterior two-thirds: lingual nerve (general sensation) and chorda tympani (taste); and posterior third: glossopharyngeal (general sensation and taste).

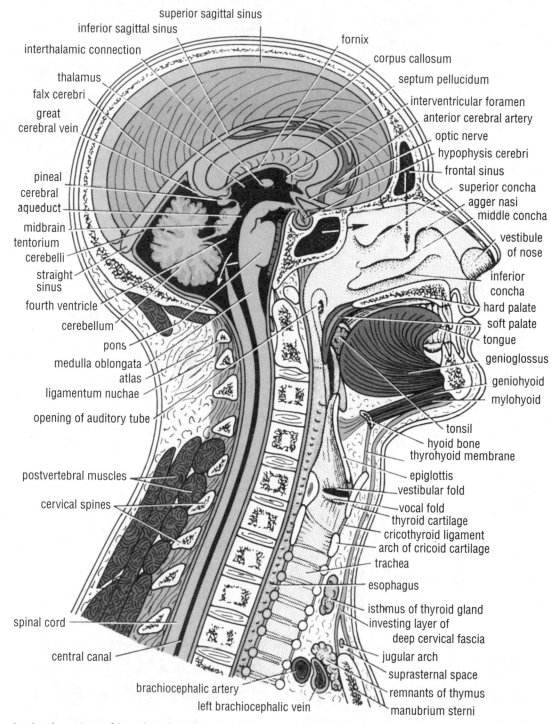

Fig. 7-20. Sagittal section of head and neck.

Movements of Tongue

Protrusion: Genioglossus muscles on both sides

Retraction: Styloglossus and hyoglossus muscles on both sides

Depression: Hyoglossus muscles on both sides

Retraction and elevation of posterior third: Styloglossus and palatoglossus muscles on both sides

Shape changes: Intrinsic muscles

Palate

The palate may be divided into two parts: hard palate in front and soft palate behind.

Hard Palate

The hard palate is formed by the palatine process of the maxillae and the horizontal plates of the palatine bones (see Fig. 7-21).

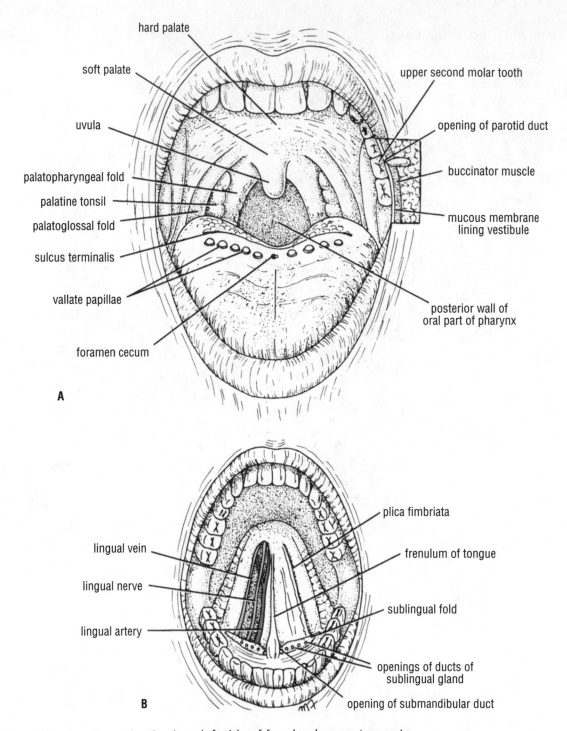

Fig. 7-21. **A.** Cavity of mouth. Cheek on left side of face has been cut away to show buccinator muscle and parotid duct. **B.** Undersurface of tongue.

Table 7-9 Muscles of Tongue

Name of muscle	Origin	Insertion	Nerve supply	Action
Intrinsic muscles				
Longitudinal	Median septum and submucosa	Mucous membrane	Hypoglossal nerve	Alters shape of tongue
Transverse				
Vertical				
Extrinsic muscles				
Genioglossus	Superior genial spine of mandible	Blends with other muscles of tongue	Hypoglossal nerve	Protrudes apex of tongue through mouth
Hyoglossus	Body and greater cornu of hyoid bone	Blends with other muscles of tongue	Hypoglossal nerve	Depresses tongue
Styloglossus	Styloid process of temporal bone	Blends with other muscles of tongue	Hypoglossal nerve	Draws tongue upward and backward
Palatoglossus	Palatine aponeurosis	Side of tongue	Pharyngeal plexus	Pulls root of tongue upward and backward, narrows oropharyngeal isthmus

Soft Palate

The soft palate is a mobile fold attached to the posterior border of the hard palate (see Fig. 7-21). Its free border has a conical projection, the **uvula,** in the midline.

Palatoglossal Arch

The palatoglossal arch is a muscular fold (see Fig. 7-21) containing the **palatoglossus muscle,** which extends from the soft palate to the side of the tongue (Fig. 7-22). **The palatoglossal arch marks the place where the mouth becomes the pharynx.**

Palatopharyngeal Arch

The palatopharyngeal arch is a muscular fold (see Fig. 7-21) that lies behind the palatoglossal arch (see Fig. 7-22). It runs downward and laterally to join the pharyngeal wall. The muscle contained within the fold is the **palatopharyngeus muscle.** The **palatine tonsils,** which are masses of lymphoid tissue, are located between the palatoglossal and palatopharyngeal arches.

The muscles of the soft palate are the **tensor veli palatini,** the **levator veli palatini,** the **palatoglossus,** the **palatopharyngeus,** and the **musculus uvulae.** The origins, insertions, nerve supply, and actions of these muscles are summarized in Table 7-10.

Blood Supply

Greater palatine branch of maxillary artery, ascending palatine branch of facial artery, and ascending pharyngeal artery.

Lymph Drainage

Deep cervical lymph nodes.

Sensory Innervation

Greater and lesser palatine nerves, and nasopalatine and glossopharyngeal nerves.

Movements of Soft Palate

Raise palate: The levator veli palatini closes off the channel between the oral pharynx and the nasal pharynx. At the same time, the upper fibers of the superior constrictor contract and pull the posterior wall of the pharynx forward. The palatopharyngeus muscles also contract on both sides, pulling in the palatopharyngeal folds, like side curtains.

Lower palate: The palatoglossus and palatopharyngeus muscles lower the soft palate.

The soft palate is stretched by the contraction of the tensor veli palatini muscle during the process of raising and lowering of the palate.

Salivary Glands
Parotid Gland

The parotid gland is the largest salivary gland and is composed almost entirely of serous acini. It lies in a deep hollow below the external auditory meatus, behind the ramus of the mandible, and in front of the sternocleidomastoid muscle. The facial nerve divides the gland into **superficial and deep lobes.** The **duct** passes forward over the masseter muscle, then turns medially and pierces the buccinator muscle to open into the mouth opposite the upper second molar tooth.

Structures Within the Parotid Gland

From lateral to medial, the structures are the facial nerve, the retromandibular vein, and the external carotid artery. Some lymph nodes are also present.

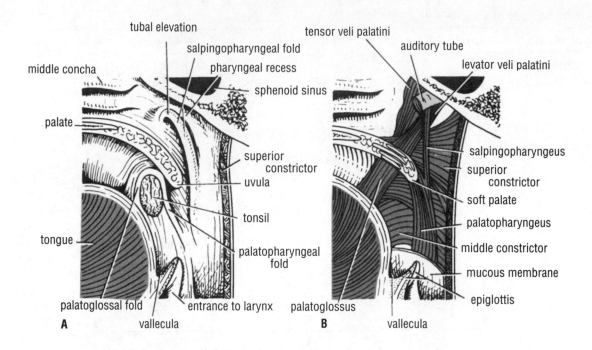

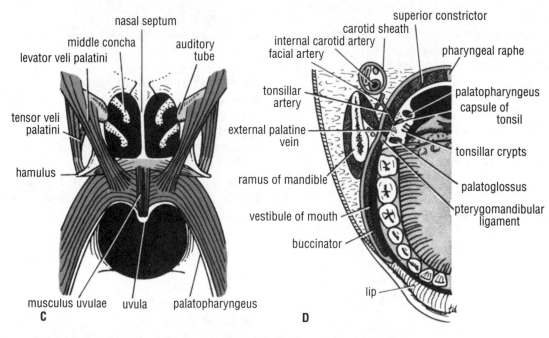

Fig. 7-22.   A. Junction of nose with nasal part of pharynx and mouth with oral part of pharynx. Note position of tonsil and opening of auditory tube. **B.** Muscles of soft palate and upper part of pharynx. **C.** Muscles of soft palate as seen from behind. **D.** Horizontal section through mouth and oral part of pharynx, showing relations of tonsil.

Table 7-10 Muscles of the Soft Palate

Name of muscle	Origin	Insertion	Nerve supply	Action
Tensor veli palatini	Spine of sphenoid, auditory tube	With muscle of other side, forms palatine aponeurosis	Nerve to medial pterygoid from mandibular nerve	Tenses soft palate
Levator veli palatini	Petrous part of temporal bone, auditory tube	Palatine aponeurosis	Pharyngeal plexus	Raises soft palate
Palatoglossus	Palatine aponeurosis	Side of tongue	Pharyngeal plexus	Pulls root of tongue upward and backward, narrows oropharyngeal isthmus
Palatopharyngeus	Palatine aponeurosis	Posterior border of thyroid cartilage	Pharyngeal plexus	Elevates wall of pharynx, pulls palatopharyngeal folds medially
Musculus uvulae	Posterior border of hard palate	Mucous membrane of uvula	Pharyngeal plexus	Elevates uvula

Nerve Supply

Parasympathetic from the glossopharyngeal nerve. The axons pass to the gland via the tympanic branch, the lesser petrosal nerve, and the otic ganglion. Postganglionic fibers pass to the gland via the auriculotemporal nerve. Postganglionic sympathetic fibers reach the gland from the superior cervical sympathetic ganglion.

Submandibular Gland

The submandibular gland is composed of a mixture of serous and mucous acini. It is located beneath the lower border of the body of the mandible. The gland is divided into superficial and deep parts by the mylohyoid muscle. The deep part of the gland lies below the mucous membrane of the mouth on the side of the tongue. The **duct** leaves the anterior end of the deep part of the gland and runs forward beneath the mucous membrane of the mouth. The duct is crossed by the lingual nerve and is covered laterally by the sublingual gland. The duct opens into the mouth on a small papilla at the side of the frenulum of the tongue (see Fig. 7-21).

Nerve Supply

Parasympathetic from the facial nerve via the chorda tympani, the nerves pass to the submandibular ganglion. Postganglionic fibers pass to the gland. Postganglionic sympathetic fibers run from the superior cervical sympathetic ganglion.

Sublingual Gland

The sublingual gland lies beneath the mucous membrane of the floor of the mouth (sublingual fold). It has both serous and mucous acini, the latter predominating. The **ducts** are 8 to 20 in number and open into the mouth on the sublingual fold (see Fig. 7-21).

Nerve Supply

Same as submandibular gland.

Pharynx

The pharynx is funnel-shaped and may be divided into **nasal, oral,** and **laryngeal parts** (see Fig. 7-20). It is continuous with the esophagus opposite the sixth cervical vertebra. The anterior wall is deficient and is replaced by the posterior opening of the nose, the opening into the mouth, and the inlet of the larynx.

The muscles of the pharyngeal wall consist of the superior, middle, and inferior constrictor muscles, whose fibers run in a more or less circular direction; and the stylopharyngeus and salpingopharyngeus muscles, whose fibers run in a more or less longitudinal direction. The origin, insertion, nerve supply, and action of these muscles are summarized in Table 7-11.

Interior of Pharynx
Nasal Pharynx

The nasal pharynx lies above the soft palate and behind the nasal cavities (see Fig. 7-20). In the roof is the collection of lymphoid tissue called the **pharyngeal tonsil.** The **pharyngeal isthmus** is the opening in the floor between the soft palate and the posterior pharyngeal wall. The **auditory tube** opens on the lateral wall (Fig. 7-23); the elevated edge of the tube is called the **tubal elevation.** The **pharyngeal recess** is a depression in the pharyngeal wall behind the tubal elevation. The **salpingopharyngeal fold** is a vertical fold of mucous membrane covering the salpingopharyngeus muscle (see Fig. 7-22).

Table 7-11   Muscles of the Pharynx

Name of muscle	Origin	Insertion	Nerve supply	Action
Superior constrictor	Medial pterygoid plate, pterygoid hamulus, ptery-gomandibular ligament, mylohyoid line of mandible	Pharyngeal tubercle of occipital bone, raphe in midline posteriorly	Pharyngeal plexus	Aids soft palate in closing off nasal pharynx, propels bolus downward
Middle constrictor	Lower part of stylo-hyoid ligament, lesser and greater cornu of hyoid bone	Pharyngeal raphe	Pharyngeal plexus	Propels bolus down-ward
Inferior constrictor	Lamina of thyroid cartilage, cricoid cartilage	Pharyngeal raphe	Pharyngeal plexus	Propels bolus down-ward
Cricopharyngeus	Lowest fibers of inferior constrictor muscle			Sphincter at lower end of pharynx
Stylopharyngeus	Styloid process of temporal bone	Posterior border of thyroid cartilage	Glossopharyngeal nerve	Elevates larynx during swallowing
Salpingopharyngeus	Auditory tube	Blends with palato-pharyngeus	Pharyngeal plexus	Elevates pharynx
Palatopharyngeus	Palatine aponeurosis	Posterior border of thyroid cartilage	Pharyngeal plexus	Elevates wall of pharynx, pulls palatopharyngeal folds medially

Oral Pharynx

The oral pharynx lies behind the oral cavity. In the floor in the midline between the tongue and the epiglottis is the **median glossoepiglottic fold;** on each side is the **lateral glossoepiglottic fold.** The **vallecula** is the depression on each side of the median glossoepiglottic fold.

On the lateral wall on each side are located the **palatine tonsils** between the palatoglossal and palatopharyngeal folds (see Fig. 7-22). The **tonsillar sinus** is a recess between these folds that is occupied by the tonsil. The **oropharyngeal isthmus** is the interval between the two palatoglossal folds that marks the boundary between the mouth and the pharynx.

Laryngeal Pharynx

The laryngeal pharynx lies behind the opening into the larynx. The lateral wall is formed by the thyroid cartilage and the thyrohyoid membrane. The **piriform fossa** is a depression in the mucous membrane on each side of the laryngeal inlet (see Fig. 7-23). It is a common site for the lodging of fish bones.

Sensory Nerve Supply of the Pharyngeal Mucous Membrane

Nasal pharynx: Maxillary nerve

Oral pharynx: Glossopharyngeal nerve

Laryngeal pharynx (around entrance into larynx): Internal laryngeal nerve, branch of the superior laryngeal branch of the vagus

Palatine Tonsil

The palatine tonsil is a mass of lymphoid tissue located on each side of the oral pharynx. The medial surface is pitted by numerous small openings, which lead into the **tonsillar crypts** (see Fig. 7-22). The lateral surface is covered by a capsule. The tonsils reach their maximum size in childhood but become smaller after puberty. The lymph drainage is into the upper deep cervical lymph nodes, just below and behind the angle of the mandible.

Swallowing

The masticated food is formed into a ball or bolus on the dorsum of the tongue and voluntarily pushed upward and backward against the undersurface of the hard palate. This is brought about by the contraction of the styloglossus muscles on both sides, which pull the root of the tongue upward and backward. The palatoglossus muscles squeeze the bolus backward into the pharynx. From this point onward, the process of swallowing becomes an involuntary act.

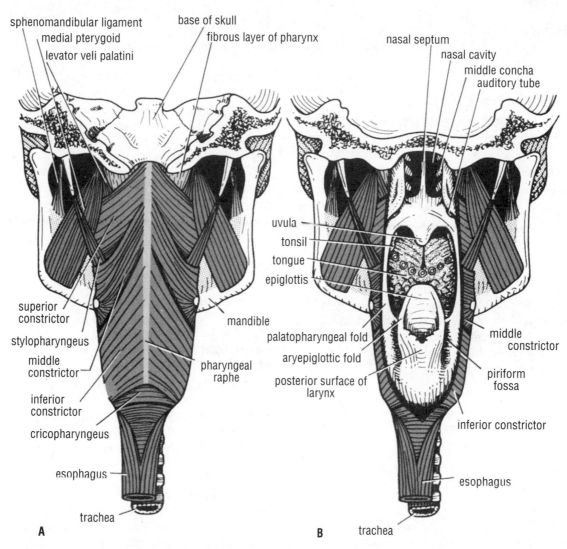

Fig. 7-23. Pharynx seen from behind. **A.** Note three constrictor muscles and position of stylopharyngeus muscles. **B.** Greater part of posterior wall of pharynx has been removed to display nasal, oral, and laryngeal parts of pharynx.

The nasal part of the pharynx is shut off from the oral part of the pharynx by the elevation of the soft palate, the pulling forward of the posterior wall of the pharynx, and the pulling medially of the palatopharyngeal folds. This prevents the passage of food and drink into the nasal cavities.

The larynx and the laryngeal part of the pharynx are then pulled upward by the contraction of the stylopharyngeus, salpingopharyngeus, thyrohyoid, and palatopharyngeus muscles. The main part of the larynx is thus elevated to the posterior surface of the epiglottis, and the entrance into the larynx is closed.

The bolus moves downward over the epiglottis, the closed entrance into the larynx, and reaches the lower part of the pharynx as the result of the successive contraction of the superior, middle, and inferior constrictor muscles. Some of the food slides down the piriform fossa on either side of the entrance into the larynx. Finally, the lower part of the pharyngeal wall relaxes and the bolus enters the esophagus.

Esophagus

The esophagus begins in the neck at the level of the cricoid cartilage as a continuation of the pharynx (see Fig. 7-23). The esophagus descends in the midline behind the trachea (see Fig. 7-5).

Important Relations

Anteriorly: Trachea and recurrent laryngeal nerves

Posteriorly: Prevertebral muscles and vertebral column

Laterally: Thyroid gland, carotid sheath, and the thoracic duct on the left side

Blood Supply

Inferior thyroid arteries.

Lymph Drainage

Deep cervical lymph nodes.

Nerve Supply

Recurrent laryngeal nerves and branches from sympathetic trunk.

Review of the Respiratory System in the Head and Neck

Nose
External Nose

The external nose is formed above by the nasal bones, the frontal processes of the maxillae, and the nasal part of the frontal bone. Below, the framework is formed of plates of hyaline cartilage.

Nasal Cavity

The nasal cavity is divided into right and left halves by the **nasal septum** (Fig. 7-24). The septum is made up of the **septal cartilage,** the **vertical plate of the ethmoid,** and the **vomer.** Each nasal cavity extends from the **nostril** in front to the **choanae** behind. The nasal cavity has a floor, a roof, and lateral and medial walls.

Floor

Palatine process of maxilla and horizontal plate of palatine bone.

Roof

Body of sphenoid, cribriform plate of ethmoid, frontal bone, and nasal bone.

Lateral Wall

This has three projections called the **superior, middle,** and **inferior nasal conchae** (see Fig. 7-24). The space below each concha is called a **meatus.**

Sphenoethmoidal Recess. A small area that lies above the superior concha. It receives the opening of the **sphenoidal air sinus.**

Superior Meatus. Lies below the superior concha. It receives the openings of the **posterior ethmoidal sinuses.**

Middle Meatus. Lies below the middle concha. It has a rounded swelling called the **bulla ethmoidalis,** caused by the underlying **middle ethmoidal sinuses,** which open on its upper border. A curved opening, the **hiatus semilunaris,** lies just below the bulla (see Fig. 7-24). The anterior end of the hiatus leads into a funnel-shaped channel called the **infundibulum.** The **maxillary sinus** opens into the middle meatus through the hiatus semilunaris. The **frontal sinus** opens into and is continuous with the infundibulum. The middle meatus is continuous in front with the **vestibule** via the **atrium.** The vestibule is lined by modified skin bearing **hairs.**

Inferior Meatus. Lies below the inferior concha and receives the opening of the **nasolacrimal duct.** The opening is guarded by a fold of mucous membrane.

Medial Wall

Formed by the nasal septum.

Mucous Membrane of Nasal Cavity

The vestibule is lined by modified skin. The lateral wall above the superior concha, the corresponding area of the septum, and the roof are lined with olfactory mucous membrane. The lower part of the nasal cavity is lined by respiratory mucous membrane.

Nerve Supply

Olfactory nerves innervate the olfactory mucous membrane; the remainder is supplied by branches of the ophthalmic and maxillary divisions of the trigeminal nerve.

Blood Supply

Branches of the maxillary artery. The sphenopalatine branch anastomoses with the septal branch of the facial artery in the vestibule, a common site for nose bleeds.

Paranasal Sinuses
Maxillary Sinus

The maxillary sinus is pyramidal in shape and lies within the maxilla on each side (Fig. 7-25). The roof is formed by the floor of the orbit, and the floor is related to the roots of the premolar and molar teeth. The maxillary sinus opens into the middle meatus of the nose through the hiatus semilunaris.

Frontal Sinus

The two frontal sinuses are contained within the frontal bone (see Fig. 7-24). Each sinus opens into the middle meatus of the nose through the infundibulum.

Sphenoidal Sinuses

The two sphenoidal sinuses lie within the body of the sphenoid bone (see Fig. 7-24). Each sinus opens into the sphenoethmoidal recess above the superior concha.

Ethmoidal Sinuses

The ethmoidal sinuses are anterior, middle, and posterior and are contained within the ethmoid bone on each side (see Fig. 7-25). The anterior sinus opens into the infundibulum of the middle meatus; the middle sinus opens into the middle meatus, above or on the bulla ethmoidalis; and the posterior sinus opens into the superior meatus.

Table 7-12 shows the various sinuses and their openings into the nose.

Larynx

The larynx is a sphincter that is responsible for voice production (Figs. 7-26 and 7-27). It opens above into the laryngeal part of the pharynx and is continuous with the trachea below.

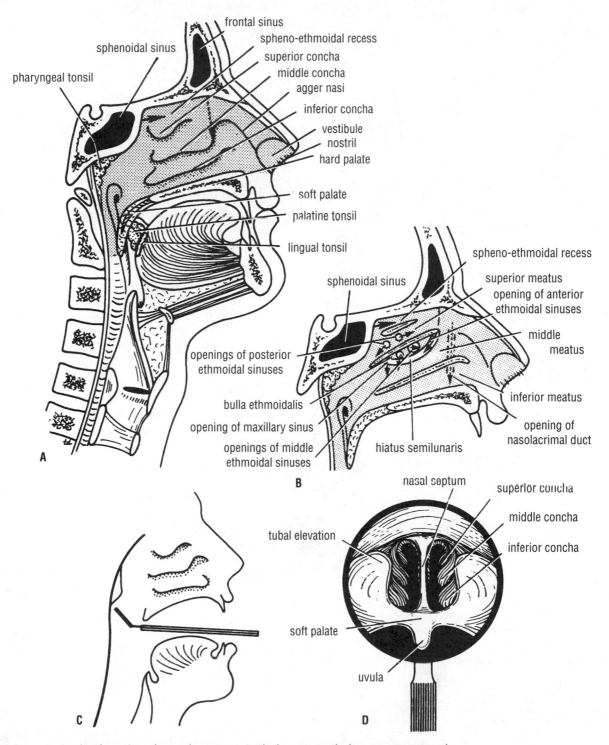

Fig. 7-24. **A.** Sagittal section through nose, mouth, larynx, and pharynx. **B.** Lateral wall of nose and nasal part of pharynx. **C.** Position of mirror in posterior rhinoscopy. **D.** Structures seen in posterior rhinoscopy.

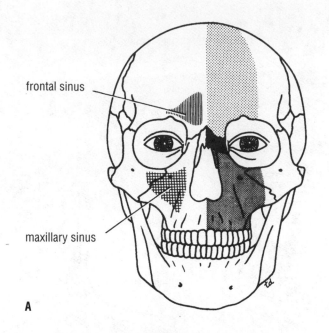

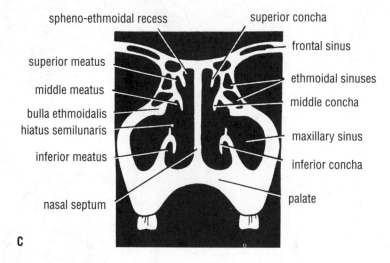

Fig. 7-25. A. Bones of face, showing positions of frontal and maxillary sinuses. **B.** Regions where pain is experienced in sinusitis. (Lightly dotted area in frontal sinusitis; solid area in sphenoethmoidal sinusitis; and heavily dotted area in maxillary sinusitis.) **C.** Coronal section through nasal cavity, showing the frontal, ethmoidal, and maxillary sinuses.

Table 7-12 The Paranasal Sinuses and Their Site of Drainage into the Nose

Name of sinus	Site of drainage
Maxillary sinus	Middle meatus through hiatus semilunaris
Frontal sinus	Middle meatus via infundibulum
Sphenoid sinuses	Sphenoethmoidal recess
Ethmoidal sinuses	
Anterior group	Infundibulum and into middle meatus
Middle group	Middle meatus on or above bulla ethmoidalis
Posterior group	Superior meatus

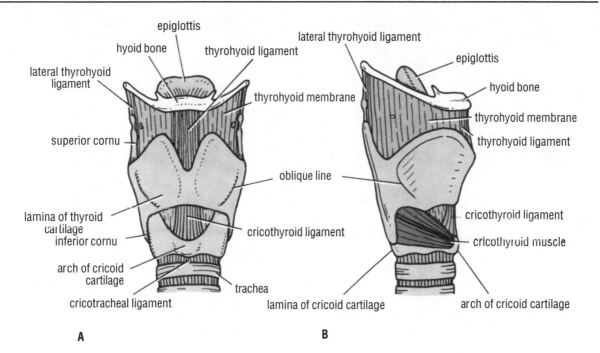

A

B

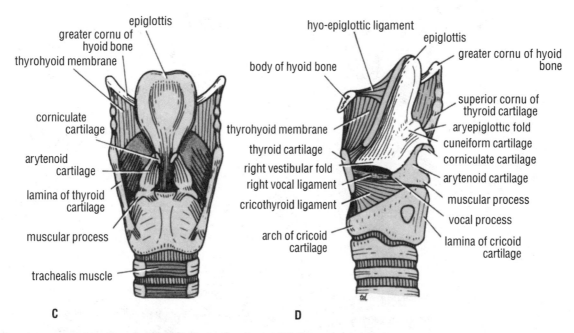

C

D

Fig. 7-26. Larynx and its ligaments: **(A)** from the front, **(B)** from the lateral aspect, **(C)** from behind. **(D)** Left lamina of thyroid cartilage has been removed to display interior of larynx.

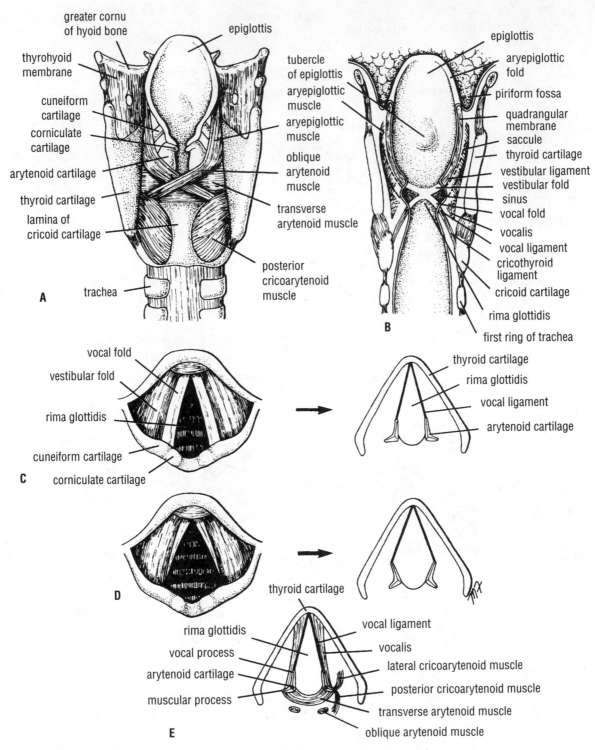

Fig. 7-27.   **A.** Muscles of larynx as seen from behind. **B.** Coronal section through larynx. **C.** Rima glottidis partially open as in quiet breathing. **D.** Rima glottidis wide open as in deep breathing. **E.** Muscles that move vocal ligaments.

Cartilages of the Larynx

Thyroid Cartilage (see Fig. 7-26)

This is the largest and consists of two laminae, each of which has a **superior cornu** and an **inferior cornu.** On the outer surface is an oblique line for the attachment of muscles.

Cricoid Cartilage (see Fig. 7-27)

This has a shallow arch in front and a broad lamina behind. It lies below the thyroid cartilage and articulates with the inferior cornu. Posteriorly, the lamina articulates on its upper edge with the arytenoid cartilages.

Arytenoid Cartilages (see Fig. 7-26)

Two in number and pyramidal in shape, they articulate with the upper border of the lamina of the cricoid cartilage. Each cartilage has an **apex,** a **base,** a **vocal process,** and a **muscular process.**

Corniculate Cartilages (see Fig. 7-26)

Two in number and conical in shape, they articulate with the arytenoid cartilages.

Cuneiform Cartilages (see Fig. 7-26)

These are rod-shaped and strengthen the aryepiglottic folds.

Epiglottis

The epiglottis is leaf-shaped and lies behind the root of the tongue (see Figs. 7-26 and 7-27). It is attached by its stalk to the thyroid cartilage.

Membranes and Ligaments

Thyrohyoid Membrane

Connects the upper margin of the thyroid cartilage to the hyoid bone (see Fig. 7-27). It is pierced by the superior laryngeal vessels and the internal laryngeal nerve.

Quadrangular Membrane

Extends between the epiglottis and the arytenoid cartilages; its inferior margin forms the **vestibular ligaments.** The vestibular ligaments form the interior of the vestibular folds.

Cricothyroid Ligament

Attached below to the upper margin of the cricoid cartilage (see Fig. 7-27). Above, instead of being attached to the lower margin of the thyroid cartilage, it ascends on the medial surface of the cartilage. The upper margin forms the important **vocal ligament** on each side (see Fig. 7-27). The vocal ligaments form the interior of the **vocal folds** (vocal cords). The anterior end of each vocal ligament is attached to the thyroid cartilage; the posterior end is attached to the vocal process of the arytenoid cartilage.

Cavity of the Larynx

The cavity has three regions: (1) the **vestibule,** situated between the inlet and the vestibular folds; (2) the middle region, situated between the vestibular folds above and the vocal folds below; and (3) the lower region, situated between the vocal folds above and the lower border of the cricoid cartilage below.

Laryngeal Folds

Vestibular Fold

This is a **fixed** fold on each side of the larynx (see Fig. 7-27). It is formed of mucous membrane covering the vestibular ligament. It is **pink** in color.

Vocal Fold (vocal cord)

This is a **mobile** fold on each side of the larynx. It is formed of mucous membrane covering the vocal ligament. It is **white** in color.

Sinus of the Larynx

The sinus of the larynx is a small recess on each side of the larynx situated between the vestibular and the vocal folds (see Fig. 7-27). It is lined with mucous membrane.

Saccule of the Larynx

The saccule of the larynx is a diverticulum that ascends from the sinus (see Fig. 7-27).

Muscles of the Larynx

Extrinsic Muscles

Elevation of Larynx.   Digastric, stylohyoid, mylohyoid, geniohyoid, stylopharyngeus, salpingopharyngeus, and palatopharyngeus muscles.

Depression of Larynx.   Sternothyroid, sternohyoid, and omohyoid muscles.

Intrinsic Muscles

Muscles Modifying the Laryngeal Inlet (see Fig. 7-27)

Narrowing the inlet: Oblique arytenoid

Widening the inlet: Thyroepiglottic

Muscles Moving the Vocal Folds (see Fig. 7-27)

Tensing the vocal folds: Cricothyroid

Relaxing the vocal folds: Thyroarytenoid (vocalis)

Adducting the vocal folds: Lateral cricoarytenoid

Abducting the vocal folds: Posterior cricoarytenoid

Approximates arytenoids: Transverse arytenoid

The muscles of the larynx are shown in Table 7-13.

Nerve Supply of Larynx

Sensory Nerves

Above vocal folds: Internal laryngeal nerve

Below vocal folds: Recurrent laryngeal nerve

Motor Nerves

All the intrinsic muscles are supplied by the recurrent laryngeal nerve except the cricothyroid muscle, which is supplied by the external laryngeal nerve.

Table 7-13   Muscles of the Larynx

Name of muscle	Origin	Insertion	Nerve supply	Action
Extrinsic muscles				
Elevator muscles	These include the digastric, the stylohyoid, the mylohyoid, the geniohyoid, the stylopharyngeus, the salpingopharyngeus, and the palatopharyngeus muscles. For details see Tables 7-4 and 7-11.			
Depressor muscles	These include the sternothyroid, the sternohyoid, and the omohyoid muscles. For details see Table 7-4.			
Intrinsic muscles				
Muscles controlling the laryngeal inlet				
Oblique arytenoid	Muscular process of arytenoid cartilage	Apex of opposite arytenoid cartilage	Recurrent laryngeal nerve	Narrows inlet by bringing the aryepiglottic folds together
Thyroepiglottic	Medial surface of thyroid cartilage	Lateral margin of epiglottis and aryepiglottic fold	Recurrent laryngeal nerve	Widens the inlet by pulling aryepiglottic folds apart
Muscles controlling the movements of the vocal cords				
Cricothyroid	Side of cricoid cartilage	Lower border and inferior cornu of thyroid cartilage	External laryngeal nerve	Tenses vocal cords
Thyroarytenoid (vocalis is part of muscle)	Inner surface of thyroid cartilage at angle between laminae	Anterior surface of arytenoid cartilage	Recurrent laryngeal nerve	Relaxes vocal cords
Lateral cricoarytenoid	Upper border of cricoid cartilage	Muscular process of arytenoid cartilage	Recurrent laryngeal nerve	Adducts the vocal cords by rotating arytenoid cartilage
Posterior cricoarytenoid	Back of cricoid cartilage	Muscular process of arytenoid cartilage	Recurrent laryngeal nerve	Abducts the vocal cords by rotating arytenoid cartilage
Transverse arytenoid	Back and medial surface of arytenoid cartilage	Back and medial surface of opposite arytenoid cartilage	Recurrent laryngeal nerve	Closes posterior part of rima glottidis by approximating arytenoid cartilages

Trachea

The trachea begins at the lower border of the larynx and descends in the midline of the neck to enter the thorax (see Figs. 7-5 and 7-20).

Important Relations in the Neck

Anteriorly: Skin, fascia, **isthmus of thyroid gland** (in front of second, third, and fourth rings), inferior thyroid veins, jugular arch, thyroidea ima artery (if present), and left brachiocephalic vein in the child, overlapped by the sternothyroid and sternohyoid muscles

Posteriorly: Right and left recurrent laryngeal nerves and esophagus

Laterally: Lobes of thyroid gland and the carotid sheath and contents

Review of the Endocrine Glands in the Head and Neck

Pituitary Gland (Hypophysis Cerebri)

The pituitary gland is situated in the hypophyseal fossa in the sella turcica of the sphenoid bone (Fig. 7-28). It is connected to the undersurface of the brain by the **stalk (infundibulum)**. The pituitary gland may be divided into an **anterior lobe (adenohypophysis)** and a **posterior lobe (neurohypophysis)**. The anterior lobe may be subdivided into the **pars anterior** and the **pars intermedia** by a cleft that is a remnant of an embryonic pouch. A projection from the pars anterior, the **pars tuberalis** extends up along the anterior and lateral surfaces of the pituitary stalk.

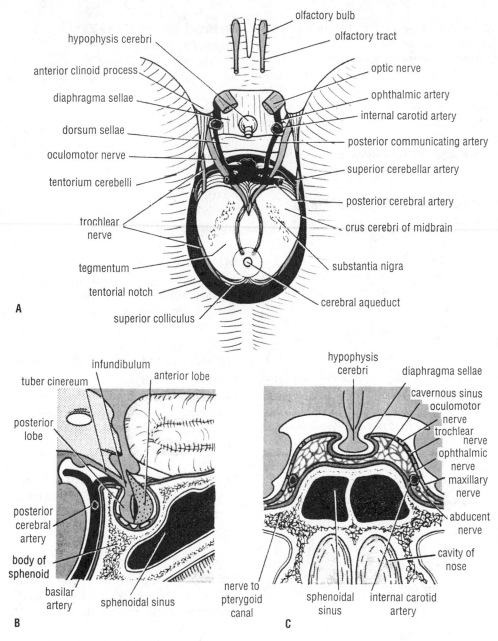

Fig. 7-28. **A.** Interior of skull as seen from above. Forebrain has been removed, leaving midbrain, hypophysis cerebri, and internal carotid and basilar arteries in position. **B.** Sagittal section through sella turcica, showing hypophysis cerebri. **C.** Coronal section through body of sphenoid, showing hypophysis cerebri and cavernous sinuses. Note position of cranial nerves.

Important Relations

Anteriorly: Sphenoid sinus

Posteriorly: Dorsum sellae, basilar artery, and pons

Superiorly: Diaphragma sellae and optic chiasma

Inferiorly: Body of sphenoid and sphenoid sinus

Laterally: Cavernous sinus and its contents (internal carotid artery and abducent nerve)

Blood Supply

Superior and inferior hypophyseal arteries, branches of the internal carotid artery.

Thyroid Gland

The thyroid gland consists of right and left lobes connected by a narrow isthmus (see Fig. 7-4). It is surrounded by a sheath formed of pretracheal fascia. The sheath attaches the gland to the larynx and the trachea.

Each **lobe** of the gland is pear-shaped with its apex directed upward along the lateral side of the thyroid cartilage; its base lies below alongside the trachea. The **isthmus** crosses the midline in front of the **second, third,** and **fourth rings of the trachea.** A **pyramidal lobe** is often present, and it projects upward from the isthmus. A muscular band, the **levator glandulae thyroideae,** often connects the pyramidal lobe to the hyoid bone.

Important Relations of the Thyroid Gland

Anterolaterally: Infrahyoid group of muscles and anterior border of sternocleidomastoid muscle

Posterolaterally: Carotid sheath and contents

Medially: Larynx, trachea, pharynx, esophagus, external laryngeal nerve, and recurrent laryngeal nerve

Posteriorly: Parathyroid glands

Blood Supply
Arteries

Superior thyroid artery (related to external laryngeal nerve) from external carotid, inferior thyroid artery (related to recurrent laryngeal nerve) from thyrocervical trunk, and thyroidea ima artery (if present) from brachiocephalic or aortic arch.

Veins

Superior and middle thyroid veins drain into internal jugular vein; inferior thyroid vein drains into left brachiocephalic vein.

Parathyroid Glands
Superior Parathyroid Glands
The two superior parathyroid glands lie behind the middle of the posterior surface of the thyroid gland.

Inferior Parathyroid Glands
The two inferior parathyroid glands lie close to the inferior poles of the thyroid gland (sometimes found some distance below in the superior mediastinum). The parathyroid glands lie within the fascial capsule of the thyroid gland.

Blood Supply
Superior and inferior thyroid arteries.

Orbit

Orbital Margin
The orbital margin is formed by the frontal, maxilla, and zygomatic bones.

Orbital Walls (Fig. 7-29)
Superior Wall
The superior wall, or roof, is formed by the orbital plate of the frontal bone and the lesser wing of the sphenoid bone.

Inferior Wall
The inferior wall, or floor, is formed by the orbital plate of the maxilla, which separates the orbital cavity from the maxillary sinus.

Lateral Wall
The lateral wall is formed by the zygomatic bone and the greater wing of the sphenoid.

Medial Wall
The medial wall is formed from before backward by the frontal process of the maxilla, the lacrimal bone, the orbital plate of the ethmoid (which separates the orbital cavity from the ethmoid sinuses), and the body of the sphenoid.

Openings into the Orbital Cavity (see Fig. 7-29)
Orbital Opening
The orbital opening lies anteriorly. About one-sixth of the eye is exposed; the remainder is protected by the walls of the orbit.

Supraorbital Groove (Foramen)
The supraorbital groove is situated on the superior orbital margin. It transmits the supraorbital nerve and blood vessels.

Infraorbital Groove and Canal
The infraorbital groove and canal is situated on the floor of the orbit in the orbital plate of the maxilla; it transmits the infraorbital nerve and blood vessels.

Nasolacrimal Canal
The nasolacrimal canal is located anteriorly on the medial wall; it communicates with the inferior meatus of the nose. It transmits the nasolacrimal duct.

Inferior Orbital Fissure
The inferior orbital fissure is located posteriorly between the maxilla and the greater wing of the sphenoid; it communicates with the pterygopalatine fossa. It transmits the maxillary nerve and its zygomatic branch, the inferior ophthalmic vein, and sympathetic nerves.

Superior Orbital Fissure
The superior orbital fissure is located posteriorly between the greater and lesser wings of the sphenoid; it communicates

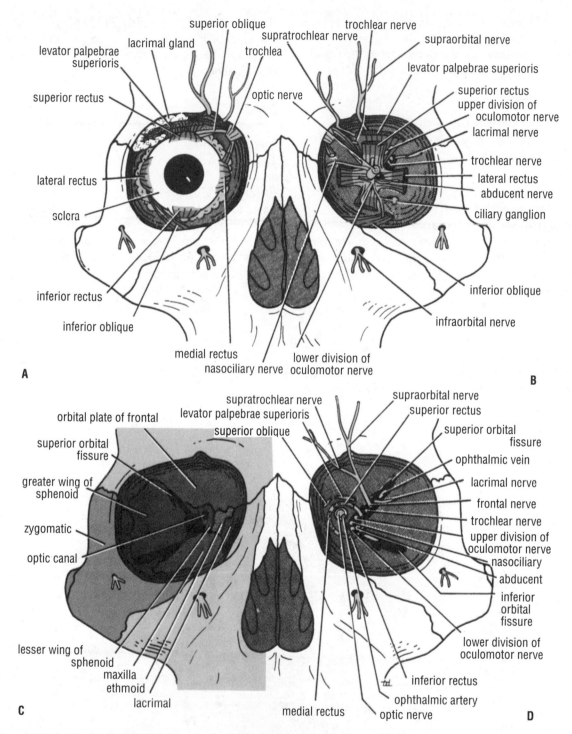

Fig. 7-29. **A.** Right eyeball exposed from in front. **B.** Muscles and nerves of left orbit seen from in front. **C.** Bones forming walls of right orbit. **D.** Optic canal and superior and inferior orbital fissures on left side.

with the middle cranial fossa. It transmits the lacrimal nerve, the frontal nerve, the trochlear nerve, the oculomotor nerve (upper and lower divisions), the abducent nerve, the nasociliary nerve, and the superior ophthalmic vein.

Optic Canal
The optic canal is located posteriorly in the lesser wing of the sphenoid (see Fig. 7-29); it communicates with the middle cranial fossa. It transmits the optic nerve and the ophthalmic artery.

Zygomaticotemporal and Zygomaticofacial Foramina
Two small openings in the lateral wall, the foramina transmit the zygomaticotemporal and the zygomaticofacial nerves respectively.

Anterior and Posterior Ethmoidal Foramina
The ethmoidal foramina are located on the medial wall in the ethmoid bone; they transmit the anterior and posterior ethmoidal nerves respectively.

Eye

Eye Muscles
There are extrinsic and intrinsic eye muscles.

Extrinsic Muscles
There are six extrinsic muscles that run from the posterior wall of the orbital cavity to the eyeball (Fig. 7-30; see also Fig. 7-29): the **superior rectus,** the **inferior rectus,** the **medial rectus,** the **lateral rectus,** and the **superior** and **inferior oblique muscles.**

Fig. 7-30. Muscles and nerves of right orbit viewed from lateral side. Maxillary nerve and pterygopalatine ganglion are also shown.

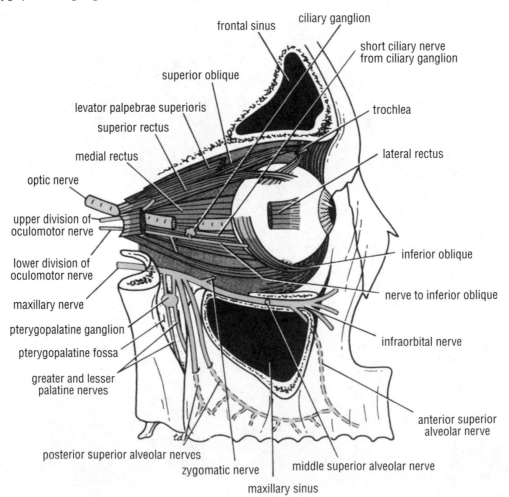

Intrinsic Muscles

The intrinsic muscles are the **ciliary muscle** and the **constrictor** and **dilator pupillae of the iris** (see below).

Because the superior and inferior recti are inserted on the medial side of the vertical axis of the eyeball, they not only raise and depress the cornea respectively, but **rotate it medially.** For the superior rectus muscle to raise the cornea directly upward, the inferior oblique must assist; for the inferior rectus to depress the cornea directly downward, the superior oblique must assist. Note that the tendon of the superior oblique muscle passes through a fibrocartilaginous pulley (trochlea) attached to the frontal bone. The tendon now turns backward and laterally and is inserted into the sclera beneath the superior rectus.

The origins, insertions, nerve supply, and actions of the muscles of the eyeball are summarized in Table 7-14.

Main Parts of the Eyeball (Fig. 7-31)

There are three **coats:** (1) fibrous, (2) vascular pigmented, and (3) nervous.

Fibrous Coat

The fibrous coat is made up of a posterior white opaque part, the **sclera,** and an anterior transparent part, the **cornea** (see Fig. 7-31). The **lamina cribrosa** is the area of the sclera that is pierced by the nerve fibers of the optic nerve. The sclera is directly continuous with the cornea at the **corneoscleral junction (limbus).**

Vascular Pigmented Coat

From back to front, the vascular pigmented coat consists of the choroid, the ciliary body, and the iris (see Fig. 7-31).

Choroid

This is composed of an outer pigmented layer and an inner vascular layer.

Ciliary Body

A complete ring that runs around the inside of the sclera (see Fig. 7-31). It is composed of the ciliary processes and the ciliary muscle. The **ciliary processes** are radially arranged folds that are connected to the **suspensory ligaments of the lens.** The **ciliary muscle,** which is responsible for changing the shape of the lens, is composed of meridional and circular fibers of smooth muscle. The meridional fibers run backward from the region of the corneoscleral junction to the ciliary processes. The circular fibers run around the eyeball within the ciliary body.

Nerve Supply of Ciliary Muscle. Parasympathetic fibers within the oculomotor nerve synapse in the ciliary ganglion. Postganglionic fibers reach the eyeball in the short ciliary nerves.

Action of Ciliary Muscle. By pulling the suspensory ligaments forward, the muscle slackens the ligaments, the elastic lens becomes more convex, and the refractive power is increased.

Iris

The iris is a thin, contractile, pigmented sheet with a central hole, the **pupil.** The iris is suspended in the aqueous humor between the cornea and the lens (see Fig. 7-31). It divides the space between the cornea and the lens into an **anterior** and a **posterior chamber.** The function of the iris is to control the amount of light entering the eye. **Sphincter pupillae** is a collection of smooth muscle fibers arranged around the pupil. **Dilator pupillae** consists of radial fibers.

Nerve Supply of Muscle of Iris. The sphincter pupillae is supplied by parasympathetic fibers from the oculomotor nerve; they synapse in the ciliary ganglion, and the postganglionic fibers reach the eyeball in the short ciliary nerves. The dilator pupillae is supplied by sympathetic fibers that reach the eye in the long ciliary nerves.

Action of Iris. The sphincter pupillae constricts the pupil in the presence of bright light and during accommodation. The dilator pupillae dilates the pupil in the presence of light of low intensity or in the presence of excessive sympathetic activity.

Nervous Coat: Retina

The nervous coat is the innermost layer of the eyeball (see Fig. 7-31). It consists of an outer **pigmented layer** and an inner **nervous layer.**

The **optic nerve** leaves the retina at the optic disc. The **optic disc** is depressed at its center and is pierced by the **central artery of the retina** (see Fig. 7-31).

Contents of the Eyeball

Aqueous Humor

The aqueous humor is a clear fluid that fills the anterior and posterior chambers.

Vitreous Body

The vitreous body fills the eyeball behind the lens; it is a transparent gel.

Lens

The lens is a transparent, biconvex disc that lies behind the iris and in front of the vitreous body (see Fig. 7-31). The circumference of the lens is attached to the ciliary processes of the ciliary body by the suspensory ligament. The pull of the radiating fibers of the suspensory ligament tends to keep the elastic lens flattened so that the eye may be focused on distant objects.

Periorbital Structures

Eyelids

The eyelids are thin movable folds having a fibrous framework, the **orbital septum.** The orbital septum is thickened at the lid margins to form the **tarsal plates.** The ends of the

Head and Neck

Table 7-14   Muscles of the Eyeball and Eyelids

Name of muscle	Origin	Insertion	Nerve supply	Action
Extrinsic muscles of eyeball (striated skeletal muscle)				
Superior rectus	Tendinous ring on posterior wall of orbital cavity	Superior surface of eyeball just posterior to corneoscleral junction	Oculomotor nerve (third cranial nerve)	Raises cornea upward and medially
Inferior rectus	Tendinous ring on posterior wall of orbital cavity	Inferior surface of eyeball just posterior to corneoscleral junction	Oculomotor nerve (third cranial nerve)	Depresses cornea downward and medially
Medial rectus	Tendinous ring on posterior wall of orbital cavity	Medial surface of eyeball just posterior to corncoscleral junction	Oculomotor nerve (third cranial nerve)	Rotates eyeball so that cornea looks medially
Lateral rectus	Tendinous ring on posterior wall of orbital cavity	Lateral surface of eyeball just posterior to corneoscleral junction	Abducent nerve (sixth cranial nerve)	Rotates eyeball so that cornea looks laterally
Superior oblique	Posterior wall of orbital cavity	Passes through pulley and is attached to superior surface of eyeball beneath superior rectus	Trochlear nerve (fourth cranial nerve)	Rotates eyeball so that cornea looks downward and laterally
Inferior oblique	Floor of orbital cavity	Lateral surface of eyeball deep to lateral rectus	Oculomotor nerve (third cranial nerve)	Rotates eyeball so that cornea looks upward and laterally
Intrinsic muscles of eyeball (smooth muscle)				
Sphincter pupillae of iris			Parasympathetic via oculomotor nerve	Constricts pupil
Dilator pupillae of iris			Sympathetic	Dilates pupil
Ciliary muscle			Parasympathetic via oculomotor nerve	Controls shape of lens; in accommodation, makes lens more globular
Muscles of eyelids				
Orbicularis oculi (see Table 7-8)				
Levator palpebrae superioris	Back of orbital cavity	Anterior surface and upper margin of superior tarsal plate	Striated muscle oculomotor nerve, smooth muscle sympathetic	Raises upper lid

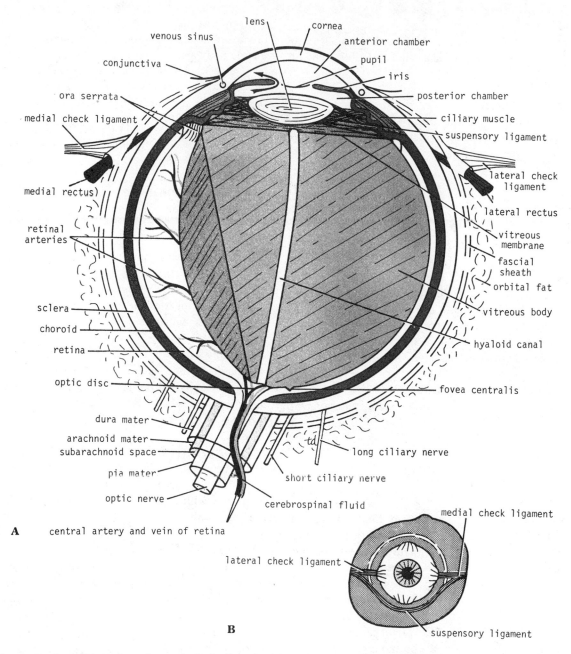

Fig. 7-31. A. Horizontal section through eyeball and optic nerve. Note that central artery and vein of retina cross subarachnoid space to reach optic nerve. **B.** Check ligaments and suspensory ligament of the eyeball.

plates are attached to the orbital margins by the **lateral and medial palpebral ligaments.**

The eyelids are closed by the contraction of the orbicularis oculi muscle. The eye is opened by the levator palpebrae superioris, which raises the upper lid.

Lacrimal Apparatus

The apparatus consists of the lacrimal gland and the ducts that drain the conjunctival sac.

Lacrimal Gland

The lacrimal gland secretes tears and is situated above the eyeball in the anterior and upper part of the orbit (see Fig. 7-29). Several ducts open from the gland into the superior part of the conjunctival sac.

Nerve Supply

Parasympathetic secretomotor nerves from the facial nerve (seventh cranial nerve).

Lacrimal Sac

Tears circulate across the cornea and enter the canaliculi to open into the lacrimal sac. The sac is the upper blind end of the nasolacrimal duct.

Nasolacrimal Duct

The nasolacrimal duct emerges from the lower end of the sac and descends downward and laterally in a bony canal to enter the inferior meatus of the nose (see Fig. 7-24).

Ear

The ear consists of the external ear, the middle ear or tympanic cavity, and the internal ear.

External Ear

The external ear has an **auricle** and an **external auditory meatus** (Fig. 7-32). The meatus is a curved tube about 1 in. (2.5 cm) long that leads from the auricle to the tympanic membrane. The outer third has a framework of elastic cartilage, and the inner two-thirds have a framework of bone.

Middle Ear (Tympanic Cavity)

The middle ear has four walls: lateral, medial, anterior, and posterior.

Lateral Wall

Tympanic membrane (see Fig. 7-32).

Medial Wall

Lateral wall of the inner ear. The **promontory** is a rounded projection due to the underlying first turn of the cochlea (Fig. 7-33). The **fenestra vestibuli (oval window),** which is oval in shape and closed by the footpiece of the stapes, lies above and behind the promontory. The **fenestra cochlea,** which is round in shape and closed by the **secondary tympanic membrane,** lies below the posterior end of the promontory. The **prominence of the facial canal,** for the facial nerve, runs backward above the promontory to reach the posterior wall. Here, it turns downward behind the pyramid.

Anterior Wall

Canal for the **tensor tympani muscle** and the opening for the **auditory tube.**

Posterior Wall

Aditus to the mastoid antrum. Below this is a hollow conical projection from whose apex emerges the tendon of the **stapedius muscle.**

Auditory Ossicles

These are the malleus, incus, and stapes (see Fig. 7-32). The **malleus** (hammer) is the largest ossicle and possesses a head, a neck, a long process or handle, an anterior process, and a lateral process. The **incus** (anvil) possesses a large body and two processes. The **stapes** (stirrup) has a head, a neck, two limbs, and a base. The edge of the base is attached to the margin of the fenestra vestibuli by a ring of fibrous tissue, the **anular ligament.**

Muscles of the Ossicles

The muscles of the tympanic cavity (ossicles) are summarized in Table 7-15.

Tympanic Membrane

The tympanic membrane is concave laterally. At the depth of the concavity is a small depression, the **umbo** (see Fig. 7-32). The membrane is composed of fibrous tissue and is covered on the outer surface with stratified squamous epithelium; the outer surface is innervated by the trigeminal and vagus nerves. The inner surface of the membrane is covered with mucous membrane and is innervated by the glossopharyngeal nerve.

Auditory Tube

The auditory tube extends from the anterior wall of the tympanic cavity to the nasal pharynx (see Fig. 7-33). Its posterior third is bony, and its anterior two-thirds are cartilaginous.

Mastoid Antrum

The mastoid antrum lies behind the tympanic cavity in the petrous part of the temporal bone (see Fig. 7-33). It communicates with the tympanic cavity through the posterior tympanic wall.

Mastoid Air Cells

The mastoid air cells are a series of communicating cavities within the mastoid process (see Fig. 7-33). The cells are continuous above with the antrum and the tympanic cavity.

Facial Nerve

The facial nerve enters the internal acoustic meatus and then the facial canal (see Fig. 7-33). It emerges through the stylomastoid foramen and enters the parotid salivary gland. The branches of this important nerve are described on page 239.

Tympanic Plexus

The tympanic plexus lies on the promontory of the medial wall of the tympanic cavity. It is formed from the tympanic branch of the glossopharyngeal nerve, sympathetic nerves, and a communicating branch from the facial nerve. The plexus supplies the mucous membrane of the tympanic cavity and gives origin to the lesser petrosal nerve, which passes to the otic ganglion (see p. 239).

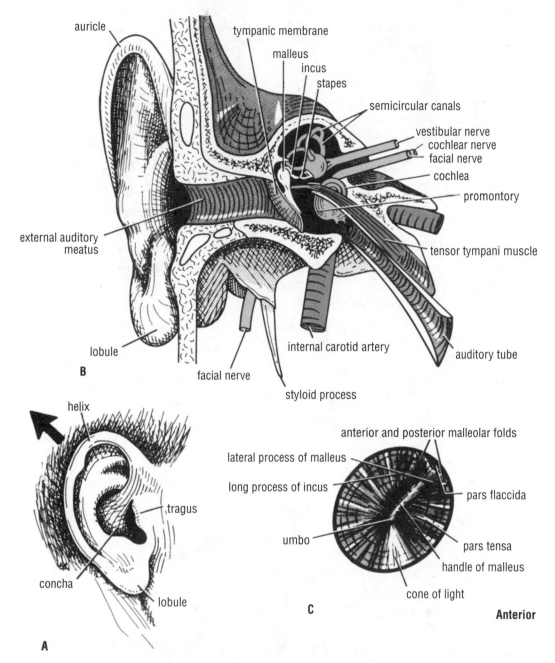

Fig. 7-32. **A.** Different parts of auricle of external ear. Arrow indicates direction that auricle should be pulled to straighten external auditory meatus prior to insertion of otoscope in the adult. **B.** External and middle portions of right ear, viewed from in front. **C.** Right tympanic membrane as seen through otoscope.

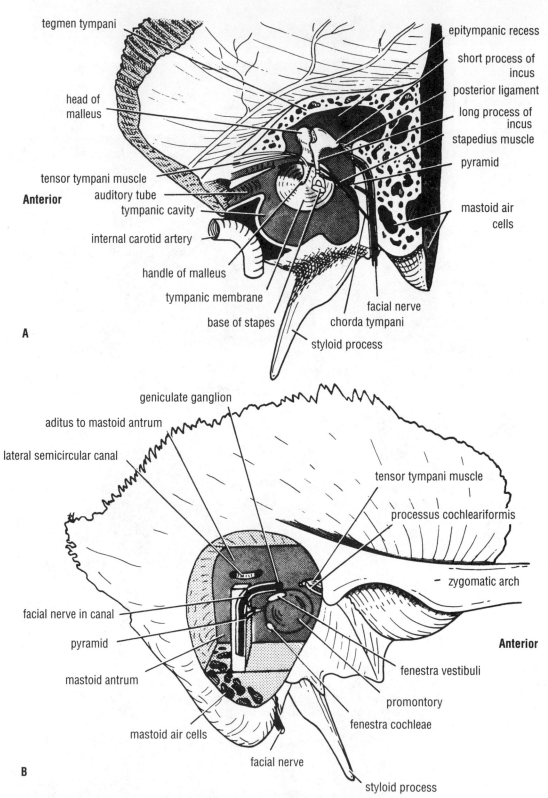

tegmen tympani

epitympanic recess

short process of
incus

posterior ligament

head of
malleus

long process of
incus

stapedius muscle

pyramid

tensor tympani muscle

auditory tube

Anterior

tympanic cavity

mastoid air
cells

internal carotid artery

handle of malleus

tympanic membrane

facial nerve

base of stapes

chorda tympani

styloid process

A

geniculate ganglion

aditus to mastoid antrum

lateral semicircular canal

tensor tympani muscle

processus cochleariformis

zygomatic arch

facial nerve in canal

Anterior

pyramid

mastoid antrum

fenestra vestibuli

promontory

fenestra cochleae

mastoid air cells

facial nerve

B

styloid process

Fig. 7-33.   A. Lateral wall of right middle ear viewed from medial side. Note position of ossicles and mastoid antrum. **B.** Medial wall of right middle ear viewed from lateral side. Note position of facial nerve in its bony canal.

Table 7-15 Muscles of the Tympanic Cavity

Name of muscle	Origin	Insertion	Nerve supply	Action
Tensor tympani	Wall of auditory tube and wall of its own canal	Handle of malleus	Mandibular division of trigeminal nerve (fifth cranial nerve)	Dampens down vibrations of tympanic membrane
Stapedius	Pyramid (bony projection on posterior wall of middle ear)	Neck of stapes	Facial nerve (seventh cranial nerve)	Dampens down vibrations of stapes

National Board Type Questions

Match the numbered structures shown in the lateral radiograph of the skull with the lettered structures (A–F) listed below.

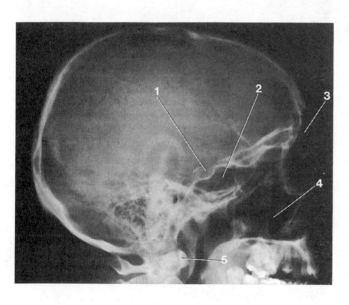

1. Structure 1.
2. Structure 2.
3. Structure 3.
4. Structure 4.
5. Structure 5.

A. Maxillary sinus.
B. Frontal sinus.
C. Anterior arch of atlas.
D. Sella turcica.
E. Sphenoid sinus.
F. Temporomandibular joint.

Match the numbered structures shown in the anteroposterior radiograph of the skull with the lettered structures (A–F) listed below.

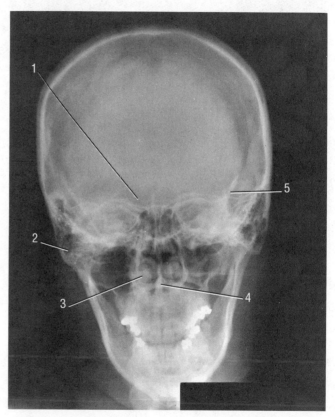

6. Structure 1.
7. Structure 2.
8. Structure 3.
9. Structure 4.
10. Structure 5.

A. Nasal septum.
B. Mastoid air cells.
C. Greater wing of sphenoid.
D. Frontal sinus.
E. Maxillary sinus.
F. Inferior concha.

Match the numbered structures shown in the axial (horizontal) computed tomographic (CT) scan of the skull with the lettered structures (A–E) listed below.

Match the numbered structures shown in the axial (horizontal) magnetic resonance imaging (MRI) of the skull with the lettered structures (A–F) listed below.

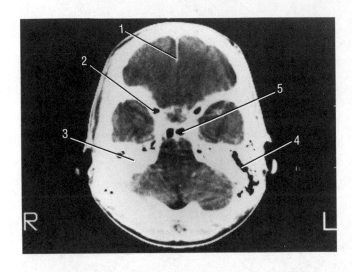

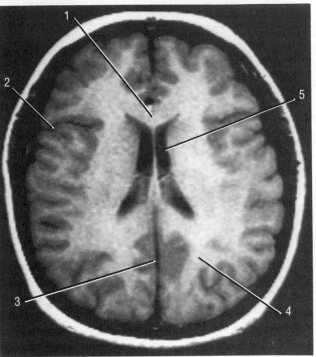

11. Structure 1.   A. Tympanic cavity.
12. Structure 2.   B. Petrous part of temporal bone.
13. Structure 3.   C. Falx cerebri.
14. Structure 4.   D. Optic canal.
15. Structure 5.   E. Sphenoid air sinus.

16. Structure 1.   A. White matter of occipital lobe.
17. Structure 2.   B. Falx cerebri.
18. Structure 3.   C. Genu of corpus callosum.
19. Structure 4.   D. Lateral ventricle.
20. Structure 5.   E. Septum pellucidum.
                       F. Lateral sulcus.

Match the openings in the skull on the left with the appropriate description on the right.

21. Foramen ovale.   A. Allows entrance of the
22. Foramen spinosum.   spinal part of the acces-
23. Internal acoustic   sory nerve into the cra-
    meatus.   nial cavity.
24. Foramen magnum.   B. Is located in the petrous
25. Jugular foramen.   part of the temporal bone.
                          C. Allows entrance of the
                             middle meningeal artery
                             into the cranial cavity.
                          D. Allows exit of the man-
                             dibular division of the
                             trigeminal nerve.
                          E. Allows exit of the glos-
                             sopharyngeal nerve.

Match the air sinuses or structures on the left with the regions into which they open on the right. Answers may be used more than once.

26. Middle ethmoid sinuses.
27. Sphenoid sinus.
28. Frontal sinus.
29. Anterior ethmoidal sinuses.
30. Nasolacrimal duct.

A. Middle meatus of nose.
B. Superior meatus of nose.
C. Sphenoethmoidal recess.
D. Inferior meatus.
E. None of the above.

Match the area of drainage on the left with the appropriate group of lymph nodes on the right.

31. Center of lower lip.
32. Tip of the nose.
33. Cheek.
34. Lateral edge of anterior two-thirds of tongue.
35. Lateral ends of eyelids.
36. Thyroid gland.

A. Submandibular.
B. Mastoid.
C. Submental.
D. Superficial cervical.
E. None of the above.

Match the area of mucous membrane on the left with the appropriate sensory nerve supply on the right.

37. Covering anterior two-thirds of tongue (taste).
38. Piriform fossa or recess (common sensation).
39. Covering posterior third of tongue (common sensation).
40. Roof of mouth (common sensation).

A. Internal laryngeal nerve.
B. Glossopharyngeal nerve.
C. Maxillary division of trigeminal nerve.
D. Recurrent laryngeal nerve.
E. Chorda tympani.

In each of the following questions, answer:

A. If only (1) is correct
B. If only (2) is correct
C. If both (1) and (2) are correct
D. If neither (1) nor (2) is correct

41. (1) The parotid salivary gland contains within its substance the facial nerve and the external carotid artery.
(2) The parotid duct opens into the mouth opposite the upper second molar tooth.

42. Concerning the trachea:
(1) As it descends in the neck, it rests posteriorly on the vertebral column.
(2) The sensory nerve supply of the mucous membrane lining the upper part of the trachea is from the recurrent laryngeal nerve.

43. The spinal part of the accessory nerve:
(1) Can be easily injured because it passes superficially across the posterior triangle of the neck.
(2) Innervates the sternocleidomastoid and the inferior belly of the omohyoid muscles.

44. Which of the following statement(s) is (are) correct?
(1) Afferent (sensory) nerve fibers for the gag reflex are contained in the glossopharyngeal nerve.
(2) Afferent (sensory) nerve fibers for the cough reflex are contained in the vagus nerve.

45. Which of the following statement(s) is (are) correct concerning the pituitary gland (hypophysis cerebri)?
(1) It is related anterosuperiorly to the optic chiasma.
(2) Receives its arterial supply from the hypophyseal branches of the middle meningeal artery.

46. The vertebral artery:
(1) Arises from the subclavian artery.
(2) Joins the artery of the opposite side to form the basilar artery.

47. The following important structure(s) pass(es) **through** the cavernous sinus.
(1) Oculomotor nerve.
(2) Internal carotid artery.

48. The tympanic membrane:
(1) Is crossed by the chorda tympani over the medial surface of the inferior part of the membrane.
(2) Has sensory innervation from the eighth cranial nerve.

49. The nerve supply to the lacrimal gland includes:
(1) Nerve fibers from the facial nerve.
(2) Postganglionic fibers from the ciliary ganglion.

50. The hypoglossal nerve:
(1) Innervates the anterior belly of the digastric muscle.
(2) Innervates the muscles of the tongue.

Select the **best** response.

51. Impaired function of which of the following muscles would result in difficulty in protruding the jaw?
A. Anterior belly of the digastric.
B. Lateral pterygoid.
C. Medial pterygoid.
D. Masseter.
E. Temporalis.

52. The falx cerebri:
A. Contains the superior sagittal sinus.
B. Is attached at one end to the anterior clinoid process.
C. Separates the pituitary gland from the cerebrum.
D. Is inferior to the tentorium cerebelli.
E. Contains the superior petrosal sinus.

53. The nasal cavity is closed off from the oropharynx during swallowing by:
A. Elevation of the tongue to the roof of the mouth.
B. Contraction of the aryepiglottic muscles.
C. Contraction of the tensor and levator veli palatini muscles.
D. Relaxation of the pharyngeal constrictor muscles.
E. Bending of the epiglottis.

54. A muscle that protrudes the tongue is the:
    A. Styloglossus.
    B. Hyoglossus.
    C. Genioglossus.
    D. Palatoglossus.
    E. Mylohyoid.

55. All the following statements concerning the palatine tonsil are correct **except:**
    A. It is related laterally to the superior constrictor muscle and the external palatine vein.
    B. The main blood supply is from the facial artery.
    C. The lymphatic drainage is into the submandibular lymph nodes.
    D. It is covered on its medial surface by mucous membrane and on its lateral surface by a fibrous capsule.
    E. The tonsil reaches its maximum size during early childhood.

56. The following statements concerning the pharynx are true **except:**
    A. The anterior wall of the nasal part of the pharynx is formed by the posterior nasal apertures.
    B. The lowest fibers of the inferior constrictor muscle serve as a sphincter at the lower end of the pharynx.
    C. The mucous membrane of the oral part of the pharynx is mainly innervated by the internal laryngeal nerve.
    D. The piriform fossa is situated on each side of the laryngeal inlet into the pharynx.
    E. All the constrictors of the pharynx are innervated by the pharyngeal plexus.

57. Sustained tension of the vocal cords (folds) is best achieved through the action of which one of the following muscles?
    A. Cricopharyngeus.
    B. Cricothyroid.
    C. Aryepiglottic.
    D. Salpingopharyngeus.
    E. Posterior cricoarytenoid.

58. The following structures are found posterior to the left scalenus anterior muscle **except:**
    A. Subclavian artery.
    B. Roots of the brachial plexus.
    C. Left phrenic nerve.
    D. Anterior ramus of the seventh cervical nerve.
    E. Suprapleural membrane.

59. Muscles or nerves that are responsible for adducting the eyeball (rotating the cornea medially) include the following **except:**
    A. Superior rectus.
    B. Medial rectus.
    C. Oculomotor nerve.
    D. Inferior oblique.
    E. Inferior rectus.

60. During operations on the thyroid gland, the following nerve may be injured when tying the inferior thyroid artery:
    A. Sympathetic trunk.
    B. Internal laryngeal nerve.
    C. Descendens cervicalis.
    D. Recurrent laryngeal nerve.
    E. Superior laryngeal nerve.

61. The following facts concerning the Horner's syndrome are true **except:**
    A. Ptosis may be present due to loss of innervation to the smooth muscle portion of the levator palpebrae superioris.
    B. Pupillary constriction may occur due to loss of innervation to the dilator pupillae muscle.
    C. Can be caused by injury to the superior cervical sympathetic ganglion.
    D. Excessive sweating may occur on one side of the face.
    E. Can be caused by injury to the sympathetic chain (trunk) in the neck.

62. A patient has lost cutaneous sensation over the tip of the nose. Which nerve is most likely to have been damaged?
    A. Facial nerve.
    B. Ophthalmic division of the trigeminal nerve.
    C. Greater auricular nerve.
    D. Mandibular division of the trigeminal nerve.
    E. Maxillary division of the trigeminal nerve.

63. Infection of the middle ear can spread along the following pathways **except:**
    A. Through the tegmen tympani to the middle cranial fossa.
    B. Through the medial wall into the labyrinth.
    C. Through the canal for the tensor tympani muscle into the internal carotid artery.
    D. Through the floor into the internal jugular vein.
    E. Through the aditus to the mastoid antrum into the mastoid air cells.

64. The superior orbital fissure transmits the following structures **except:**
    A. Frontal nerve.
    B. Ophthalmic veins.
    C. Abducent nerve.
    D. Trochlear nerve.
    E. Optic nerve.

65. Compression of the facial nerve in the facial canal in the posterior wall of the tympanic cavity could result in the following **except:**
    A. A cessation of lacrimal secretion.
    B. Paralysis of the posterior belly of the digastric muscle.
    C. Inability to whistle.

D. Decreased saliva in the mouth.

E. Loss of taste sensation to the anterior two-thirds of the tongue.

In the next section, study each case history. Select the best answer to each question following it.

A 45-year-old man was riding his bicycle when he swerved to avoid a pot hole and lost his balance. He crashed, hitting his head against a rock. When he regained consciousness in the emergency department of a neighboring hospital, it was immediately noted that he had medial strabismus (squint) of his right eye. On questioning, the cyclist admitted that he had not been wearing a crash helmet.

66. Which eye muscle was paralyzed in this injury?
    A. Medial rectus.
    B. Inferior rectus.
    C. Superior rectus.
    D. Lateral rectus.
    E. Superior oblique.
67. Which nerve was damaged by the crash?
    A. Facial nerve.
    B. Oculomotor nerve.
    C. Abducent nerve.
    D. Trochlear nerve.
    E. Ophthalmic division of the trigeminal nerve.
68. Which anatomical structure is most likely to have caused damage to the nerve?
    A. Petrous part of the temporal bone.
    B. Greater wing of the sphenoid bone.
    C. Lesser wing of the sphenoid bone.
    D. Tentorium cerebelli.
    E. Falx cerebelli.

A 59-year-old woman was seen by her physician with a small swelling below her chin. On examination, a single hard swelling could be palpated in the submental triangle. It was mobile on the deep tissues and not tethered to the skin. A diagnosis of a malignant secondary deposit in a submental lymph node was considered.

69. Using your knowledge of anatomy, where would you look for the primary carcinoma?
    A. Posterior one-third of the tongue.
    B. Maxillary sinus.
    C. Angle of mouth.
    D. Center of the lower lip.
    E. Lining of the cheek.
70. Where do the submental lymph nodes drain?
    A. Superficial cervical nodes.
    B. Deep cervical nodes.
    C. Submandibular nodes.
    D. Tracheobronchial nodes.
    E. Anterior cervical nodes.

Answers and Explanations

1. D
2. E
3. B
4. A
5. C
6. D
7. B
8. F
9. A
10. C
11. C
12. D
13. B
14. A
15. E
16. C
17. F
18. B
19. A
20. D
21. D
22. C
23. B
24. A
25. E
26. A   On the bulla ethmoidalis.
27. C
28. A   Via the infundibulum.
29. A   Via the infundibulum.
30. D   Guarded by a valve.
31. C
32. A
33. A   Via the buccal nodes.
34. A
35. E   Parotid lymph nodes.
36. E   Tracheal and deep cervical group.
37. E
38. A
39. B
40. C
41. C
42. B   The esophagus, the prevertebral layer of deep cervical fascia, and the prevertebral muscles separate the trachea from the vertebral column.
43. A   The spinal part of the accessory nerve innervates the sternocleidomastoid and the trapezius muscles. The inferior belly of the omohyoid muscle is supplied by the ansa cervicalis (C1, 2, and 3).
44. C
45. A   The blood supply of the hypophysis cerebri is from the superior and inferior hypophyseal branches of the internal carotid artery.

46. C
47. B
48. D   The chorda tympani crosses the medial surface of the superior part of the tympanic membrane. The sensory innervation of the membrane is from the tympanic branch of the glossopharyngeal nerve, the auriculotemporal branch of the mandibular division of the trigeminal nerve, and the auricular branch of the vagus nerve.
49. A
50. B   The innervation of the anterior belly of the digastric muscle is from the mylohyoid nerve, from the inferior alveolar branch of the mandibular branch of the trigeminal nerve.
51. B   The anterior belly of the digastric muscle depresses the jaw, the medial pterygoid muscle elevates the jaw, the masseter muscle elevates the jaw, the temporalis muscle (anterior fibers) elevates the jaw, and the posterior fibers retract the jaw.
52. A
53. C   The tensor and levator veli palatini muscles raise the soft palate, which closes off the nasal from the oral part of the pharynx. The tensor veli palatini serves to stretch the soft palate to allow it to be elevated.
54. C   Both genioglossus muscles pull the tongue forward.
55. C   The lymphatic drainage of the tonsil is into the jugulodigastric lymph node.
56. C   The mucous membrane of the oral part of the pharynx is innervated mainly by the glossopharyngeal nerve via its pharyngeal branch to the pharyngeal plexus.
57. B   Cricothyroid muscle. This muscle tilts the cricoid cartilage and the arytenoid cartilages backward and thus tenses the vocal cords.
58. C   The phrenic nerve is located on the anterior surface of the scalenus anterior muscle in the neck.
59. D   The oblique muscles turn the eyeball laterally. (In addition, the superior oblique turns the eye downward and the inferior oblique turns the eye upward.) The superior rectus turns the eye medially as well as upward and the inferior rectus turns the eye medially as well as downward (because these muscles take origin from the back of the orbit medial to the vertical axis of the eyeball). The oculomotor nerve supplies the medial, superior, and inferior recti and the inferior oblique muscles.
60. D   Remember that the superior thyroid artery is closely related to the external laryngeal nerve.
61. D   In Horner's syndrome sweating is reduced due to loss of the sympathetic innervation of the sweat glands of the facial skin.
62. B   The skin of the tip of the nose is innervated by the external nasal branch of the nasociliary branch of the ophthalmic division of the trigeminal nerve.
63. C   The canal for the tensor tympani muscle is closed at its deep end and is filled by the origin of the tensor tympani muscle.
64. E   The optic nerve enters the orbital cavity through the optic canal in the lesser wing of the sphenoid bone along with the ophthalmic artery.
65. A   Lacrimal secretion is controlled by the lacrimal nucleus of the facial nerve; the fibers leave the facial nerve as the greater petrosal nerve on the medial wall of the tympanic cavity before the facial nerve reaches the posterior wall of the tympanic cavity.
66. D
67. C
68. A   The long slender abducent nerve (sixth cranial nerve) is commonly damaged in severe head injuries. Sudden movement of the head may result in injury to the nerve as it leaves the posterior cranial fossa by passing over the superior border of the petrous part of the temporal bone to enter the cavernous sinus.
69. D   The lymph from the posterior one-third of the tongue drains into the deep cervical nodes. The lymph from the lining of the cheek, the angle of the mouth, and the maxillary sinus drains into the submandibular nodes.
70. B

Back

The back extends from the skull to the tip of the coccyx. Superimposed on the upper part are the scapulae and the muscles that connect the scapulae to the trunk.

It is suggested that the back be reviewed in the following order with the help of an articulated skeleton:

1. A review of the vertebral column and a brief study of the regional differences between vertebrae.
2. A brief overview of the joints of the vertebral column, including the atlanto-occipital and atlantoaxial joints.
3. A review of the structure and function of the intervertebral disc.
4. A brief review of the muscles of the back (the detailed attachments of the muscles are not required).
5. A brief review of the arteries, veins, and lymphatic drainage of the back.
6. A review of the nerves of the back.
7. Finally, an overview of the gross anatomy of the spinal cord and its meninges, including the anatomy of the lumbar puncture.

Vertebral Column

The vertebral column is the central pillar of the body (Fig. 8-1). The head is balanced on the upper end of the pillar and the ribs; the thoracic and abdominal viscera are suspended from the front. The vertebral column protects the spinal cord and supports the weight of the head and the trunk, which it transmits to the hip bones and the lower limbs. It is a flexible structure because it is segmented and made up of **vertebrae,** their joints, and pads of fibrocartilage called **intervertebral discs.** The intervertebral discs form a fourth of the length of the column.

The vertebrae are grouped as follows:

Cervical (7)

Thoracic (12)

Lumbar (5)

Sacral (5 fused to form the sacrum)

Coccygeal (4, the lower 3 are commonly fused)

Intervertebral Foramina

Intervertebral foramina are present between the vertebrae to allow the spinal nerves, which leave the spinal cord, to be distributed to the different parts of the body.

Curves of the Vertebral Column

On side view, there are four curves of the vertebral column (Fig. 8-2).

Cervical region: Posterior concavity

Thoracic region: Posterior convexity

Lumbar region: Posterior concavity

Sacral region: Posterior convexity

Pregnant women, with the weight of the fetus, show an increase in the posterior lumbar concavity **(lordosis). Old age,** with atrophy of the intervertebral discs, tends to produce a continuous posterior convexity of the vertebral column so that the individual has a bent forward appearance **(kyphosis).**

General Characteristics of a Vertebra

A typical vertebra has a rounded **body** anteriorly and a **vertebral arch** posteriorly (see Fig. 8-2). The **vertebral foramen** is the space enclosed by the body and the arch. The vertebral arch consists of a pair of **cylindrical pedicles,** which form the sides of the arch, and a pair of flattened **laminae,** which complete the arch posteriorly. The vertebral arch gives rise to seven processes: one spinous, two transverse, and four articular (see Fig. 8-2).

Spinous Process (Spine)

The spinous process is directed posteriorly from the junction of the two laminae. **Transverse processes** are directed laterally from the junction of the laminae and the pedicles. Both processes serve as levers and receive attachments of muscles and ligaments.

Articular Processes

Articular processes are arranged vertically and consist of two superior and two inferior processes. They arise from

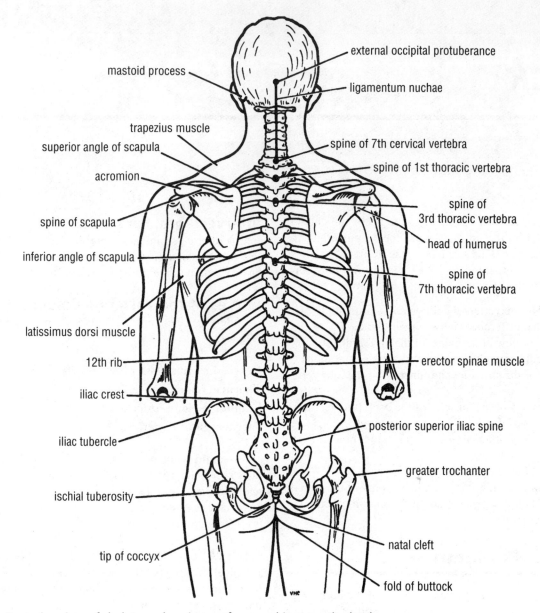

mastoid process

external occipital protuberance

ligamentum nuchae

trapezius muscle

superior angle of scapula

spine of 7th cervical vertebra

acromion

spine of 1st thoracic vertebra

spine of
3rd thoracic vertebra

spine of scapula

head of humerus

inferior angle of scapula

spine of
7th thoracic vertebra

latissimus dorsi muscle

12th rib

erector spinae muscle

iliac crest

posterior superior iliac spine

iliac tubercle

greater trochanter

ischial tuberosity

tip of coccyx

natal cleft

fold of buttock

Fig. 8-1.   Posterior view of skeleton, showing surface markings on the back.

the junction of the laminae and the pedicles. The two superior articular processes of one vertebra articulate with the two inferior articular processes of the vertebra above it.

The pedicles are notched on their upper and lower borders, forming the **superior and inferior vertebral notches.** The superior notch of one vertebra and the inferior notch of an adjacent vertebra together form an **intervertebral foramen** (see Fig. 8-4). These foramina transmit the spinal nerves and blood vessels.

Cervical Vertebra

A typical cervical vertebra has the following characteristics (see Fig. 8-2):

1. **Foramen transversarium** in the transverse process for the passage of the vertebral artery and veins.
2. **Spines** are small and bifid.
3. **Body** is small and broad and has two small synovial joints on each side.
4. **Vertebral foramen** is large and triangular in shape.
5. **Superior articular processes** have small, flat articular facets that face backward and upward.
6. **Inferior articular processes** have facets that face downward and forward.

Atypical Cervical Vertebrae

The **first cervical vertebra (atlas)** has no body and no spinous process. It has an **anterior arch,** a **posterior arch,**

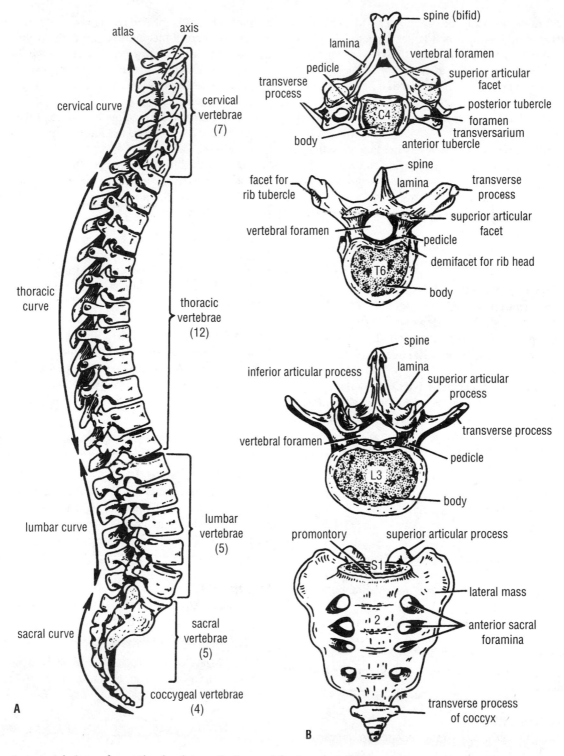

Fig. 8-2. **A.** Lateral view of vertebral column. **B.** General features of different kinds of vertebrae.

and a **lateral mass** on each side. Each lateral mass has **articular surfaces** on its upper and lower aspects. The atlas articulates above with the occipital condyles of the skull and below with the axis.

The **second cervical vertebra (axis)** has a peglike **odontoid process,** which projects upward from the superior surface of the body and represents the body of the atlas that has fused with the axis.

The **seventh cervical vertebra** has the longest spinous process. The spinous process is **not** bifid. The **transverse process** is large, but the foramen transversarium is small and does not transmit the vertebral artery.

Thoracic Vertebra

A typical thoracic vertebra has the following characteristics (see Fig. 8-2):

1. **Body** is medium-sized and heart-shaped.
2. **Vertebral foramen** is small and circular.
3. **Spines** are long and inclined downward.
4. **Costal facets** are present on sides of bodies and on transverse processes. (T11 and 12 have no facets on transverse processes.)
5. **Superior articular processes** have facets that face backward and laterally, whereas the facets on the inferior articular processes face forward and medially.

Lumbar Vertebrae

A typical lumbar vertebra has the following characteristics (see Fig. 8-2):

1. **Body** is massive and kidney-shaped.
2. **Pedicles** are strong.
3. **Laminae** are thick.
4. **Vertebral foramina** are triangular in shape.
5. **Transverse processes** are long and slender.
6. **Spinous processes** are short, flat, and quadrangular in shape and project directly backward.
7. **Superior articular processes** have facets that face medially, whereas the facets for the inferior articular processes face laterally.
8. **There are no rib facets** and **no foramina in the transverse processes.**

Sacral Vertebrae

Sacral vertebrae consist of five rudimentary vertebrae that are fused together to form a single wedge-shaped bone, the **sacrum.** The sacrum has the following characteristics (see Fig. 8-2):

1. **Sacral promontory** is the anterior and upper margin of the first sacral vertebra that bulges forward into the pelvic cavity; **it is an important obstetric landmark.**

2. **Sacral canal** is formed from the sacral foramina. It contains part of the **cauda equina,** the **filum** terminale, and the **meninges** down as far as the lower border of the second sacral vertebra. The lower part of the canal contains the lower sacral and coccygeal nerve roots and the filum terminale.
3. **Sacral hiatus** is formed by the failure of the laminae of the fifth and sometimes the fourth sacral vertebra to meet in the midline.
4. **Anterior and posterior sacral foramina** on the anterior and posterior surfaces of the sacrum for the passage of the anterior and posterior rami of the upper four sacral spinal nerves.

The upper border, or base, of the sacrum articulates with the fifth lumbar vertebra (see Fig. 8-2). The narrow inferior end articulates with the coccyx. Laterally, the sacrum articulates with the two hip bones to form the sacroiliac joints (see Fig. 8-1).

Coccygeal Vertebrae

There are four coccygeal vertebrae that are fused together to form a single small triangular bone called the **coccyx** (see Fig. 8-2). The first coccygeal vertebra is commonly not fused with the second vertebra.

Joints of the Vertebral Column

Atlanto-occipital Joints

Atlanto-occipital joints are synovial joints that are formed between the occipital condyles, which are found on either side of the foramen magnum of the skull, and the facets on the superior surfaces of the lateral masses of the atlas below (Fig. 8-3). They are enclosed by a capsule.

Ligaments
Anterior Atlanto-occipital Membrane
Connects the anterior arch of the atlas to the anterior margin of the foramen magnum (see Fig. 8-3).

Posterior Atlanto-occipital Membrane
Connects the posterior arch of the atlas to the posterior margin of the foramen magnum (see Fig. 8-3).

Movements
Flexion, extension, and lateral flexion. No rotation is possible.

Atlantoaxial Joints

Atlantoaxial joints are synovial joints, one of which is between the odontoid process and the anterior arch of the atlas, whereas the other two are between the lateral masses of the bones (see Fig. 8-3). The joints are enclosed by capsules.

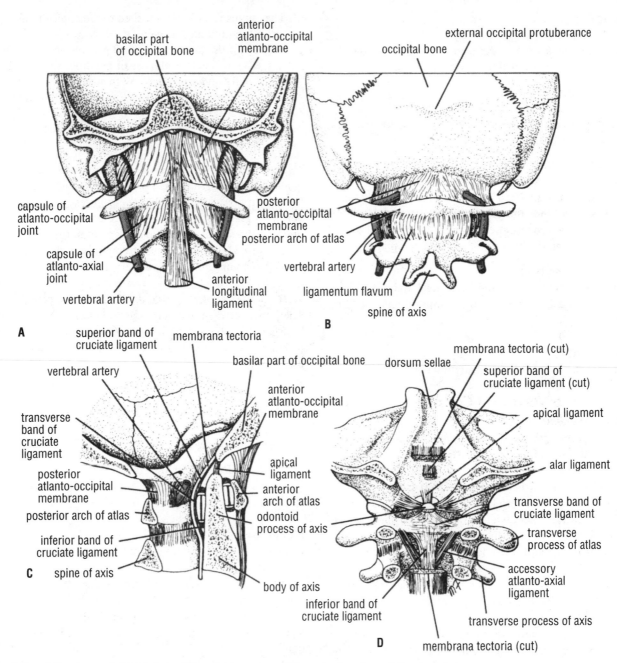

Fig. 8-3. Atlanto-occipital joints: (**A**) anterior view; (**B**) posterior view. Atlantoaxial joints: (**C**) sagittal section; (**D**) posterior view; note that the posterior arch of the atlas and the laminae and spine of the axis have been removed.

Ligaments
Apical Ligament
Connects the apex of the odontoid process to the anterior margin of the foramen magnum (see Fig. 8-3).

Alar Ligaments
Lie on each side of the apical ligament and connect the odontoid process to the medial sides of the occipital condyles (see Fig. 8-3).

Cruciate Ligament
Consists of transverse and vertical parts (see Fig. 8-3). The **transverse part** is attached on each side to the lateral mass of the atlas and binds the odontoid process to the anterior arch of the atlas. The **vertical part** connects the body of the axis to the anterior margin of the foramen magnum.

Membrana Tectoria
An upward continuation of the posterior longitudinal ligament that is attached above to the occipital bone (see Fig. 8-3).

Movements
Rotation of the atlas with the head on the axis.

Joints of the Vertebral Column Below the Axis
1. **Cartilaginous joints.** Between the vertebral bodies.
2. **Synovial joints.** Between the articular processes.

Joints Between Two Vertebral Bodies
The bodies of adjacent vertebrae are covered by a thin plate of hyaline cartilage. Sandwiched between the plates of cartilage is an intervertebral disc of fibrocartilage (Fig. 8-4). The disc strongly unites the bodies of the two vertebrae.

Ligaments
Anterior and Posterior Longitudinal Ligaments
Run as continuous bands down the anterior and posterior surfaces of the vertebral column from the skull to the sacrum (see Fig. 8-4).

Intervertebral Discs
The intervertebral discs have the following important characteristics:

1. **Responsible** for a fourth of the length of the vertebral column. They are thicker in the cervical and lumbar regions.
2. **Anulus fibrosus** forms periphery of disc (see Fig. 8-4). It is composed of fibrocartilage, in which the collagen fibers are arranged in concentric layers or sheets. The fibers pass obliquely between adjacent vertebral bodies, and their inclination is reversed in alternate sheets.

The anulus is strongly attached to the anterior and posterior longitudinal ligaments.

3. **Nucleus pulposus** is the central part of the disc (see Fig. 8-4). It is an ovoid mass of gelatinous material and is normally under pressure. With advancing age, the water content of the nucleus diminishes and is replaced by fibrocartilage.
4. **No discs** are present between the first two cervical vertebrae or in the sacrum or coccyx.

Herniation of the nucleus pulposus may occur if there is a sudden increase in the compression load on the vertebral column. The anulus fibrosus ruptures and the nucleus herniates posteriorly into the vertebral canal where it may press on the spinal nerve roots, a spinal nerve, or even the spinal cord. Herniation occurs frequently in the lower lumbar (most common) and lower cervical regions.

Joints Between Two Vertebral Arches
The joints between two vertebral arches are synovial joints between the superior and inferior articular processes of adjacent vertebrae (see Fig. 8-4). The joints are surrounded by a capsule.

Ligaments
1. **Supraspinous ligament.** Connects adjacent spines.
2. **Interspinous ligament.** Connects adjacent spines.
3. **Ligamentum flavum.** Connects adjacent laminae.

In the cervical region, the supraspinous and interspinous ligaments are greatly thickened to form the **ligamentum nuchae**.

Nerve Supply
Branches of corresponding spinal nerves.

Movements of the Vertebral Column
The type and range of movements possible in each region of the column are dependent on the (1) thickness of the intervertebral discs and (2) the shape and direction of the articular processes.

Flexion: A forward movement.
Extension: A backward movement.
Both flexion and extension are extensive in the cervical and lumbar regions but restricted in the thoracic region due to the presence of ribs.
Lateral flexion: The bending of the body to one or the other side. It is extensive in the cervical and lumbar regions, but restricted in the thoracic region due to the presence of ribs.
Rotation: A twisting of the vertebral column. It is least extensive in the lumbar region.
Circumduction: A combination of all these movements.

Back

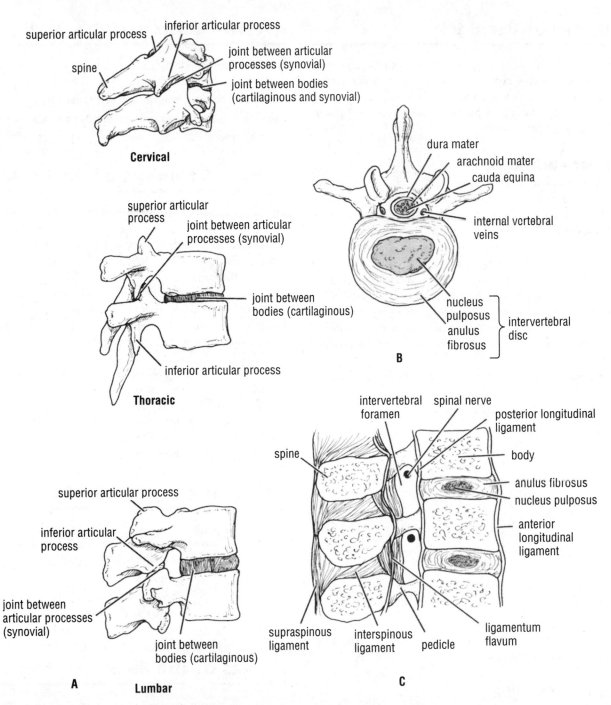

Fig. 8-4. **A.** Joints in cervical, thoracic, and lumbar regions of vertebral column. **B.** Third lumbar vertebra seen from above, showing relationship between intervertebral disc and cauda equina. **C.** Sagittal section through three lumbar vertebrae, showing ligaments and intervertebral discs. Note relationship between emerging spinal nerve in an intervertebral foramen and the intervertebral disc.

Muscles of the Back

The muscles of the back may be divided into three groups: **superficial muscles** connected with the shoulder girdle, **intermediate muscles** involved with movements of the thoracic cage, and **deep or postvertebral muscles** belonging to the vertebral column.

Postvertebral Muscles

The postvertebral muscles are very well developed in man. They form a broad, thick column of muscle tissue, which occupies the hollow on each side of the spinous processes (Fig. 8-5). The spines and transverse processes of the vertebrae serve as levers that assist the muscle actions. The muscles of longest length lie superficially and run from the sacrum to the rib angles, the transverse processes, and the upper vertebral spines. The muscles of intermediate length run obliquely from transverse processes to spines. The shortest and deepest muscle fibers run between spines and between transverse processes of adjacent vertebrae.

The postvertebral muscles may be classified as follows:

Superficial vertically running muscles
Erector spinae: Iliocostalis
                        Longissimus
                        Spinalis
Intermediate oblique running muscles
Transversospinalis: Semispinalis
                        Multifidus
                        Rotatores
Deepest muscles
                        Interspinales
                        Intertransversarii

Students are not required to learn the detailed attachments of these muscles. Figure 8-5 shows the arrangement of the deep muscles of the back.

Muscular Triangles
Auscultatory Triangle

The auscultatory triangle is the site on the back where breath sounds may be most easily heard with a stethoscope. The boundaries are latissimus dorsi, trapezius, and the medial border of the scapula; the rhomboid major forms the floor.

Lumbar Triangle

The lumbar triangle is the site where pus may emerge from the abdominal wall. The boundaries are the latissimus dorsi, the posterior border of the external oblique muscle of the abdomen, and the iliac crest.

Arterial Supply of the Back

Cervical region: Occipital artery, vertebral artery, deep cervical artery, and the ascending cervical artery
Thoracic region: Posterior intercostal arteries
Lumbar region: Subcostal and lumbar arteries
Sacral region: Iliolumbar and lateral sacral arteries

Venous Drainage of the Back

The veins form complicated plexuses that extend along the vertebral column from the skull to the coccyx. The **external vertebral venous plexus** lies external to the vertebral column; the **internal vertebral venous plexus** lies within the vertebral canal (Fig. 8-6).

There is free communication between the plexuses and the veins in the neck, thorax, abdomen, and pelvis. They communicate above through the foramen magnum with the venous sinuses in the cranial cavity.

The internal plexus also communicates with the veins draining the vertebral bodies (**basivertebral veins**) and the veins of the meninges and spinal cord.

The vertebral plexuses are drained into the vertebral, intercostal, lumbar, and lateral sacral veins. **The plexuses provide a pathway for the spread of malignant disease from the pelvis to the skull.**

Lymphatic Drainage of the Back

Superficial lymph: Above the iliac crest drains into the axillary nodes; below the iliac crest drains into the superficial inguinal nodes
Deep lymph: Drains into the deep cervical, posterior mediastinal, lateral aortic, and sacral nodes

Nerves of the Back

Segmental supply to the skin and muscles arises from the posterior rami of the 31 pairs of spinal nerves. Note that the posterior rami of C1, 6, 7, and 8 and L4 and 5 spinal nerves supply the deep muscles of the back and **do not supply the skin.**

A dermatomal chart for the posterior surface of the body is shown in Figure 1-8.

Spinal Cord

The spinal cord is cylindrical in shape and begins superiorly at the foramen magnum where it is continuous with

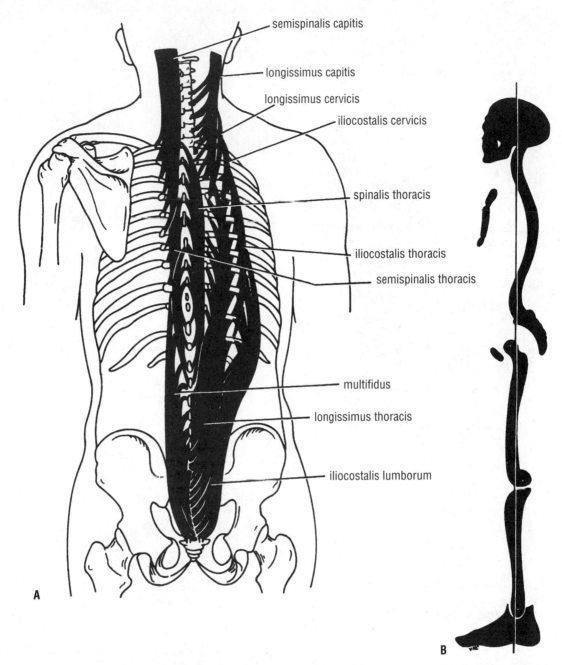

semispinalis capitis

longissimus capitis

longissimus cervicis

iliocostalis cervicis

spinalis thoracis

iliocostalis thoracis

semispinalis thoracis

multifidus

longissimus thoracis

iliocostalis lumborum

A

B

Fig. 8-5. **A.** Arrangement of deep muscles of back. **B.** Lateral view of skeleton, showing line of gravity. Since greater part of body weight lies anteriorly to vertebral column, deep muscles of back are important in maintaining normal postural curves of vertebral column in standing position.

the medulla oblongata of the brain. It terminates below in the adult at the level of the lower border of the first lumbar vertebra. The spinal cord thus occupies the upper two-thirds of the vertebral canal and is surrounded by the three meninges: the **dura mater,** the **arachnoid mater,** and the **pia mater** (Fig. 8-7; see also Fig. 8-6). Additional protection is provided by the **cerebrospinal fluid** in the **subarachnoid space.** The spinal cord has **cervical** and **lumbar enlargements** where it gives origin to the brachial and lumbar plexuses. The spinal cord is tapered below to form the **conus medullaris** (see Fig. 8-7). The **filum terminale** is a prolongation of the pia mater that extends from the conus to the back of the coccyx (see Fig. 8-7).

Fissures of the Spinal Cord
Anterior Median Fissure
The anterior median fissure is located in the midline on the anterior surface of the cord.

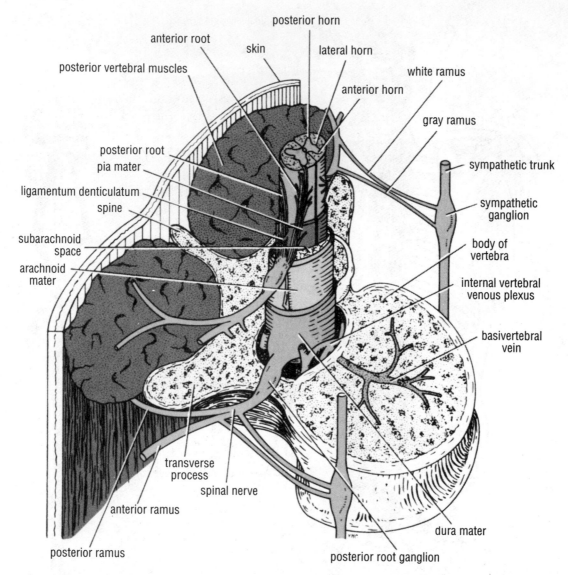

Fig. 8-6.   Oblique section through first lumbar vertebra, showing spinal cord and its covering membranes. Note relationship between spinal nerve and sympathetic trunk on each side. Note, also, important internal vertebral venous plexus.

Posterior Median Fissure

The posterior median fissure is located in the midline on the posterior surface of the cord.

Roots of the Spinal Nerves

Thirty-one pairs of spinal nerves are attached to the spinal cord by **anterior and posterior roots** (see Fig. 8-7). The anterior roots are motor and the posterior roots are sensory. Each posterior root possesses a **posterior root ganglion** (see Fig. 8-7).

In the upper cervical region, the spinal nerve roots are short and run almost horizontally, but the roots of the lumbar and sacral nerves below the level of the termination of the cord (lower border of the first lumbar vertebra in the adult) form a vertical leash of nerves around the filum terminale called the **cauda equina.**

As the spinal nerve roots pass through the intervertebral foramina, they unite to form a **spinal nerve** (see Fig. 8-6). After emerging from the intervertebral foramen, each spinal nerve divides into a large **anterior ramus** and a smaller **posterior ramus,** each containing both motor and sensory fibers (see Fig. 8-6).

Blood Supply of the Spinal Cord
Arteries
Posterior Spinal Arteries

Arise directly or indirectly from the vertebral arteries.

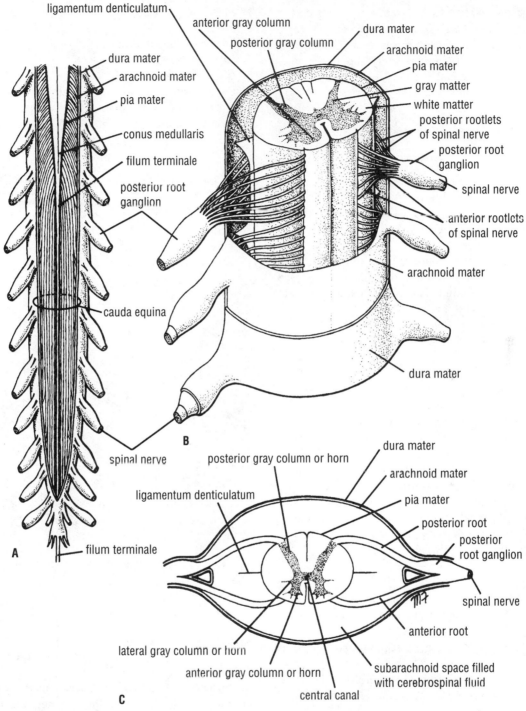

Fig. 8-7. **A.** Lower end of spinal cord and cauda equina. **B.** Section through thoracic part of spinal cord, showing anterior and posterior roots of spinal nerves and meninges. **C.** Transverse section through spinal cord, showing meninges and position of cerebrospinal fluid.

Anterior Spinal Arteries

Arise from the vertebral arteries and unite to form a single artery, which descends in the anterior median fissure.

Branches of the anterior spinal artery supply the anterior two-thirds of the spinal cord, whereas the posterior third is supplied by the posterior spinal arteries.

Radicular Arteries

Branches of regional arteries that reinforce the anterior and posterior spinal arteries.

Veins

Drain into the internal vertebral venous plexus.

Meninges of the Spinal Cord
Dura Mater

An external membrane of dense fibrous tissue, the dura mater encloses the spinal cord and the cauda equina (see Fig. 8-7). The dura is continuous above with the meningeal layer of dura covering the brain. Below, it ends on the filum terminale at the level of the lower border of the second sacral vertebra. The dura gives sheaths to all the spinal nerve roots.

Arachnoid Mater

The arachnoid mater is a delicate impermeable membrane that lies within the dura and outside the pia (see Fig. 8-7). It is separated from the pia mater by a wide space, the **subarachnoid space,** which is filled with **cerebrospinal fluid.** The arachnoid is continuous above through the foramen magnum with the arachnoid covering the brain. Inferiorly, it ends on the filum terminale at the level of the **lower border of the second sacral vertebra.** The arachnoid continues along the spinal nerve roots, forming small lateral extensions of the subarachnoid space.

Pia Mater

The pia mater is a vascular membrane that closely covers the spinal cord. It is thickened on either side between the nerve roots to form the **ligamentum denticulatum,** which passes laterally to adhere to the arachnoid and dura (see Fig. 8-7). The pia mater extends along each nerve root as far as the spinal nerve. Inferiorly, it is prolonged off the lower end of the spinal cord as the **filum terminale.**

Lumbar Puncture (Spinal Tap)

The fourth lumbar spine is identified by passing an imaginary line between the highest points on the iliac crests; the spine lies on that line. With the patient lying on the side with the vertebral column well flexed, a lumbar puncture needle is passed into the vertebral canal above or below the fourth lumbar spine. The needle passes through the following anatomical structures before entering the subarachnoid space:

1. **Skin**
2. **Fascia**
3. **Supraspinous ligament**
4. **Interspinous ligament**
5. **Ligamentum flavum**
6. **Fatty tissue and internal vertebral venous plexus**
7. **Dura mater**
8. **Arachnoid mater**

National Board Type Questions

Match the numbered structures shown in the anteroposterior radiograph of the cervical part of the vertebral column with the appropriate lettered structures listed below.

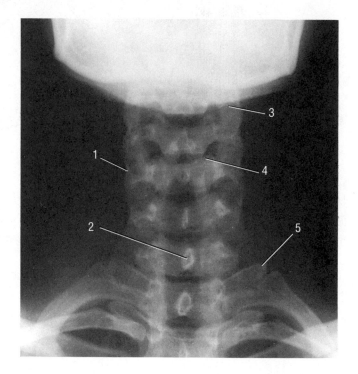

1. Structure 1.
2. Structure 2.
3. Structure 3.
4. Structure 4.
5. Structure 5.

A. Transverse process of first thoracic vertebra.
B. Joint between articular processes of the third and fourth cervical vertebrae.
C. Transverse process of fifth cervical vertebra.
D. Spinous process of seventh cervical vertebra.
E. Lateral synovial joint between adjacent vertebral bodies.
F. None of the above.

Match the numbered structures shown in the axial (transverse) computed tomographic (CT) scan through the lumbar part of the vertebral column with the appropriate lettered structures listed below.

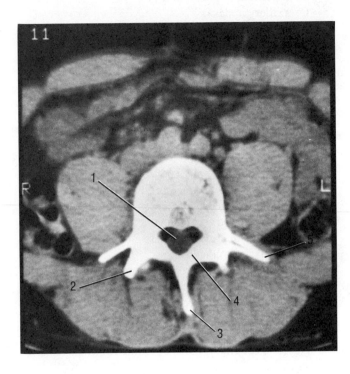

6. Structure 1.   A. Lamina.
7. Structure 2.   B. Articular process.
8. Structure 3.   C. Spinous process.
9. Structure 4.   D. Cauda equina.
10. Structure 5.   E. Transverse process.
                    F. None of the above.

In each of the following questions, answer:

A. If only (1) is correct
B. If only (2) is correct
C. If both (1) and (2) are correct
D. If neither (1) nor (2) is correct

11. (1) There are seven cervical spinal nerves and eight cervical vertebrae.
(2) The posterior ramus of the first cervical spinal nerve supplies the skin over the occipital bone.

12. (1) With an individual in the standing position, the line of gravity passes anterior to the cervical part of the vertebral column and posterior to the thoracic and lumbar regions of the column.
(2) The intervertebral disc makes up about a fourth of the length of the vertebral column.

13. (1) In old age, atrophy of the intervertebral discs tends to produce a continuous posterior convexity of the vertebral column.

(2) In pregnancy, the weight of the developing fetus causes an increase in the posterior lumbar concavity of the vertebral column.

14. (1) In the cervical region, the most prominent spinous process is that of the sixth cervical vertebra.
(2) The foramen transversarium of the sixth cervical vertebra contains the vertebral artery and vein.

15. (1) The odontoid process of the axis represents developmentally the body of the atlas.
(2) The vertebral canal of the third cervical vertebra contains the rootlets of the spinal part of the accessory nerve.

16. (1) The vertebral canal in the thoracic region is relatively large so that the spinal cord has plenty of room.
(2) The tip of a spine of a thoracic vertebra lies directly behind the vertebral body of the vertebra below.

17. (1) The sacral hiatus is formed by the failure of the laminae of the fifth sacral vertebra to meet in the midline.
(2) The anterior sacral foramina permit the entrance of the sacral spinal nerves into the pelvis.

18. (1) The atlanto-occipital joints permit the flexion and extension of the head on the vertebral column, but not lateral flexion.
(2) The atlantoaxial joints permit rotation of the atlas with the head on the axis."

19. (1) The joint surfaces between two vertebral bodies are covered with a plate of hyaline cartilage.
(2) There is usually a fibrocartilaginous disc between the lower end of the sacrum and the coccyx.

20. (1) The anterior longitudinal ligament of the vertebral column is weak and narrow and is attached to the anterior margins of the intervertebral discs.
(2) The intervertebral disc is innervated by a recurrent branch of a spinal nerve that enters the vertebral canal through the intervertebral foramen.

Select the **best** response.

21. The subarachnoid space ends inferiorly at the level of:
A. L5 vertebra.
B. L3 vertebra.
C. S2–3 vertebrae.
D. T12 vertebra.
E. L1 vertebra.

22. The spinal cord in the adult usually ends inferiorly at the following vertebral level:
A. At the lower border of S2.
B. At the upper border of S1.
C. At the lower border of S4.
D. At the upper border of the coccyx.
E. At the lower border of L1.

23. The cervical spinal nerves lack:
A. Motor fibers to striated muscles.
B. Connective tissue sheaths.

C. Motor fibers to smooth muscle.

D. Postganglionic sympathetic fibers.

E. Preganglionic sympathetic fibers.

24. The least serious congenital abnormality involving the neural arch and the neural tube is:

A. Rachischisis.

B. Meningohydroencephalocele.

C. Meningomyelocele.

D. Spina bifida occulta.

E. Meningoencephalocele.

25. The following ligaments are **not** attached to the odontoid process of the axis **except:**

A. Cruciate ligament.

B. Alar ligaments.

C. Membrana tectora.

D. Ligamentum flavum.

E. Ligamentum nuchae.

26. When performing a lumbar puncture (spinal tap), the following structures are pierced by the needle **except:**

A. Supraspinous ligament.

B. Posterior longitudinal ligament.

C. Arachnoid mater.

D. Ligamentum flavum.

E. Dura mater.

27. All the following characteristics are present in a lumbar vertebra **except:**

A. A massive kidney-shaped body.

B. The transverse processes are short and thick.

C. The spinous processes are short, flat, and quadrangular in shape.

D. The transverse processes have no foramen.

E. The articular surfaces of the superior articular processes face medially.

28. In order to perform a lumbar puncture (spinal tap) in the adult, the needle is introduced between which of the following spinous processes?

A. T12 and L1.

B. L2 and 3.

C. S1 and 2.

D. L4 and 5.

E. S3 and 4.

29. The following joints associated with the sacrum and atlas are synovial **except:**

A. Sacroiliac.

B. Atlanto-occipital.

C. Between the superior articular process of the sacrum and the inferior articular process of the fifth lumbar vertebra.

D. Between the body of the first sacral vertebra and the body of the fifth lumbar vertebra.

E. Between the odontoid process and the anterior arch of the atlas.

30. The lymphatic drainage of the skin of the back in the region of the spinous process of the tenth thoracic vertebra is into which of the following lymph nodes?

A. Posterior mediastinal nodes.

B. Superficial inguinal nodes.

C. Axillary nodes.

D. Lateral aortic nodes.

E. Sacral nodes.

31. Which of the following structures pass through the intervertebral foramen between the sixth and seventh cervical vertebrae?

A. Vertebral artery.

B. Sixth cervical spinal nerve.

C. Seventh cervical spinal nerve.

D. Vertebral vein.

E. Eighth cervical spinal nerve.

32. The following facts concerning the internal vertebral venous plexus are true **except:**

A. It drains blood from the vertebral bodies.

B. It permits malignant cells from the prostate to metastasize to the skull.

C. It does not possess competent valves.

D. The venous flow is indirectly influenced by changes in the intra-abdominal pressure.

E. It does not communicate with the veins within the thorax.

33. The cauda equina is made up of the following structures:

A. The spinal nerves of S1, 2, and 3.

B. The anterior rami of spinal nerves L2–5.

C. The posterior rami of spinal nerves L1–Cocc 1.

D. The anterior and posterior nerve roots of spinal nerves below the first lumbar segment of the spinal cord.

E. The spinal nerves of L1–5.

34. The following facts concerning the herniation of an intervertebral disc are true **except:**

A. It generally herniates posteriorly.

B. In the lumbar region, it will usually affect the spinal nerve whose number corresponds to the vertebra below.

C. It is a portion of the nucleus pulposus that actually herniates.

D. A contributing factor to the herniation may be excessive compression of the posterior region of the disc.

E. It may press on the spinal cord.

35. The strength of the flexor muscles of the vertebral column can be assessed by asking the patient to:

A. Sit up from the supine position while keeping the hips and knees extended.

B. Lift the shoulders from the examining table while in the prone position.

C. Extend the head while lying in the prone position.

D. Sit up from the supine position while keeping the hips and knees flexed.

E. Performing the movement of forced expiration.

36. Which of the following facts concerning the vertebral column is **not** true?

A. Injuries to the intervertebral discs are most common in the lower lumbar and lower cervical regions.

B. Rotatory movements are not possible in the thoracic region.

C. In the lumbar region, flexion and extension movements are possible.

D. The amount of movement possible in any particular region of the vertebral column is dependent on the thickness of the intervertebral discs and the shape and direction of the articular processes.

E. The vertebral column protects the spinal cord but a vertebral fracture may cause irreversible nerve cell damage in the cord.

37. Which of the following facts concerning an intervertebral disc is **not** true?

A. During aging, the fluid within the nucleus pulposus is diminished and the amount of fibrocartilage is increased.

B. The discs are thickest in the lumbar region.

C. The discs play a major role in the development of the curvatures of the vertebral column.

D. The atlantoaxial joint possesses a small disc.

E. The discs are innervated by adjacent spinal nerves.

38. A herniated disc that causes sensory changes in a specific dermatome is pressing on a:

A. Anterior primary ramus.

B. Spinal nerve or posterior root.

C. Anterior root.

D. Posterior primary ramus.

E. Anterior gray horn of spinal cord.

39. Which of the following facts is true of the blood supply to the spinal cord?

A. The anterior spinal arteries are two in number and run down the anterior surface of the spinal cord close to the anterior nerve roots.

B. The posterior spinal arteries supply the posterior one-third of the spinal cord.

C. The veins of the spinal cord drain into the external vertebral venous plexus.

D. The anterior and posterior spinal arteries do not anastomose with the radicular arteries.

E. The spinal cord has a profuse blood supply.

In the next section, study each case history. Select the **best** answer to each question following it.

A 69-year-old woman was seen by her physician complaining of a burning pain over her left shoulder and the upper part of the left arm. The pain had started about 2 weeks previously and had progressively worsened. The pain was made worse by moving the neck. On examination, the patient showed hyperesthesia of the skin over the lower part of the left deltoid muscle and down the lateral side of the arm; her left deltoid and biceps brachii muscles were weaker than those on the right side. Radiologic examination showed extensive osteoarthritic changes of the vertebrae with spur formation on the bodies of the fourth, fifth, and sixth cervical vertebrae.

40. Assuming that the osteoarthritic changes had caused narrowing of the cervical intervertebral foramina, which spinal nerve roots were most likely to have been pressed on to cause the burning pain and hyperesthesia?

A. C3 and C4.

B. C6.

C. C5 and C6.

D. C7.

E. C8 and T1.

41. The weakness of the left deltoid and biceps brachii muscles could have been due to pressure on which nerve roots?

A. Posterior nerve roots of C5 and C6.

B. Anterior nerve roots of C5 and C6.

C. Anterior nerve roots of C7 and C8.

D. Anterior nerve roots of C8 and T1.

E. None of the above nerve roots.

42. The sensory and motor changes in this patient were due to pressure on the cervical spinal nerves or their roots as they passed through the intervertebral foramina. The following structures formed the boundaries of each foramen **except:**

A. Intervertebral disc.

B. Bodies of cervical vertebrae.

C. Posterior longitudinal ligament.

D. Articular processes of vertebrae.

E. Pedicles of vertebrae.

Answers and Explanations

1. C
2. D
3. B
4. E
5. A
6. D
7. B
8. C
9. A
10. E
11. D   There are eight cervical spinal nerves and seven cervical vertebrae. The posterior ramus of the second

cervical nerve gives rise to the greater occipital nerve that supplies the skin over the back of the head.

12. B   In the standing position, the line of gravity passes through the odontoid process of the axis, in front of the thoracic vertebrae, and through the lumbar vertebrae, but anterior to the sacrum (see Fig. 8-5).

13. C

14. B   The most prominent spinous process in the cervical region is that of the seventh cervical vertebra.

15. C

16. B   The vertebral canal in the thoracic region is small and circular. Fracture dislocation in the thoracic part of the vertebral column almost invariably causes damage to the spinal cord.

17. A   The anterior sacral foramina permit the passage of the anterior rami into the pelvis and **not** the spinal nerves.

18. B   Lateral flexion is possible at the atlanto-occipital joints.

19. C

20. B   The anterior longitudinal ligament is strong and wide.

21. C

22. E

23. E   The sympathetic outflow is confined to the T1–L2[3] segments of the spinal cord.

24. D

25. B   The alar ligaments are attachments to the tip of the odontoid process on either side of the apical ligament.

26. B

27. B   The transverse processes are long and slender.

28. D

29. D   This is a cartilaginous joint.

30. C   The subscapular group of axillary nodes.

31. C

32. E   The internal vertebral venous plexus communicates with the veins of the neck, the thorax, the abdomen, and the pelvis, via radicular veins.

33. D   The cauda equina is not made up of spinal nerves or the rami of spinal nerves.

34. D   Excessive compression of the anterior part of the disc, as in excessive forced flexion, may contribute to the herniation.

35. D   Sitting up from the supine position while keeping the hip and knee joints extended results in the use of the iliacus and psoas muscles, in addition to the main flexor muscles of the vertebral column.

36. B

37. D   The atlantoaxial joint has no intervertebral disc.

38. B   Anterior and posterior primary rami arise from a spinal nerve outside the vertebral canal. An anterior root does not contain sensory nerve fibers. The anterior gray horn of the spinal cord contains motor neurons.

39. B   There is only one anterior spinal artery formed by the union of branches from each vertebral artery. The anterior spinal artery is small and runs downward in the anterior median sulcus of the spinal cord. The veins of the spinal cord drain into the internal vertebral venous plexus. The spinal cord has a relatively poor blood supply.

40. C   The burning pain and hyperesthesia occurred in the fifth and sixth cervical dermatomes on the left side.

41. B   The deltoid muscle is innervated by the axillary nerve (C5 and C6) and the biceps brachii muscle is innervated by the musculocutaneous nerve (C5, C6, and C7). The majority of the motor fibers emerge from the spinal cord in the anterior nerve roots of C5 and C6 spinal nerves.

42. C

Index

Note: Page numbers followed by *f* indicate figures; page numbers followed by *t* indicate tables.